AF726515

Lithography: The Science of Microfabrication and Nanofabrication

Lithography: The Science of Microfabrication and Nanofabrication

Edited by Burton Kohler

CLANRYE INTERNATIONAL
www.clanryeinternational.com

Clanrye International,
750 Third Avenue, 9th Floor,
New York, NY 10017, USA

Copyright © 2023 Clanrye International

This book contains information obtained from authentic and highly regarded sources. Copyright for all individual chapters remain with the respective authors as indicated. All chapters are published with permission under the Creative Commons Attribution License or equivalent. A wide variety of references are listed. Permission and sources are indicated; for detailed attributions, please refer to the permissions page and list of contributors. Reasonable efforts have been made to publish reliable data and information, but the authors, editors and publisher cannot assume any responsibility for the validity of all materials or the consequences of their use.

Trademark Notice: Registered trademark of products or corporate names are used only for explanation and identification without intent to infringe.

ISBN: 978-1-64726-602-8

Cataloging-in-publication Data

Lithography : the science of microfabrication and nanofabrication / edited by Burton Kohler.
p. cm.
Includes bibliographical references and index.
ISBN 978-1-64726-602-8
1. Nanolithography. 2. Microfabrication. 3. Microlithography. 4. Lithography.
5. Chemical engineering. I. Kohler, Burton.

TK7874.843 .L58 2023
621.381--dc23

For information on all Clanrye International publications
visit our website at www.clanryeinternational.com

CLANRYE INTERNATIONAL

Contents

Preface VII

Chapter 1 **Design of Circular Dammann Grating: Fabrication and Analysis** 1
Fung Jacky Wen and Po Sheun Chung

Chapter 2 **Fabrication of Binary Diffractive Lens on Optical Films by Electron Beam Lithography** 21
Atsushi Motogaito and Kazumasa Hiramatsu

Chapter 3 **Application of Soft Lithography and Micro-Fabrication on Neurobiology** 31
Gao Kan, Chen Haifeng, Liu Bing-Fang and Xu Qun-Yuan

Chapter 4 **Photocontrolled Reversible Dimensional Changes of Microstructured Photochromic Polymers** 49
Despina Fragouli, Roberto Cingolani and Athanassia Athanassiou

Chapter 5 **Fabrication of Surfaces with Bimodal Roughness Through Polyelectrolyte/Colloid Assembly** 65
Christine C. Dupont-Gillain, Cristèle J. Nonckreman, Yasmine Adriaensen and Paul G. Rouxhet

Chapter 6 **Approach to EUV Lithography Simulation** 83
Atsushi Sekiguchi

Chapter 7 **Physical Deposition Assisted Colloidal Lithography: A Technique to Ordered Micro/Nanostructured Arrays** 103
Yue Li, Shuyan Gao, Guotao Duan, Naoto Koshizaki and Weiping Cai

Chapter 8 **Application of Nanoimprint Lithography to Distributed Feedback Laser Diodes** 149
Masaki Yanagisawa

Chapter 9 **High-Index Immersion Lithography** 163
Keita Sakai

Chapter 10 **Scanning Probe Lithography on Organic Monolayers** 183
SunHyung Lee, Takahiro Ishizaki, Katsuya Teshima, Nagahiro Saito and Osamu Takai

Chapter 11 **Extremely Wetting Pattern by Photocatalytic Lithography and its Application** 213
Yuekun Lai, Changjian Lin and Zhong Chen

Permissions

List of Contributors

Index

Preface

This book has been a concerted effort by a group of academicians, researchers and scientists, who have contributed their research works for the realization of the book. This book has materialized in the wake of emerging advancements and innovations in this field. Therefore, the need of the hour was to compile all the required researches and disseminate the knowledge to a broad spectrum of people comprising of students, researchers and specialists of the field.

Lithography refers to a planographic technique of printing that was formerly based on the immiscibility of water and oil. The printing is done on a smooth-surfaced stone or metal plate. Due to the requirement of high functional miniaturized systems in diverse fields, techniques of lithography have been developed and implemented for their intended goals in numerous fields. These include biomedical researches, semiconductor development, and chemical and biological analysis. Lithography plays an important role in nanofabrication and microfabrication techniques. Microfabrication is the technique of fabricating microscopic structures on the micrometer and smaller scales. Nanofabrication refers to a technique utilized for creating one, two or three-dimensional nanostructures in the size range of 1-100 nanometers. Microfabrication and nanofabrication are critical to the growth of miniaturized systems. This book elucidates the concepts and innovative models around prospective developments with respect to lithography and its application in microfabrication and nanofabrication. It will help the readers in keeping pace with the rapid changes in this field of study.

At the end of the preface, I would like to thank the authors for their brilliant chapters and the publisher for guiding us all-through the making of the book till its final stage. Also, I would like to thank my family for providing the support and encouragement throughout my academic career and research projects.

Editor

1

Design of Circular Dammann Grating: Fabrication and Analysis

Fung Jacky Wen and Po Sheun Chung
Department of Electronic Engineering, City University of Hong Kong
China

1. Introduction

The term "diffraction" has been defined by Sommerfeld as follows (E. Hecht, 2002): any deviation of light rays from rectilinear paths which cannot be interpreted as reflection or refraction. Diffraction is caused by the confinement of the lateral extent of a wave and is most importantly when that confinement is comparable in size to the wavelength of the radiation being used. The first accurate report and description of such a phenomenon was made by Grimaldi and was published in the year 1665. Later, in 1678, Christian Huygens expressed the intuitive conviction that of each point on the wavefront of a disturbance were considered to be a new source of a secondary spherical disturbance. This technique, however, ignores most of each secondary wavelet and retains only that portion common to the envelope. As a result of this inadequacy, this principle is unable to account for the details of the diffraction process. The difficulty was resolved by Fresnel and Kirchhoff with his addition of the concept of interference in late 18th century. These types of diffraction are known for two centuries in the form of diffraction gratings which periodically modulate the incident wave-front. An ideal grating generates a set of waves, called diffraction orders that propagate into discrete directions. The diffraction angles θ_m are given by the well-known grating equation $\sin\theta_m = m\lambda/D$, where λ is the wavelength of light and D is the grating period, is shown in Fig. 1. The amplitudes of the diffraction orders are determined by the structure of the periodic modulation. The demand for electromagnetic analysis is arising together with the advance of the fabrication technology. A significant step was taken as well since the development of computers. It can compute such a complicated wave field numerical analysis from which the design a desired grating is much easier.

The DOE is designed for splitting the input beam into M diffraction orders. For array illuminators, equal power intensity with high uniformity is necessary. To achieve this special feature, a periodic nature together with binary phase structure, which was proposed by Dammann in the early 70's, is one of the solutions (H. Dammann & K. Gortler, 1971; H. Dammann & E. Klotz, 1977). To analyze the performance of the Dammann grating, we employ the Burckhardt, Kaspar and Knop (BKK) method and TE-polarized dependent mode is normally assumed (C. B. Burckhardt, 1966). In order to have an easier understanding, the entire Dammann grating has the identities in terms of periodic, symmetric and binary structure. The total normalized diffraction efficiency is just the sum of all required normalized diffraction power. Normally, for reducing the zero-order intensity and polarization dependent loss, the ratio of period/wavelength should be as large as possible. We have also employed Dammann Grating as a beam splitter (J. F. Wen & P. S. Chung, 2007; J. F. Wen & P. S. Chung, 2007).

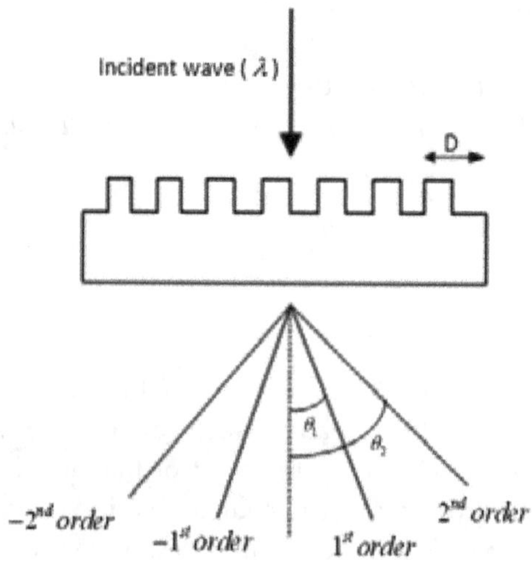

Fig. 1. Diffraction Angle for DOE.

2. Introduction to Circular Dammann Grating (CDG)

In previous section, we know that light could be diffracted into multiple orders with equal intensities and high uniformity with the theory of Dammann grating. The diffracted spots could be one-dimensional or two dimensional depending on the applications. For some applications, optical systems need circular images, e.g. laser free space communication system (J. Jia, C. Zhou & L. Liu, 2003), fast focal length measurement (S. Zhao, J. F. Wen & P. S. Chung, 2007; S. Zhao & P. S. Chung, 2007) and usage in DFB laser (C. Wu *et al,* 1991; T. Erdogan *et al,* 1992). We have then further extended the Dammann grating into CDG. The CDG is one of the possible candidates which can produce circular beams in ring-shape at the image plane.

Zhou et al (C. Zhou, J. Jia & L. Liu, 2003) first proposed the concept of CDG in 2003 based on the modulation of the Bessel function using a binary phase annulus mask. The phase and radius of each annulus can be modified so that the intensity at the far field can be manipulated. However the CDG does not have the periodic nature as most of the gratings required and therefore, it is only a DOE and equal separation cannot also be achieved. Recently Zhao and Chung (S. Zhao & P. S. Chung, 2007) proposed a new design method for the periodic CDG using the coefficients of the circular sine series for generating equal-intensity and equal-spacing of optical rings, which means those infinite circular periods can be repeated. We have also presented another two novel approaches based on the concept of circular spot rotation (J. F. Wen, S. Y. Law & P. S. Chung, 2007), Hankel transform (J. F. Wen & P. S. Chung, 2008) to achieve the same objectives as mentioned above with higher efficiency and uniformity. In our research, we have employed the Circular Dammann Grating into angle, area and distance measurements respectively (J. F. Wen & P. S. Chung, 2008; J. F. Wen, Z. Y. Chen & P. S. Chung, 2008; J. F. Wen, Z. Y. Chen, & P. S. Chung, 2010).

2.1 Circular spot rotation method

The concept of generating CDG using this method is based on the theory of conventional Dammann Grating (C. Zhou & L. Liu, 1995). The cross section and first order spectrum of CDG is shown in Fig. 2 and 3 respectively. We assumed that if the diffraction spots rotate 360 degrees continuously, circular rings will be formed. Fig. 4 and 5 illustrate these ideas. Table 1 shows some numerical results. For the formula of Circular Dammann Grating, the intensities among different orders will then be as follows:

$$n^{th}\,order:\; P_n = |M(n)|^2 = \frac{1}{n^2\pi^2}\{[\sum_{k=1}^{N}(-1)^k \cos(2\pi n r_k)]^2 + [\sum_{k=1}^{N}(-1)^k \sin(2\pi n r_k)]^2\} \tag{2.1}$$

$$0^{th}\,order: \quad P_0 = \left[2\sum_{k=1}^{N}(-1)^k r_k + 1\right]^2 \tag{2.2}$$

The overall normalized efficiency is therefore given as,

$$\eta = \sum_{n=-\infty}^{\infty} nP_{2n-1} \tag{2.3}$$

The feature size, which is the minimum distance between two different continuous transition points, is

$$\Delta = \min|r_{k+1} - r_k| \tag{2.4}$$

The uniformity is defined as,

$$uni = \frac{\max(I_n) - \min(I_n)}{\max(I_n) + \min(I_n)} \tag{2.5}$$

The radius (R) of each ring with the focal length (f) of converging lens will then be

$$R_n = \frac{n\lambda}{D} f \tag{2.6}$$

The numerical solutions with near optimum efficiency and uniformity of CDG are listed in Table 1.

Circle number	Normalized Transition Points in Half of Period*	Efficiency	Uniformity	Feature Size
1	0.5	0.81	0	0.5
2	0.20525 0.29067 0.5	0.72	0.00006	0.08542
3	0.11649 0.24024 0.26741 0.38396 0.5	0.74	0.0002	0.02717
4	0.099104 0.18382 0.26295 0.32925 0.4196 0.49492 0.5	0.8	0.00014	0.00508

*The other half period could be calculated using $x_{k+i} = x_i + 0.5$

Table 1. Some Numerical Solutions of CDG by Spot Rotation Method.

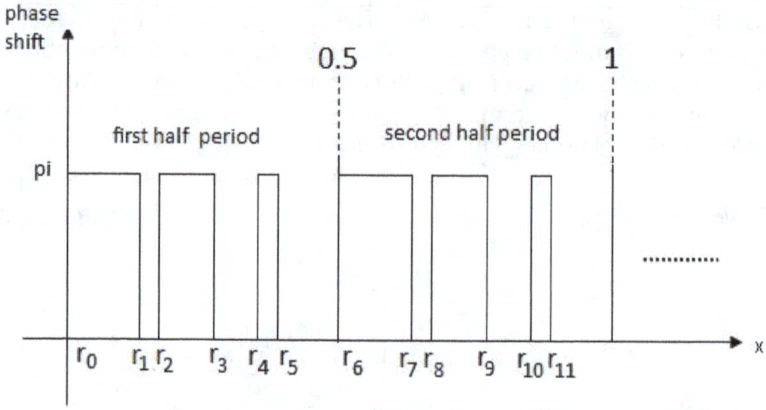

Fig. 2. Cross Section of CDG.

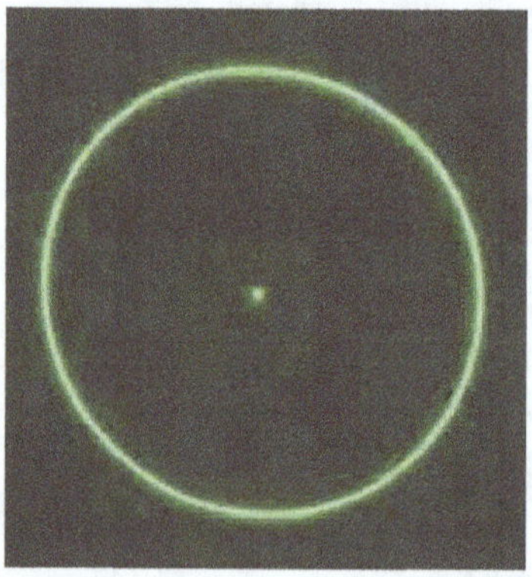

Fig. 3. First order CDG spectrum.

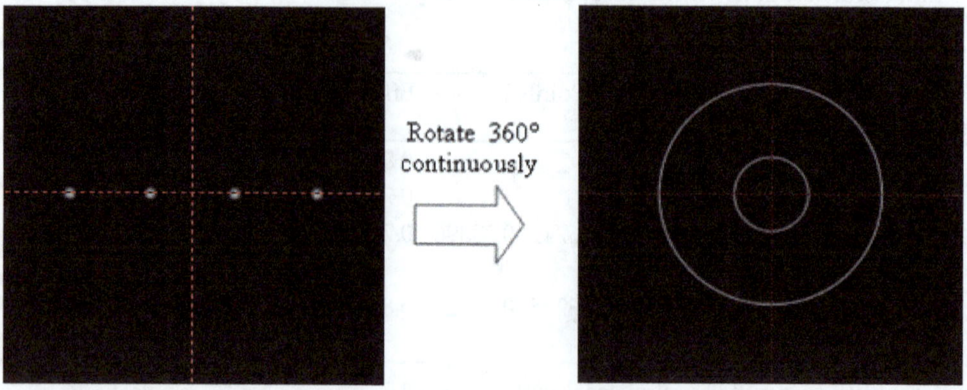

Fig. 4. Concept of Circular Spot Rotation Method's type CDG.

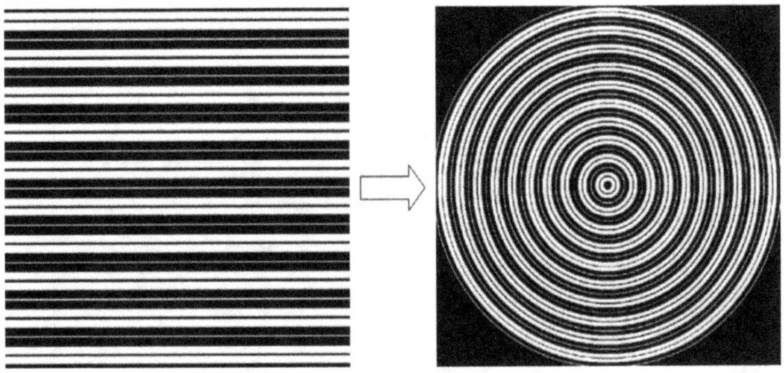

Fig. 5. Circular Spot Rotation Method's type CDG Profile Design.

2.2 Hankel transform

The concept of this grating is based on the theory of Hankel transform. It is a two dimensional Fourier transform with a radially symmetric integral and is also called the Fourier-Bessel transform (F. Bowman, 1958). The cross section of CDG is still same as Fig. 2. The problem of side lobe effect is one of the major concerns in both Zhao and our circular rotation methods. The effect is shown in Fig. 6. In this section, we present another novel approach based on the Bessel function together with Hankel transform in symmetry structure to achieve the same objectives with better performance in terms of efficiency and uniformity while the lower side lobe power could be obtained. Table 2 shows some numerical results. The comparisons are shown in Fig. 7-9.
The intensities equations for each diffraction order are given as follows:

$$n^{th} order: \; I_q = |M(q)|^2 = \frac{1}{q^2}\left|2\sum_{k=1}^{N}(-1)^{k+1} r_k J_1(2\pi q r_k) + r_N J_1(2\pi q r_N)\right|^2 \tag{2.7}$$

$$0^{th} order: \; I_0 = \left[2\sum_{k=1}^{N}(-1)^k r_k + 1\right]^2 \tag{2.8}$$

Circle Number	Normalized Transition Points in Half of Period*			Feature[1]	Efficiency[2] (Main Lobe)	Overall Efficiency[3]	Uniformity
1	0.39763	0.4907	0.5	0.0093	0.665	0.931	
2	0.091023	0.3266	0.5	0.0610	0.612	0.855	0.043
3	0.10036	0.23569	0.26741	0.0353	0.60	0.866	0.05
	0.38396	0.5					
4	0.099572	0.1838	0.26314	0.0038	0.69	0.99	0.036
	0.32894	0.41929	0.49616				
	0.5						

Note:
1: Feature is the minimum distance between two subsequent transition points
2: It describes the sum of powers in terms of main lobes only.
3: It describes the sum of all powers including the main and side lobes.
*: The other half period could be calculated using $x_{k+i} = x_i + 0.5$

Table 2. Some Numerical Solutions of CDG by Hankel Transform.

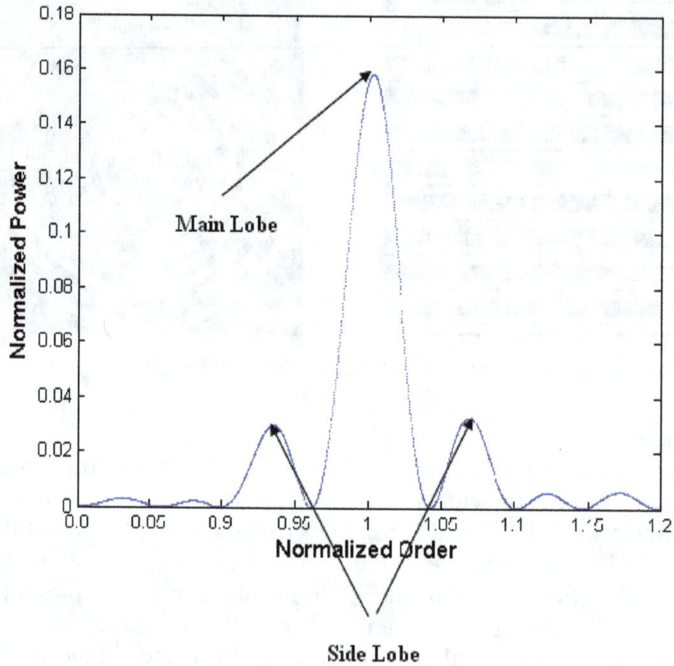

Fig. 6. Side Lobe Effect of CDG.

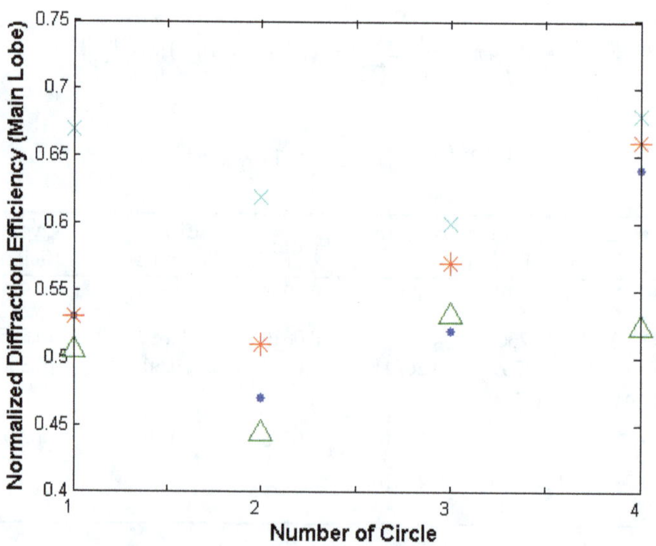

Fig. 7. Main Lobe Intensities Comparisons among Four Different Methods (Triangle: Zhao's symmetric method, Dot: Zhao's asymmetric method, Star: Spot Rotation Method, Cross: Hankel Transform).

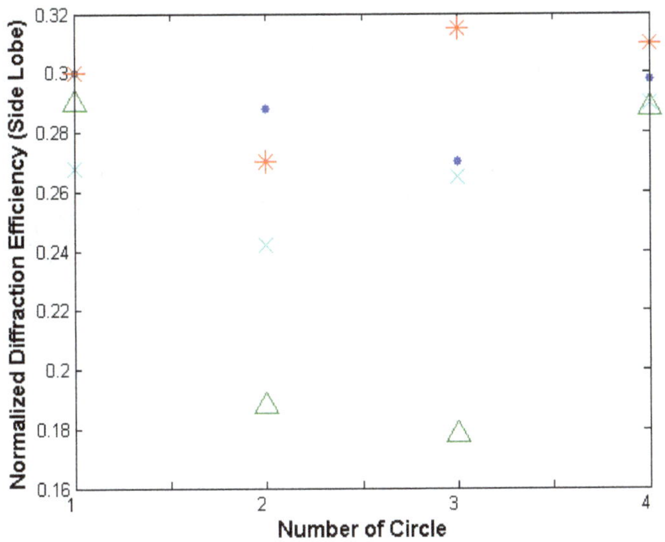

Fig. 8. Side Lobe Intensities Comparisons among Three Different Methods (Triangle: Zhao's symmetric method, Dot: Zhao's asymmetric method, Star: Spot Rotation Method, Cross: Hankel Transform).

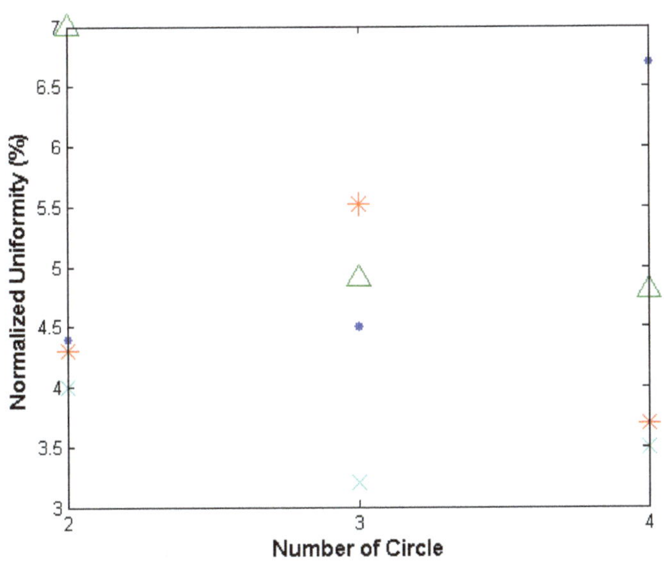

Fig. 9. Uniformities Comparisons among Three Different Methods (Triangle: Zhao's symmetric method, Dot: Zhao's asymmetric method, Star: Spot Rotation Method, Cross: Hankel Transform).

The overall normalized efficiency, feature size, uniformity and radius are defined same as equations 2.3-2.6.

Side lobe is existed around the main lobe in every diffracted order, is shown in Fig. 6, no matter which method is applied. It is because our design is not able to fully express the concept of circular phase modulation, as the circular profile can not be completely decomposed into a square pixel representation and also finite number of periods happened in practical case. Thus side lobe existed. The lobe separation is not only governed by number of period (*ND*), but also controlled by the input wavelength and the focal length of the lens. With the assistance of diffraction theory, the final equation is then defined, i.e.

$$s = \frac{2\lambda}{3ND} f \tag{2.9}$$

Fig. 10-12 shows this relationship. From these figures, we can conclude that 100 periods is the optimum solution.

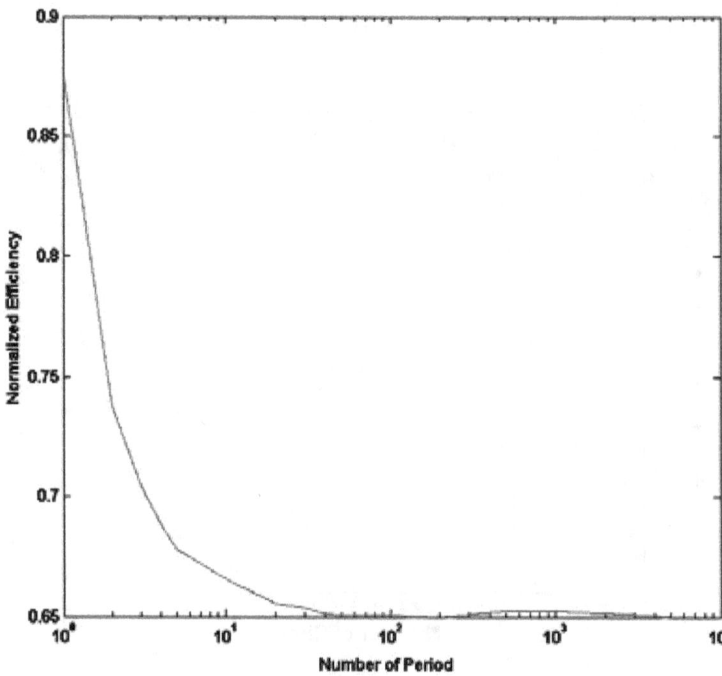

Fig. 10. Number of Period against efficiency.

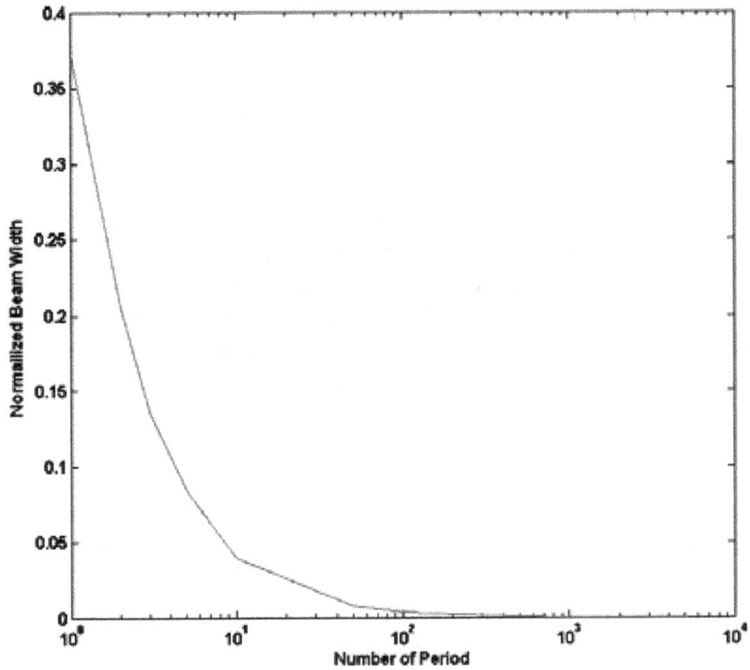

Fig. 11. Number of period against beam width (FWHM).

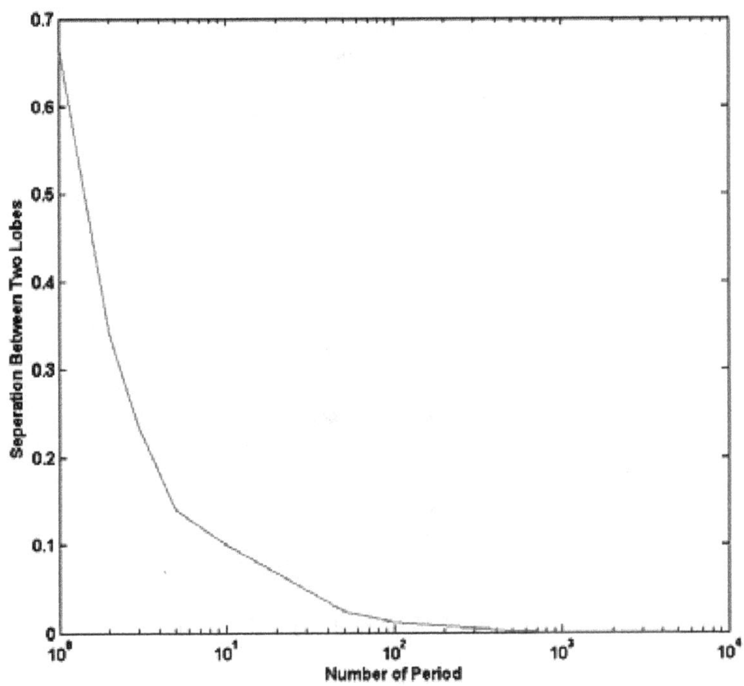

Fig. 12. Number of Period against separation between two lobes.

3. Fabrication

Many different fabrication methods exist for diffraction grating. Most of these techniques can be grouped into two main categories: lithographic techniques and electron beam writing. Lithographic techniques (J. Turunen, A Vasara, J Westerholm, G Jin and A Salin, 1990) use light sensitive polymers at the top of the substrate along with controlled etching or deposition methods. For standard e-beam lithography (Masato Okano, Hisao Kikuta, Yoshihiko Hirai, Kazuya Yamamoto, and Tsutom Yotsuya, 2004), an e-beam exposure contains enough of an electron dose that the exposed regions of e-beam resist are fully cleared during the development process. This can be used to produce different thicknesses of e-beam resist simply by varying the dose. The e-beam approach can generate the finest features, a serval tens of nm. However, because of the small size of the electron beam, it is extremely time consuming to expose a pattern of large size of sample.

3.1 Chemical etching fabrication

In previous sections, we have understood the fundamental theories of diffraction and a number of design approaches for CDG. To fabricate these gratings, we have to maintain the desired grating shape which is critical to guarantee high transmission of the order of interest. An optimal depth is very important since existing zero order transmission will happen afterwards. Other defeats, such as mask misalignment, sharpness of the profile, will scatter light to higher orders. Because of the size of the features and the need for flexibility of fabrications, lithographic method is an optimal for many types of DOE fabrication [M. T. Gale, 1997; J. M. Miller et al, 1993; C. Zheng, 2005). A simplified version of the photolithographic processing sequence is shown in Fig. 13.

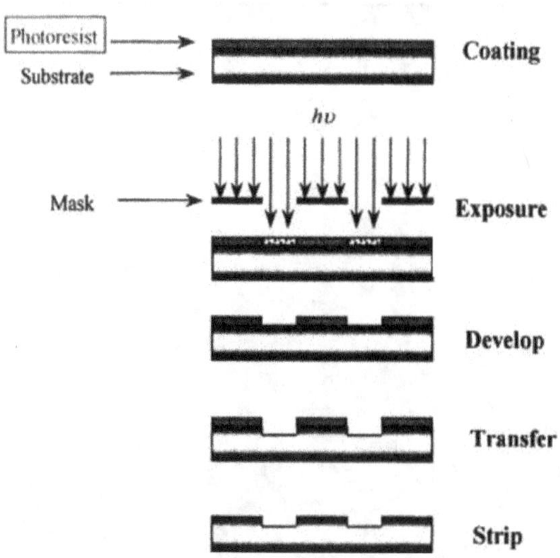

Fig. 13. Flow Diagram of chemical etching.

Once we confirm the pattern which is determined from the design process, the depth of the diffractive phase structure, which is determined by the wavelength of the incident light and the refractive indices of the substrate and surrounding material, can be calculated. The depth can be calculated by

$$h = \frac{\lambda}{2(n_1 - n_0)} \tag{3.1}$$

where λ is the input wavelength and n_1, n_0 are the indices of refraction of the substrate material and the surrounding medium at the operating wavelength respectively. The substrate in using is quartz and the refractive index is shown in Fig. 14.

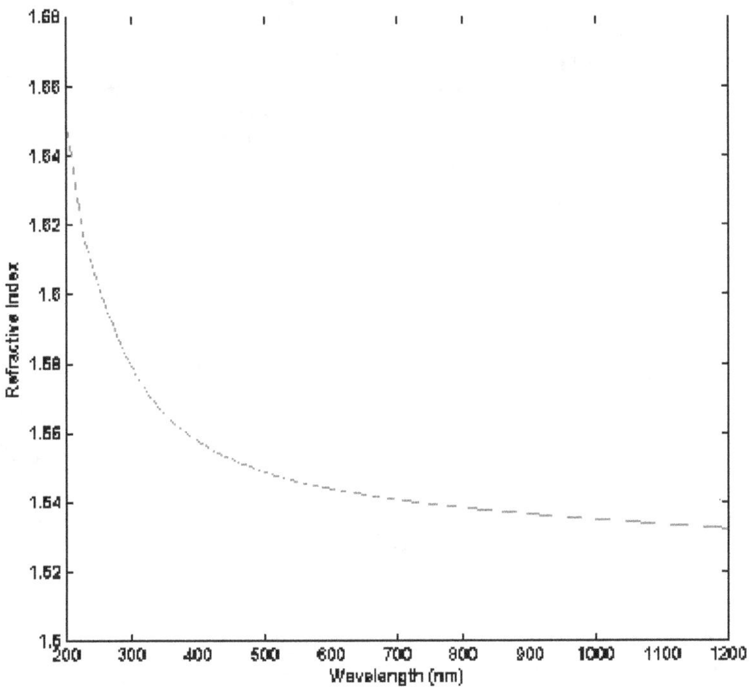

Fig. 14. Refractive Index of quartz.

Most lithographic masks are binary transmission masks. That is, they contain alternating clear and opaque areas. The opaque areas mean the Chromium remains on top of the substrate. These masks are usually made by forming the pattern in a light-sensitive photoresist on top of thin chrome later on the glass mask. Once the photoresist is developed, the chrome, where the photoresist has been removed, was protected. The mask pattern is exposed using optical pattern generators with controllable beam size. A variety of file formats, e.g. GDSII, CIF and BMP, can be used. The machine in our lab is "Microtech LW405". The positioning accuracy is 1um and the minimum linewidth is 0.8um. These patterns are transformed into pixel forms with a given dimension and the magnified mask pattern is shown in Fig. 15. The accuracy of patterning curve is related to the wavefront error introduced by the required shape approximation. Increasing the number of pixels can help to improve the accuracy. However, as we expect, the higher degree of accuracy can result of more time spending and the amount of data is also increased. Once the mask is fabricated, which is shown in Fig. 16, we can move on to the next step.

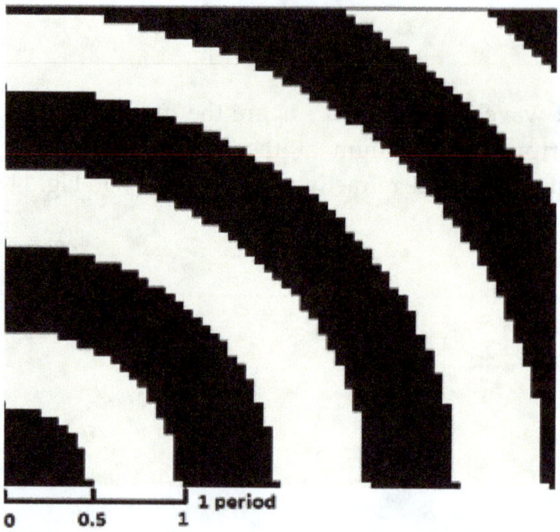

Fig. 15. Magnified CDG pattern.

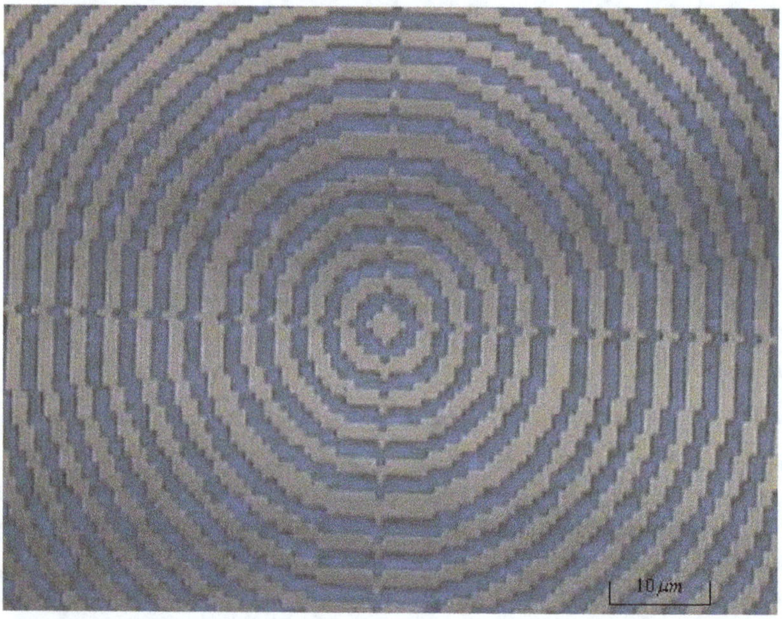

Fig. 16. CDG Pattern with the number of periods 100.

Photolithographic methods are based on the use of photoresist create relief structures on substrate surfaces. This structure is used to protect the underlying substrate during subsequent processing steps. Photoresists, which is light sensitive polymer, can be either positive, where the exposed resist dissolves, or they can be negative, where the exposed resist remains after development. Positive photoresists (e.g. SPR6112B) is used in the following processing example. After substrate is cleaned, the first step is to coat the substrate with a thin (typically microns) layer of photoresist. It can be generated by spin coat the wafer. During spin coating, liquid photoresist is distributed uniformly around the wafer as it rotates at high rates. The spin curve of this photoresist is shown in Fig. 17.

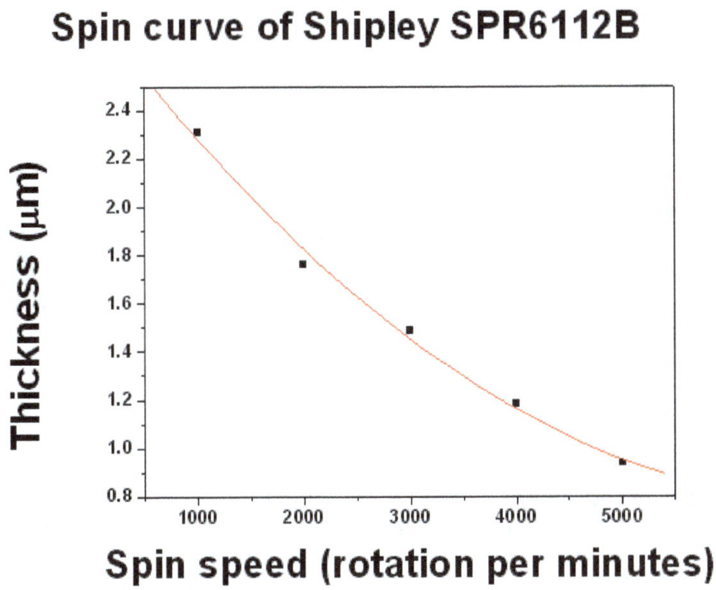

Fig. 17. Spin Curve for SPR6112B.

In order to have a better evaporation of phtoresist, it is often to be heated at a hundred degree Celsius. The final thickness of the photoresist layer is controlled by a combination of the viscosity of the photoresist and the spin speeds used during the coating process. The next step is the exposure of the photoresist. Patterns will be formed at the photoresist layer using Aligner, a uniform ultraviolet light source. The mask can be placed in contact with the photorsist layer for a high resolution, 1:1 transfer of the image scale, this process is referred to as contact printing. The exposed photoresist is washed away after exposure. Developing solution and developing time also affect the fidelity of the resulting lithographic image. Over, or underdevelopment will decrease the fidelity of the image pattern. For example, while exposed resist dissolves much faster than unexposed areas, the un-irradiated areas will also begin to lose photoresist if the development time is too long. The simplified illustration is shown in Fig. 18.

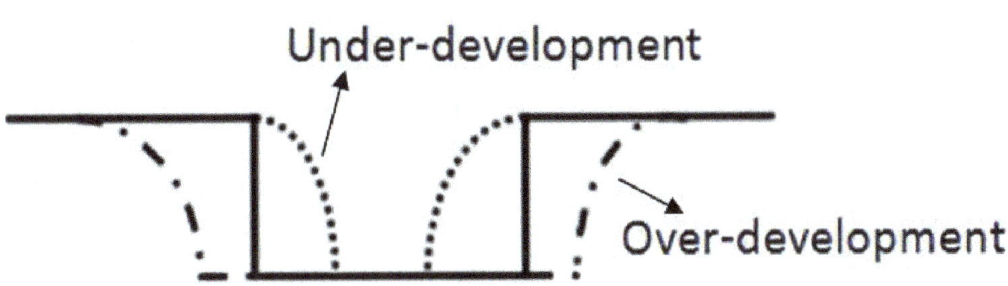

Fig. 18. Diagram of different types of development.

The wafer has been normally spin coated with Chromium with SiO_2 as the substrate. This material can be functioned from the ultraviolet through the near infrared due to its own transmission properties and low coefficient of thermal expansion. Normally, a photoresist pattern serves as a wall that protects the area under the photoresist during etching. In our experience, a single photoresist layer without any other protection cannot stand for a long period of time. In this case, we add one more chromium layer which the purpose is same as photoresist. As a result, only the areas not covered by the chromium are removed during the etching process.
The wet etching technique is isotropic, which its definition is the etching rate is equal in all directions. This desires for our grating case, particularly for applications requires sharply defined and vertical features. Standard diffractive etch chemistries for silica, usually HF acid, can act as an etching solution. In our experiment, we use another solution, named as Fluorosilicic acid. This etching solution offers a straightforward way to create smooth features in glass materials. In addition fluorosilicic acid etching is considerably safer than conventional hydrofluoric acid (HF) etching. Normally, the temperature controlled heating stage was set at 60°C, which gave an etching solution temperature of 30°C (±0.5°C). Fluorosilicic acid, 100 ml (20% concentration) was allowed to obtain an etch rate 56 nm/min in our experience.

3.2 Electron beam lithography

Electron beam lithography (EBL) is based on the principle that some polymers are sensitive to electrons and can be patterned by electron exposure, which is very much like the other lithography. PMMA (Poly Methyl Methacrylate) is normally used in EBL. Its own low exposure time and low resolution are the limitations for our case. We therefore shifted to another resist named "ZEP 520A" by ZEON corporation. ZEP520A is high performance positive EB resists which show high resolution and dry etch resistance. The spin curve and refractive index of this resist are shown in Fig. 19 and 20. The schematic of electron beam system is shown in Fig. 21.

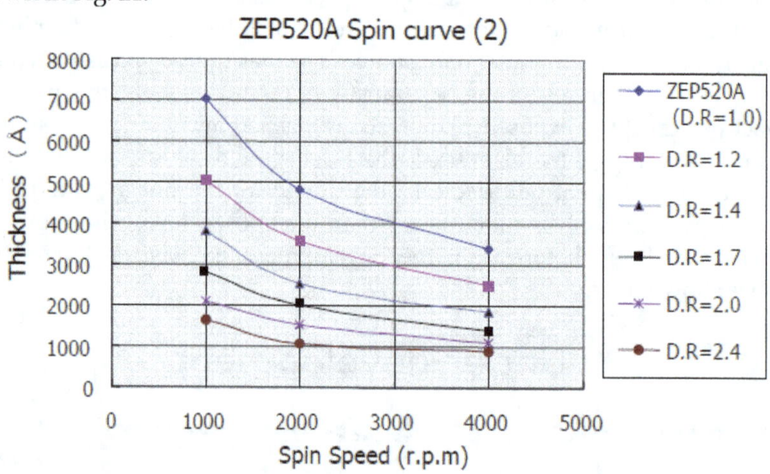

D.R(Dilution Rate)
={Original Resist(g)+Solvent(g)}/original Resist(g)
(Weight Ratio)

Fig. 19. Spin Curve for ZEP520A with Dilution Rate (provided by ZEON).

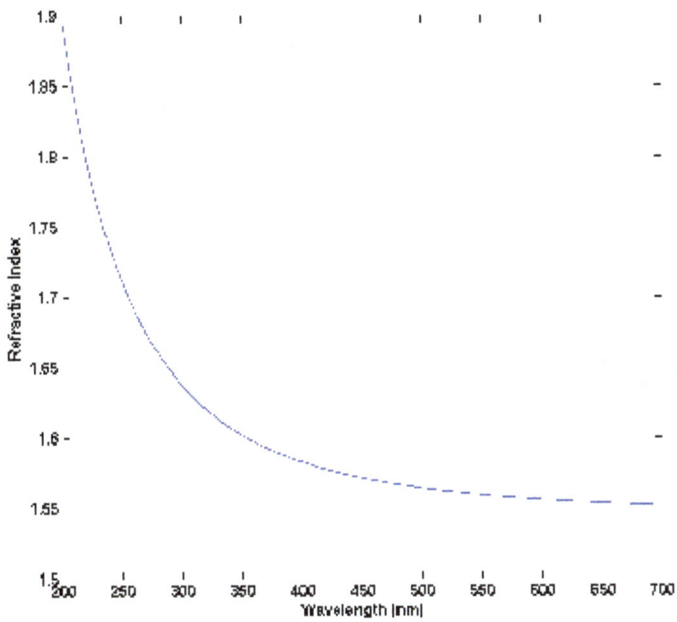

Fig. 20. Refractive Index of ZEP520A (provided by ZEON).

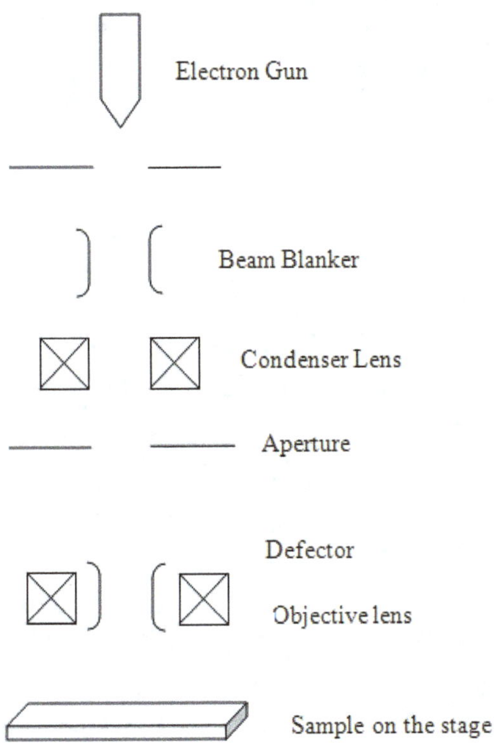

Fig. 21. Simplified System of EBL.

At the top, it consists of an electron gun, a condenser lens to allow changes in the current and corresponding beam diameter, an objective lens to focus the beam on the wafer and a deflector to scan the e-beam around within the field. The sample is placed below on a motorized stage so that it can be patterned by the desired profile. More detailed descriptions of the different types of sources, lenses and the various other components can be found in many textbooks (C. Zheng, 2005). The simplified flow diagram is shown in Fig. 22. The CDG sample is also shown in Fig. 23. The machine in our lab is "Crestec CABL-9510C".

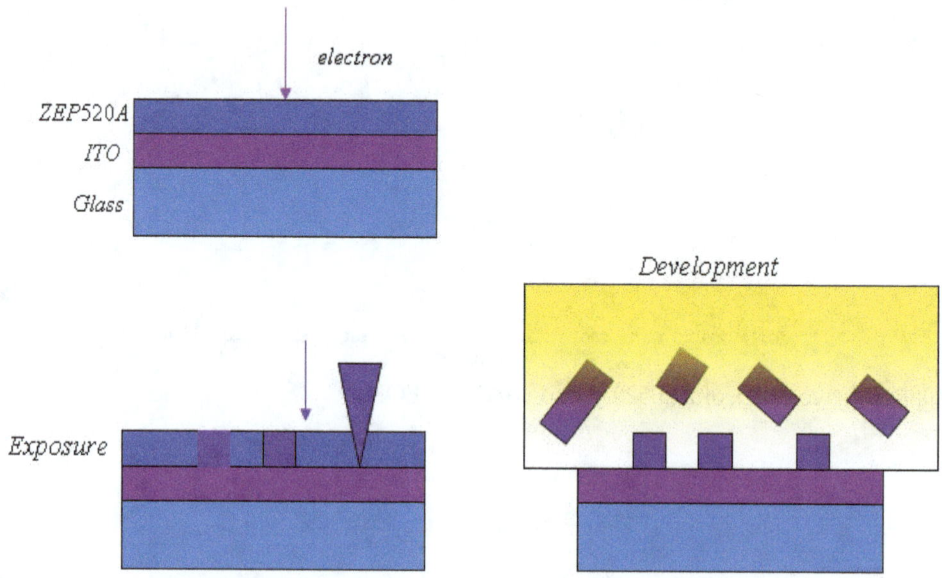

Fig. 22. Flow diagram of EBL.

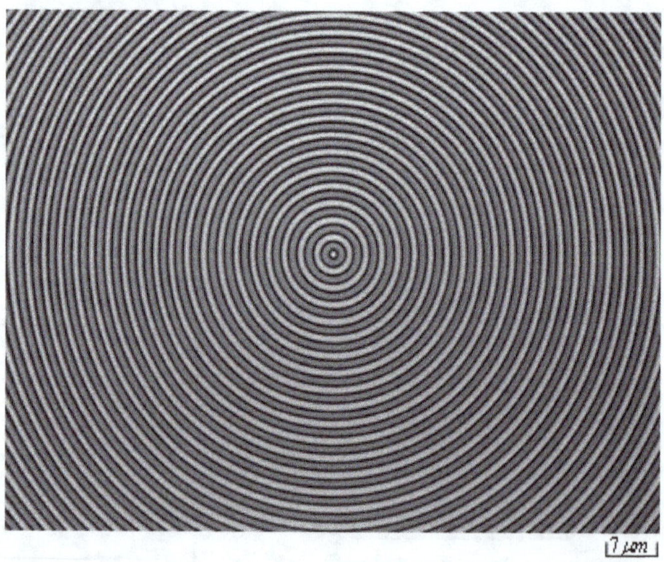

Fig. 23. CDG sample using EBL.

4. Conclusion

CDG is affected by the different errors, such as etched depth errors, feature errors, grating period deviations and non vertical side wall angle. Most of these errors will affect the distribution of light into diffracted order. The effects of variation in phase depth and grating duty cycle for a grating are shown in Fig. 24-26 respectively. In general, from the following figures, we can conclude that for getting over 60% efficiency, 100nm variation is allowed in both techniques.

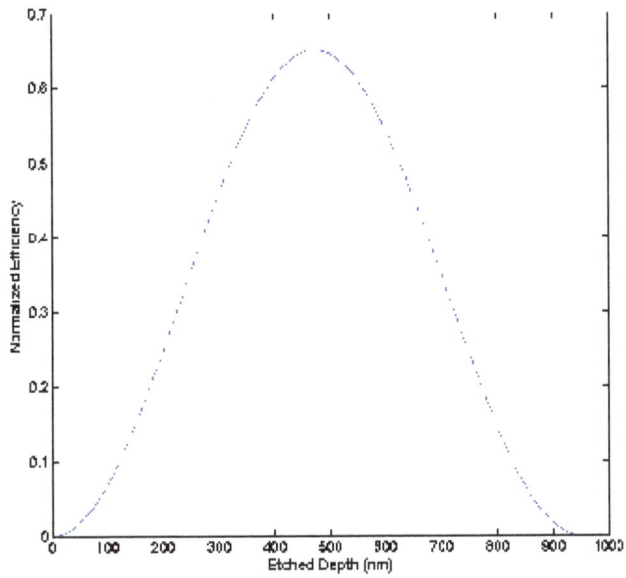

Fig. 24. Etched Depth against efficiency with photolithography.

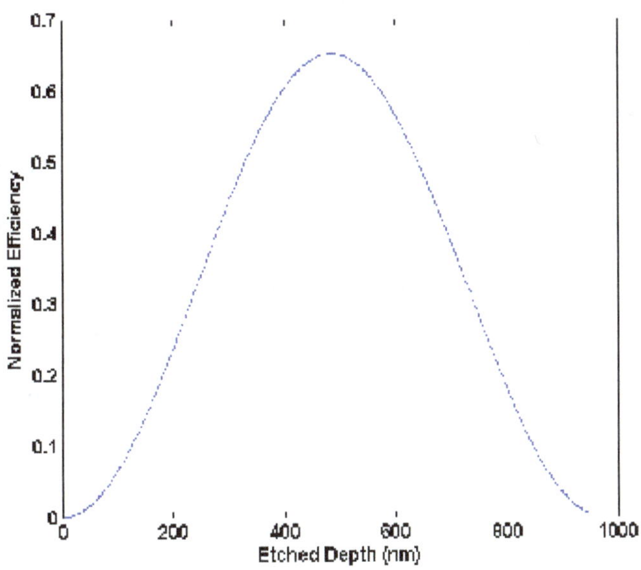

Fig. 25. Etched Depth against efficiency with EBL.

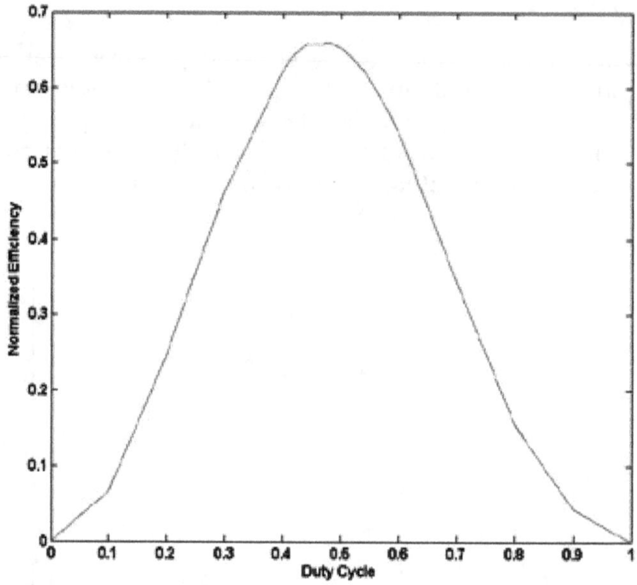

Fig. 26. Duty Cycle against efficiency.

Although fabrication errors on these structures will redistribute the energy among diffraction orders and increase the efficiency of central order, the total energy remains the same. It is also noted that the choice of fabrication method introduce different errors. For example, EBL tends to have variations in exposure dosage which has excess etch depth and shape errors. But it can provide very fine width and vertical shape walls and hence it can support high density structure. Conversely, chemical etching technique can provide very sharp side walls and accurate etch depth with carful timing. But, it never supports high density structure as the linewidth it can sustain is in order of um. The users must to balance between the cost and performance requirements.

5. Acknowledgement

This project is supported by GRF (No. 112710) of the Research Grant Council in Hong Kong.

6. References

E. Hecht (2002). *Optics*, Addison Wesley, ISBN 0-80-538566-5, US

C. B. Burckhardt (1966). Diffraction of a plane wave at a sinusoidally stratified dielectric grating. *J. Opt. Soc. Am.* Vol. 56 pp.1502–1509, ISSN 0030-3941

M. G. Moharam & T. K. Gaylord (1982). Diffraction Analysis of Dielectric Surface Relief Gratings. *Journal of the Optical Society of America,* Vol. 72, pp.1385-1392, ISSN 0030-3941

H. Dammann & K. Gortler (1971). High-efficiency in line multiple imaging by means of multiple phase holograms. *Optics Communications,* Vol.3, pp.312-315, ISSN 0030-4018

H. Dammann & E. Klotz (1977). Coherent optical generation and inspection of two- dimensional periodic structures. *Optica Acta,* Vol.4, pp.505-515, ISSN 0950-0340

J. Jahns *et al* (1989). Dammann gratings for laser beam shaping. *Optical Engineering,* Vol.28, pp.1267-1275, ISSN 0091-3286

J. Turunen *et al* (1990). Optimization and fabrication of grating beam splitters. *Journal of Physics D: Applied Physics,* Vol.21, pp.102-105, ISSN 0022-3727

J. F. Wen & P. S. Chung (2007). 2D optical splitters with Polymer Optical Fiber Arrays. *Journal of Optics A: Pure and Applied Optics,* Vol.9, pp.723-727 , ISSN 1464-4258

F. J. Wen & P. S. Chung (2007). Two-dimensional optical splitters with polymer optical fibre arrays. *Journal of Optics A: Pure and Applied Optics,* Vol.9, pp.723-727, ISSN 1464-4258

J. Jia, C. Zhou & L. Liu (2003). Superresolution technology for reduction of the far-field diffraction spot size in the laser free space communication system. *Optics Communications,* Vol.228, pp.271-278, ISSN 0030-4018

S. Zhao, J. F. Wen & P. S. Chung (2007). Simple focal length measurement technique with a circular Dammann Grating. *Applied Optics,* Vol.46, pp.44-49, ISSN 0003-6935

S. Zhao & P. S. Chung (2007). Collimation testing using a circular Dammann grating. *Optics Communications,*Vol.279, pp.1-6, ISSN 0030-4018

C. Wu *et al* (1991). Optically pumped surface-emitting DFB GalnAsP/InP lasers with circular grating. *Electronics Letters,* Vol.27, pp.1819-1821, ISSN 0013-5194

T. Erdogan *et al* (1992). Circularly-symmetric operation of a concentric-circle grating surface-emitting AlGaAs/GaAs quantum-well semiconductor laser. *Applied Physics Letters,* Vol.60, pp.1921-1923, ISSN 0003-6935

C. Zhou, J. Jia & L. Liu (2003). Circular Dammann Grating. *Optics Letters,* Vol.28, pp.2174-2176, ISSN 0146-9592

S. Zhao & P. S. Chung (2007). Design of Circular Dammann Grating, *Optics Letters,* J. F. Wen, S. Y. Law & P. S. Chung (2007). Design of Circular Dammann Grating by employing the Circular Spot Rotation Method. *Applied Optics,* Vol.46, pp.5452-5455, ISSN 0003-6935

J. F. Wen & P. S. Chung (2008). A new Circular Dammann Grating Using Hankel Transform. *Journal Optics A: Pure and Applied Optics,* Vol.10, 075306, ISSN 1464-4258

J. F. Wen & P. S. Chung (2008). The use of Circular Dammann Grating for angle measurement. *Applied Optics,* Vol.47, pp.5197-5200, ISSN 0003-6935

J. F. Wen, Z. Y. Chen, & P. S. Chung (2010). Area measurement at long-distance using a circular Dammann grating. *Applied Optics,* Vol.49, pp.648-652, ISSN 0003-6935

J. F. Wen, Z. Y. Chen & P. S. Chung (2008). A Novel Distance Measurement Technique based on Optical Fractional Fourier Transform, *Proceedings of 13th Opto-Electronics and Communications Conference,* P-42, ISBN 978-0-85825-807-5, Syndey, Australia, July 7-10, 2008

C. Zhou & L. Liu (1995). Numerical study of Dammann array illuminators. *Applied Optics,* Vol.34, pp.5961-5969, ISSN 0003-6935

F. Bowman (1958). *Introduction to Bessel functions.* Dover, ISBN 0486604624, New York, USA

J. Turunen et al (1990). Optimization and fabrication of grating beam splitters. *Journal of Physics D,* Vol. 21, pp.102-105, ISSN 0022-3727

Masato Okano et al (2004). Optimization of Diffraction Grating Profiles in Fabrication by Electron-Beam Lithography. *Appl. Opt.*, Vol. 43, pp. 5137-5142, ISSN 0003-6935

M. T. Gale (1997). Replication, In: *Binary optics fabrication in Micro-optics: Elements, Systems and Applications,* H. P. Herzig (Ed.), Taylor and Francis, ISBN 0748404813, London, England

J. M. Miller et al (1993). Multilevel grating array generators: fabrication error analysis and experiments. *Applied Optics*, Vol.32, pp.2519-2525, ISSN 0003-6935

C. Zheng (2005). *Micro-Nanofabricaton Technologies and Applications*, Springer, ISBN 7-04-017663-7, China

2

Fabrication of Binary Diffractive Lens on Optical Films by Electron Beam Lithography

Atsushi Motogaito and Kazumasa Hiramatsu
Graduate School of Engineering, Mie University
The Center of Ultimate Technology on nano-Electronics, Mie University
Japan

1. Introduction

Two types of lenses can focus light: an optical lens using refraction phenomenon and a diffractive lens using diffraction phenomena. Table 1 shows the characteristics of each lens. The focal length of the diffractive lens is controlled by the structures of the lens, as mentioned in detail in Section 2.2. This suggests that the focal length of the diffractive lens is independent of refractive index and curvature. Thus, application of diffractive lenses to UV optical elements or thin optical elements is possible.

	Refractive lens	Diffractive lens
Principle	Refraction	Diffraction
Structure	 Radius of curvature R	Close to wavelength  Radius r_{m}
Focal length f	$f = \dfrac{R}{2(n-1)}$	$f = \dfrac{r_m^{\ 2} - (m\lambda)^2}{2m\lambda}$
Control of f	n, R (depending on material)	r_m (m=1,2,···)
Short wavelength light such as UV and X-ray	Difficult	Possible (depending on material)
Thin structure	Difficult	Possible (wavelength order)
Controlling light distribution	Difficult	Possible

Table 1. Comparison of characters between refractive lens and diffractive lens

Recently, the emitting efficiency of light emitting diodes (LEDs) has improved; thus, they are used in lighting devices. To this end, miniaturizing the LEDs for smaller lighting devices and controlling the luminosity of LEDs are required. The conventional oval lamp-type LEDs cannot realize these requirements because the lens height of such LEDs is approximately 5 mm and its distribution of luminosity is determined by its shape. In this study, instead of the oval lamp-type lens, we used the diffractive lens on the optical films, as shown in Fig. 1. If the diffractive lens with short focal length (order of micrometer) can be fabricated, miniaturization of the lens system and consequently the LED lighting devices can be achieved. In order to realize the refractive lenses with the short focal length, large curvature radius is needed, thus making it difficult to realize it easily. Therefore, the diffractive lenses are suitable for realizing the short focal length lenses. Furthermore, by modifying the structure of the diffractive lens, it is easy to control the luminosity of the LEDs and the far-field pattern. Therefore, in this study, we focused on the diffractive lens because it enabled us to reduce the thickness of the lens, control the luminosity distribution of LEDs, and facilitate the realization of the binary structure.

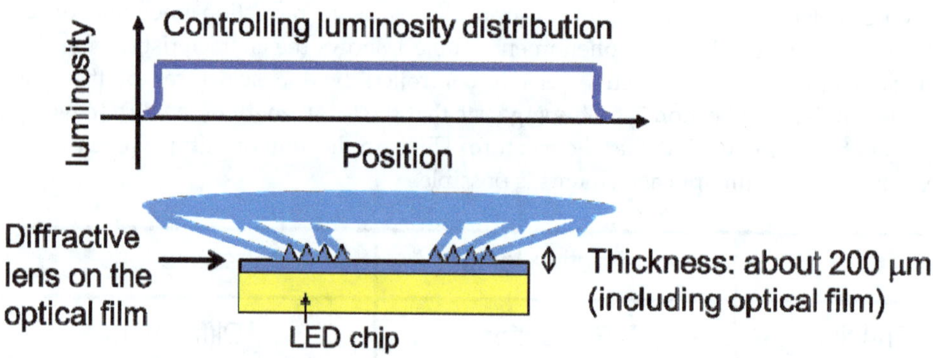

Fig. 1. Schematic representation of diffractive lens on the optical film

The zone plate was the first diffractive lens invented by I. L. Solet in 1875. To improve light efficiency, kinoform was invented by J. A. Jordan (Jordan et al., 1970). Recently, binary optics technology was developed using CAD design and VLSI technology (Swason and Veldkamp, 1989). The diffractive optical elements with multi-level grating having step-like cross-section have been developed. By controlling thestructure of the multi-level gratings, an optical effect almost same as that of the kinoform can be obtained (Orihara et al., 2001 & Yamada et al., 2004).

On the other hand, subwavelength structures (SWSs), which are equivalent to a blazed structure, were suggested by P. Lalanne (Lalanne et al., 1999 & Mait et al., 1999). These structures are fabricated binary SWSs converted from Fresnel lenses. These structures are fabricated easily than those of the multi-level gratings because they can be fabricated by electron beam lithography (EBL) or nanoimprint lithography (NIL). Furthermore, in the case of photolithography, combining some masks is not necessary. By using these structures, achromatized diffractive lenses were reported (Kleemann et al., 2008).

We aim to realize a highly effective short focal length diffractive lens using the binary diffractive lens fabricated by EBL, and expect the equivalent effect with the diffractive lens of the saw-like structure. NIL or photolithography can easily fabricate these structures at low cost and over a large area. However, EBL does not use any molds or masks. Therefore, it is convenient to examine EBL in detail to obtain optimum structures.

In this study, we carry out the design and fabrication of the binary diffractive lens with 2-mm focal lengths for controlling the luminosity distribution and the binary diffractive lens with the 100-μm-order focal length. Furthermore, to improve the diffraction efficiency, we characterize the detailed relationship between the lens structure and the light intensity.

2. Experimental procedure

In this section, the methods of design, fabrication, and characterization of the binary diffractive lens are described.

2.1 Basic optical characteristics of materials related to binary diffractive lens

The binary diffractive lenses, on which this study is focused, were fabricated on the poly(ethylene terephthalate) (PET) films. The PET films are often used as optical sheets for liquid crystal displays. There are many types of optical films such as polycarbonate (PC) and poly(methyl methacrylate) (PMMA). In this study, the EBL process was used for fabricating the binary diffractive lenses; this process required the optical films to endure high temperature and chemicals, making them more suitable than PC or PMMA.

In this study, the binary diffractive lens was fabricated by developing the resist for EBL (ZEP-520A, ZEON Co.) on the PET films (Teijin® Tetoron® Film, Teijin DuPont Films, Japan). If the refractive indexes of both materials are almost same, the binary diffractive lens can be fabricated by developing the resist instead of etching the PET films. Therefore, the refractive indexes of the PET films and the resist are evaluated by ellipsometry (M-2000DI, J.A. Woollam Co., Inc.). Fig. 2 shows the wavelength dispersion of the PET film and the resist on the PET film, including the data from the catalog of ZEP-520A. D2 and halogen lamps were used for this measurement. The refractive index of the PET film is relatively

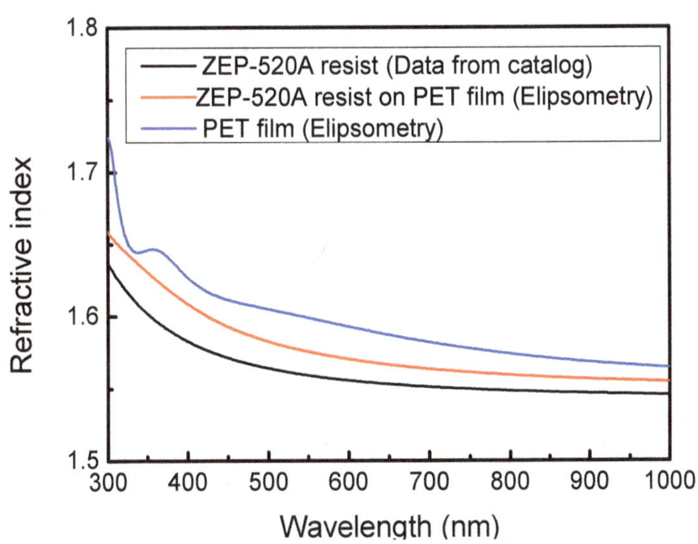

Fig. 2. Wavelength dispersion of the PET film and the resist on the PET film, including the data from the catalog of ZEP-520A

higher than that of the resist; however, in the visible region, their refractive indexes are between 1.58 and 1.60. Thus, in this study, these values are considered to be almost same. Furthermore, by using ellipsometry, the thickness of the resist is estimated using the multilayer model. Fig. 3 shows the relationship between the thickness of the resist and the number of rotations of the spin coater. The thickness of the resist varies from 760 to 460 nm and increases with the number of rotations. Thus, in this study, the binary diffractive lens structures of the electron beam (EB) resist were fabricated by developing an EB resist on the PET films. The development of the EB resist can be regarded as processing the surface of a PET film. The thickness of the resist is equivalent to the height of the binary diffractive lens.

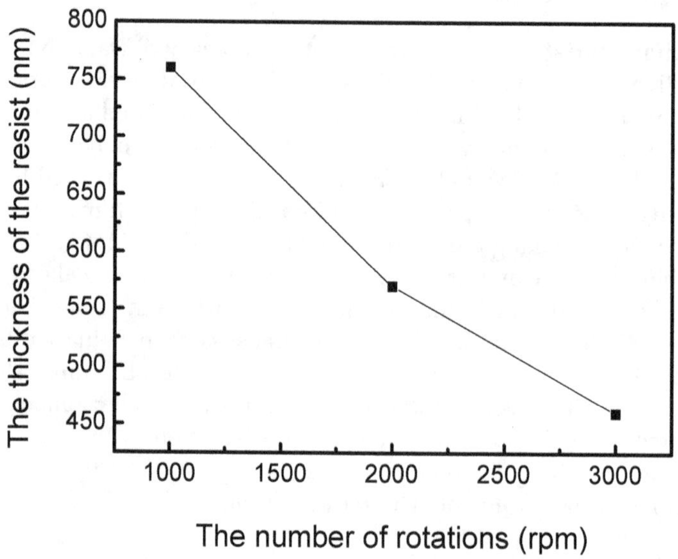

Fig. 3. Relationship between the thickness of the resist and the number of rotations of the spin coater

2.2 Design of diffractive lens

The fabricated binary diffractive lens was based on the micro-Fresnel lens. In this study, a part of two-level zone plates with a pattern of lines and spaces was fabricated. Radius of the mth zone r_m is

$$r_m = 2mf\lambda + (m\lambda)^2, \tag{1}$$

where f is the focal length of the designed lens and λ is the dominant wavelength. Equation 1 is based on the imaging theory of the diffractive lens (Buralli et al., 1989).

Then, mth period of this lens d_m is determined by $r_m - r_{m-1}$. In period d_m, the blazed structure is approximated to a step-like structure with n steps and then the step-like structures is converted to the relief structures by duty ratio of height $t_i = 1 - h(x_i)/h_{max}$ in each interval g, as shown in Fig. 4 (a). In the interval, the width of the air part is given by g^*t_i. In the binary diffractive lens, N is the number of the relief structures in a period. Examples of the structures are shown in Fig. 4 (b) and the complete structure of this lens is shown in Fig. 4 (c).

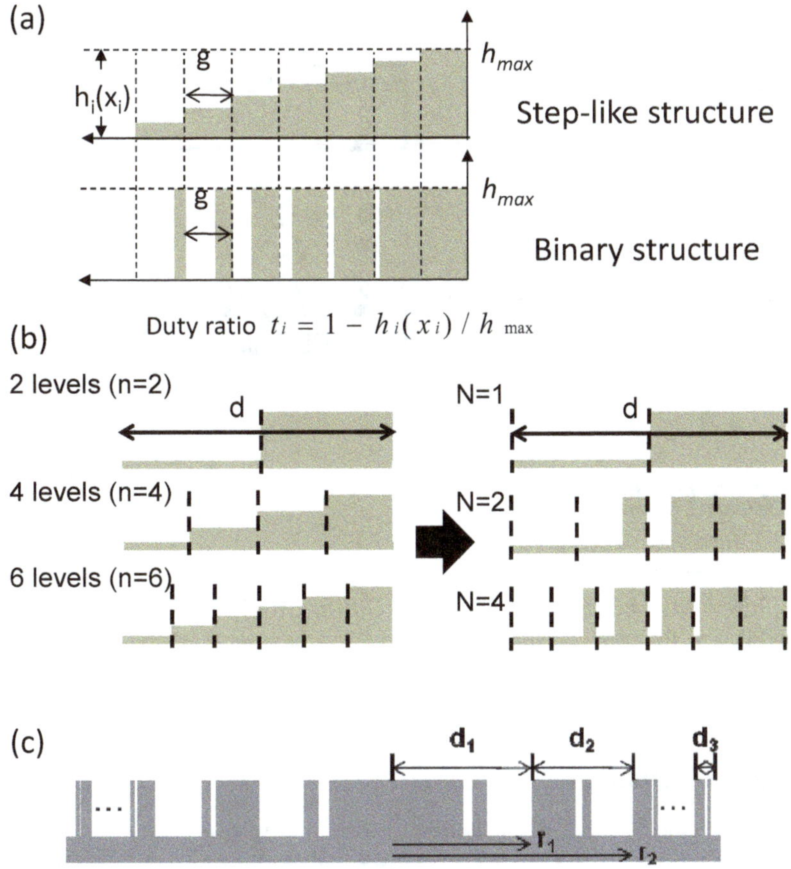

Fig. 4. Structure of the binary diffractive lens, (a) the conversion of the step-like structure to the binary structure, (b) the examples of the structures, (c) the complete structure

2.3 Fabrication of binary diffractive lens on optical films by EBL

The 125-μm-thick PET films were used as the substrate. Fig. 5 shows the procedure for the fabrication of the binary diffractive lens on the optical films by EBL. Before spin coating the EB resist, hexamethyldisilazane (HDMS) was spin coated on the surface of the PET film to improve the adherence between the PET film and the EB resist (Fig. 5 (a)). The surface was spin coated with an EB positive resist followed by pre-baking (Fig. 5 (b)). Then, the charge-up prevention was spin coated on the EB resist (Fig. 5 (c)).

The EBL system (Crestec CABL-8000) was equipped with a ZrO/W thermal field emission cathode. The acceleration voltage was 30 kV; the electrons accelerated by this voltage were able to penetrate the resist (Fig. 5 (d)). After exposure, the resist was developed and the binary diffractive lens could be obtained from these procedures (Fig. 5 (e)). The size of the patterns for the binary diffractive lens ranged from 100 × 100 μm^2 to 2 × 2 mm^2.

The optimum results obtained using the diffractive lenses fabricated by EBL, such as period, width, and height of the fabricated binary diffractive lenses, are useful for fabricating the molds of the thermal-type nanoimprint.

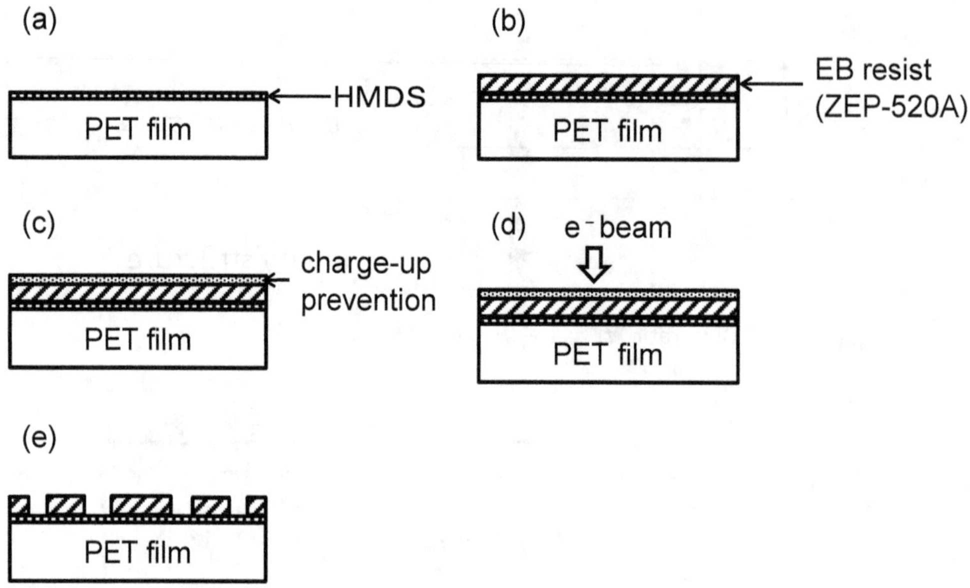

Fig. 5. Procedure for the fabrication of the binary diffractive lens on optical films by EBL, (a) spin coating HMDS, (b) spin coating EB resist and pre-baking, (c) spin coating charge-up prevention, (d) exposing e^- beam, (e) developing the resist and obtaining the binary diffractive lens

3. Results and discussion

In this section, we describe and discuss the experimental results. There are two types of the binary diffractive lenses: (1) the binary diffractive convex lens with a 2-mm focal length for controlling the luminosity of LED light and (2) the binary diffractive convex lens with a 150-μm focal length.

3.1 Binary diffractive convex lens with 2-mm focal length for controlling luminosity of LED light

The binary diffractive convex lens with 2-mm focal length was fabricated on the PET film. Fig. 6 shows the scanning electron microscopy (SEM) image of the fabricated binary diffractive lens on the PET film. The diffractive lens having width almost same as that of the designed lens was obtained.

Optical characterization of the fabricated binary diffractive lens was carried out. The luminous intensity distribution of the LED (λ = 566 nm) for the binary diffractive lens was characterized using a luminous intensity distribution system (Asahi Spectra IMS5000-LED).

The fabricated lens was then mounted on the LED chip and spectral irradiance in the vertical direction was measured; Fig. 7 shows the distribution of the irradiance. Most of the LED light was focused, as shown in Fig. 7 (a); the light distribution angle became narrow (30°) using the binary diffractive lens. As shown in Fig. 7 (b), spectral irradiance around 0° with this lens was 1.5 times higher than that without the lens. On the other hand, two side peaks in these data were observed and believed to be due to light escaping from the fabricated binary diffractive lens. From these results, it is clear that the luminous intensity distribution can be controlled using this type of lens.

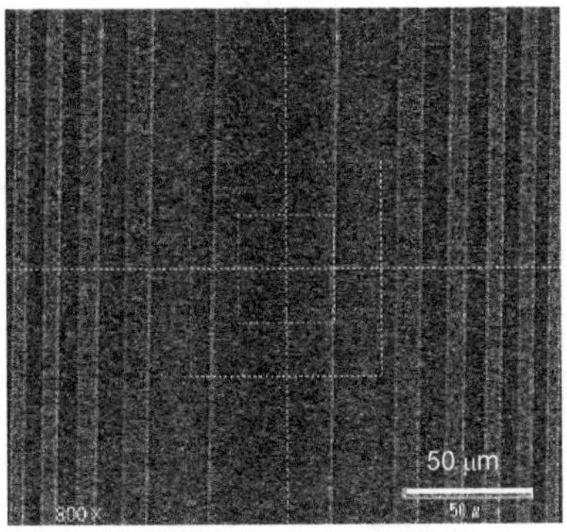

Fig. 6. SEM image of the fabricated binary diffractive convex lens with 2-mm focal length on the PET film.

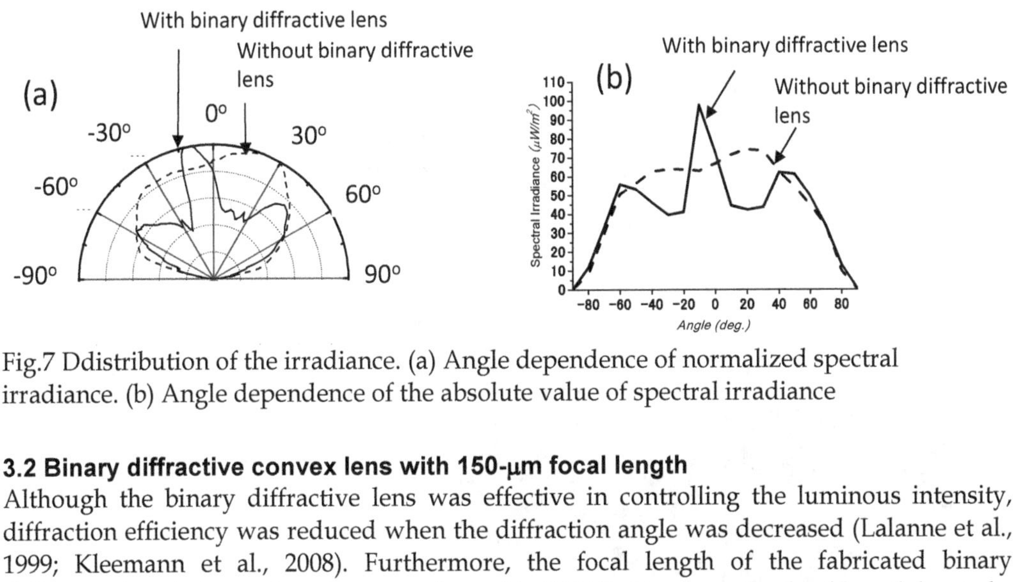

Fig.7 Ddistribution of the irradiance. (a) Angle dependence of normalized spectral irradiance. (b) Angle dependence of the absolute value of spectral irradiance

3.2 Binary diffractive convex lens with 150-μm focal length

Although the binary diffractive lens was effective in controlling the luminous intensity, diffraction efficiency was reduced when the diffraction angle was decreased (Lalanne et al., 1999; Kleemann et al., 2008). Furthermore, the focal length of the fabricated binary diffractive lens is 2 mm. In order to realize a thin LED light source, the focal length has to be shorter. In this section, to improve the diffraction efficiency and shorten the focal length, we designed the binary diffractive convex lens with 150-μm focal length.

In this study, a binary diffractive lens with a focal length of approximately 150 μm was designed and light propagation of the plane wave was simulated by the finite domain time difference (FDTD) method. Fig. 8 shows the field intensity distributions for TE polarization of the binary diffractive lens. The simulation parameters were λ = 632 nm, n = 1.575 (refractive index of the PET film), and n_0 = 1.0 (refractive index of air). The value of the period in part of the fringe was smaller than that in the center. The designed lens was placed along the x-axis (z = 0). The light was incident from z = 0 to the +z direction, resulting in the light being focused at x = 0 μm and z = 140 μm. After focusing, the light was spread with time because of diffraction. Therefore, a binary diffractive lens with a micrometer-order focal wavelength is expected to provide a small and thin light source for controlling the luminous intensity distribution. On the basis of the results of section 3.1, we speculated that the LED light can be focused at 140 μm.

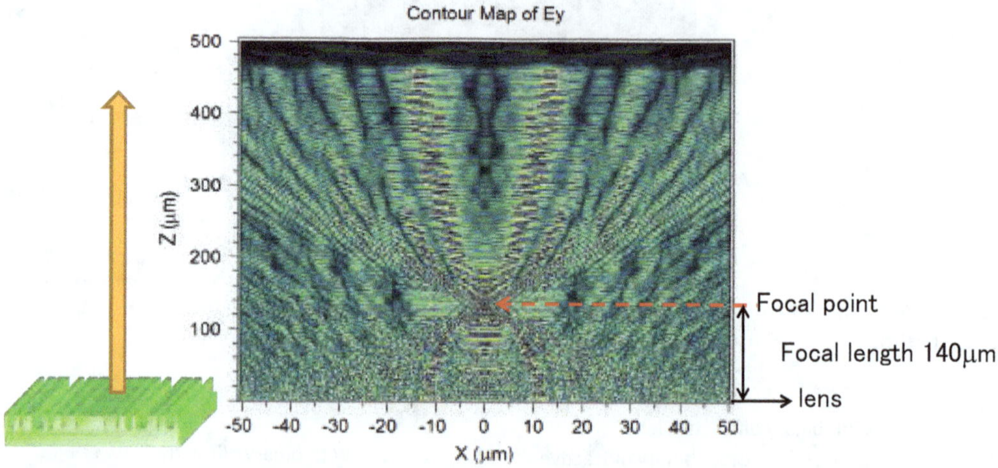

Fig. 8. Field intensity distributions for TE polarization of the binary diffractive lens

The binary diffractive lens with a 150-μm focal length was fabricated; its size was 100 × 100 μm² and thickness was 570 nm, as measured by ellipsometry. Fig. 9 shows the SEM image of the fabricated binary diffractive lens (N = 4) on the PET film. The diffractive lens, whose width was almost the same as the designed lens, was obtained.

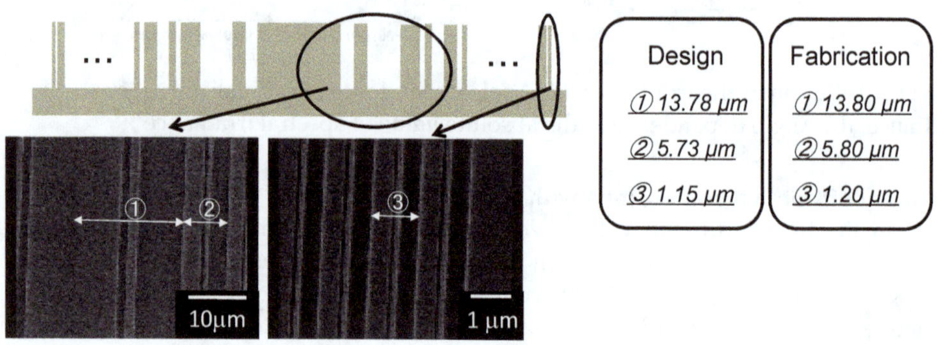

Fig. 9. SEM image of the fabricated binary convex diffractive lens with a 150-μm focal length (N = 4) on the PET film

The far-field transmitted intensity distribution of the fabricated lens is characterized by red laser light (λ = 635 nm). The aperture with a diameter of 100 μm was used for eliminating the light escaping from the edge of the lens. Fig. 10 shows the far-field transmitted intensity distribution of the fabricated lens with different N values (1, 2, 4). The focal length of this lens, which is estimated from this distribution, is approximately 160 μm, which is almost same as that in the FDTD simulation. For higher N values, the intensity of first-order diffraction decreases.

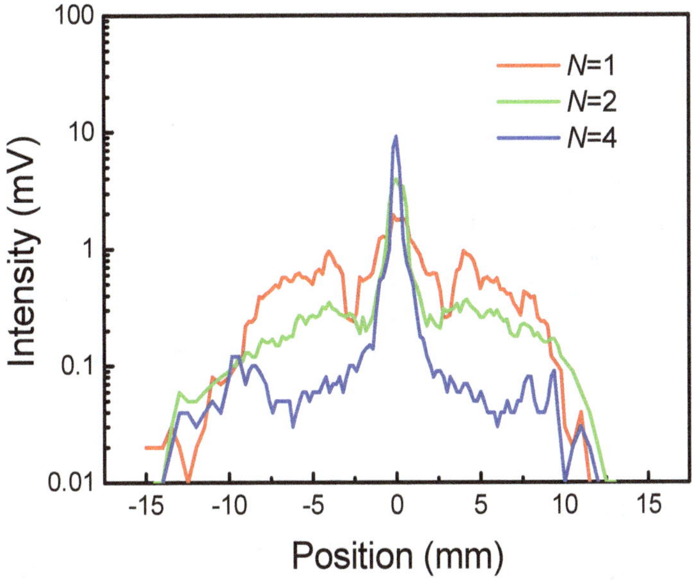

Fig. 10. Far-field transmitted intensity distribution of the fabricated lens

To determine the reason for these results, the binary diffractive lenses with only first period (d_1 = 13.78 μm) and 12th period (d_{12} = 2.12 μm) were fabricated. Fig. 11 shows the far-field light distribution of both lenses. In the case of d_1 = 13.78 μm, the first-order diffraction is observed when N = 4. Because d_1 is considerably larger than the wavelength of light, the first-order diffraction cannot be observed when N is small. On the other hand, in the case of d_{12} = 2.12 μm, the first-order diffraction is observed when N = 1, and it disappears by increasing the number of N. Therefore, in order to improve the diffraction efficiency of the diffractive lens, it is necessary to control the intensities of the zero- and first-order diffractions by choosing the binary structures.

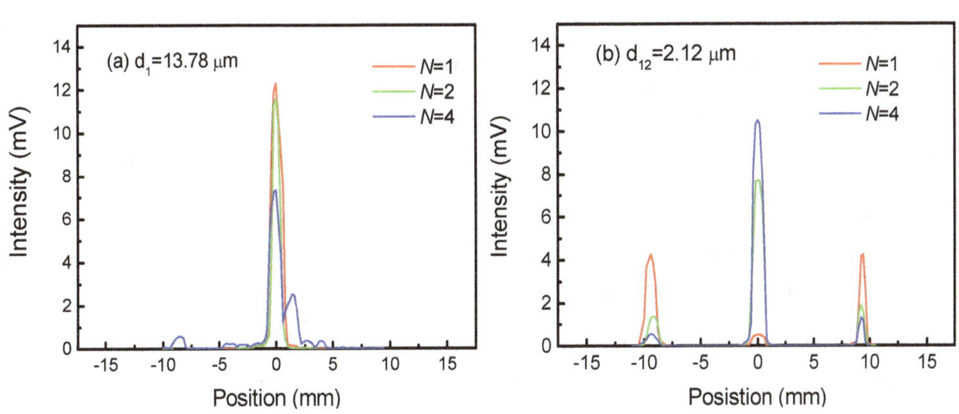

Fig. 11. Far-field light distribution of the binary diffractive lenses in the case of (a) d_1 = 13.78 μm and (b) d_{12} = 2.12 μm

4. Conclusion

In summary, we designed and fabricated two types of binary diffractive convex lenses using EBL on a PET film. In the case of the binary diffractive convex lens with 2-mm focal length, it is possible to control the luminous intensity distribution. To improve the diffraction efficiency and realize a thin LED light source, we designed a binary diffractive lens with 140-μm focal length. This type of lens with focal wavelengths in the micrometer range can produce a thin LED light source to control the luminous intensity distribution.

To realize the binary diffractive lens with the 100-μm-order focal lengths, we characterize the relationship between the diffractive lens structure and its light intensity. It is clear that the intensities of the zero- and first-order diffractions are controlled by the structure of the binary diffractive lens. By using this lens, wide luminous intensity distribution can be obtained.

5. Acknowledgment

This work was partly supported by a Grant-in-Aid for Scientific Research from the Japan Society for the Promotion of Science (No. 18360008 and 21360007), Mie University COE Projects, a Start-up Grant Program for Mie Venture Business from Mie Industry Enterprise Support Center, the Kinki Invention Center, and the Knowledge Cluster Initiative from the Ministry of Education, Culture, Sports, Science and Technology.

The authors thank Mr. T. Sato (JA Woollam Japan Co., Inc.) for ellipsometry measurement.

The authors also thank Dr. H. Miyake, Mr. K. Manabe, Mr. N. Machida, Mr. Y. Nakayama, Mr. K. Arakawa, and Mr. Y. Seriguchi for their help in the experiments and valuable discussions.

6. References

Jordan, J. A. ; Hirsch, Jr. P. M. ; Lesem, L. B. & Van Rooy D. L. (1970). Kinoform Lenses. *Applied Optics*, 9, 8, (1883-1887), ISSN0003-6935

Swason, G. J. & Veldkamp, W. B. (1989). Diffractive optical elements for use in infrared systems. *Optical Engineering*, 28, 6, (605-608) ISSN0091-3286

Orihara, Y. ; Klaus, W. ; Fujino, M. & Kodate, K. (2001). Optimization and Application of Hybrid-Level Binary Zone Plates. *Applied Optics*, 40, 32, (5877-5885), ISSN0003-6935

Yamada, K. ; Watanabe, W. ; Li, Y. ; Itoh, K. & Nishii, J. (2004). Multilevel Phase-type Diffractive Lenses in Silica Glass Induced by Filamentation of Femtosecond Laser Pulses. *Optics Letters*, 29, 16, (1846-1848), ISSN0146-9592

Lalanne, P. ; Astilean, S. ; Chavel, P. ; Cambril E. ; Cambril, E. & Launois, H. (1999). Design and Fabrication of Blazed Binary Diffractive Elements with Sampling Periods Smaller than the Structural Cutoff. *Journal of Optical Society of America. A*, 16, 5, (1143-1156) ISSN0740-3232

Lalanne, P. (1999). Waveguiding in Blazed-Binary Diffractive Elements. *Journal of Optical Society of America. A*, 16, 10, (2517-2520) ISSN0740-3232

Mait, J. M. ; Prather, D. W. & Mirotznik, M. S. (1999). Design of Binary Subwavelength Diffractive Lenses by Use of Zeroth-order effective-medium Theory. *Journal of Optical Society of America. A*, 16, 5, (1157-1167) ISSN0740-3232

Kleemann, B. H. ; Seesselberg, M. & Ruoff, J. (2008). Design Concepts for Broadband High-efficiency DOEs. *Journal of the European Optical Society*, 3, (08015-1–08015-16) ISSN 1990-2573

3

Application of Soft Lithography and Micro-Fabrication on Neurobiology

Gao Kan, Chen Haifeng, Liu Bing-Fang and Xu Qun-Yuan
Beijing Institute for Neuroscience Beijing Center of Neural Regeneration and Repairing Key Laboratory for Neurodegenerative Diseases of The Ministry Education Capital Medical University Beijing, China

1. Introduction

Soft lithography should be regarded as a complement to common lithography, providing a low-expertise route toward micro/nanofabrication and playing an important role in microfluidics (YN Xia & Whitesides, 1998). The resolution ranges from 5 to 100 nanometer (Pilnam Kim, et al. 2008). Patterns generated by the soft lithography are transfered repeatedly to the soft flexible materials, and then are printed on the medium substrates. In this field, micro contact printing (μCP) is the most widely used technique, especially in bioscience research. Combined with microfluidic patterns technology, several kinds of the extracellular matrix proteins like polymers can be printed to make cells grow according to the designed patterns (Tai Hyun Park, et al. 2003). The cell growth, differentiation in vitro can be regulated in the respect of spatial structure of extracellular matrix(Y. Nam, et al. 2004). So the morphology of neural cells and the influence of spatial structure can be investigated on the micron or even nano-scale level substrates. The closed loop of neural cells can be constructed in order to simulate the complex neural network in vivo. Finally, communication with the specific environment in vitro will be achieved by multi-electrode arrays (MEA). Our previous work used μCP technique can build more solid patterns. By comparing three different extracellular matrixes, PEI can obtain much better results, which adhering more neural cells to form reliable design. However, it is not perfect for the specific neural network construction and the patterned neural cell culture on MEA. In subsequent research, we improved the parameters of the template and achieved a big progress on microfluidic patterning technique to microfabricate patterns. Patterns of biomaterials were constructed with the help of the advanced soft lithography to do the primary cell culture, such as dopaminergic neurons in the substantial nigra and GABAergic neurons in the striatum. Finally, the biocompatibility of MEA was validated initially by primary striatal neuronal culture. Meanwhile, new strategy of structural microfabrication on MEA surface was further explored.

2. Soft lithography materials

Why soft lithography is called "soft"? One of the reasons is the use of soft organic materials such as polydimethyl siloxane(PDMS), polymethyl methacrylate (PMMA), and polycarbonate (PC). PDMS is most widely used in bioscience research, because it has good biocompatibility, chemical stability, optical transparency, air permeability, elasticity. Moreover, the polymer precursors can be aggregated into a mold by UV radiation. PDMS polymerization is shown in Figure 1.

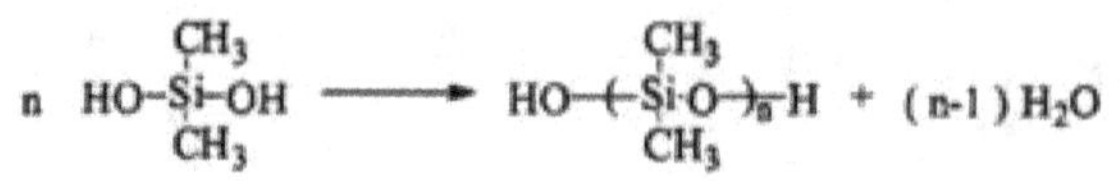

Fig. 1. PDMS polymerization

To cure the PDMS prepolymer in general, a mixture of silicon elastomer and a curing agent (10:1, Sylgard 184 silicone elastomer kit, Dow Corning Corp.) is poured onto the master and placed at 70-80°C for 1 h. The character of the PDMS is closely related to the mixture ratio, curing temperature, and vacuum. Silicon, quartz or glass, and some photoresist are the most common materials to fabricate the masters by standard lithography, transferring the patterns to the PDMS stamp.

3. Soft lithography fabrication methods and it's application in patterning

Various soft lithographic technologies have been applied to fabricate high-quality microstructures and nanostructures including micro contact printing (μCP), replica molding (REM), microtransfer molding (μTM), micromolding in capillaries (MIMIC), and solvent-assisted micromolding (SAMIM). Here, three soft lithographic methods are introduced to fabricate micropatterns onto a surface or MEA: μCP, microfluidic patterning technique and microstencil. The former two can be achieved using the same PDMS stamps and molds. A novel technology was applied to get the high depth-to-width ratio silicon-based mold to fabricate the topographic PDMS microstencil with microfluidic channel (Y.Nam, et al, 2006). Finally, microchannels on MEA with polyimide (PI) guiding the cell growing were also introduced.

3.1 Microcontact printing (μCP)

μCP is a direct method for pattern transfer, generating a non-structured, chemically modified surface. The process of μCP is shown in figure 2. Photolithography was used for the fabrication of silicon-based masters in preparing PDMS stamps. Multi-layer molds were made of thick photoresist like SU-8 on silicon or glass wafers by standard lithography techniques. It was subsequently placed at least 30min in an oven at 160°C to make the photoresist adhere to the substrate closely. Release agent DC20 or OTS were always spin-coated and drying on the master before pouring PDMS. Liquid PDMS (Sylgard-184 from Dow Corning) was poured onto the mold and clamped by the foil, so that the shape of the mold microstructure was transferred to PDMS membrane. It was subsequently placed at least 2 h in an oven at 80°C. The molded PDMS slab was then peeled off and placed onto a glass slide for handling. After curing, PDMS stamps are soaked in a protein "ink", such as PEI, PLL, or LN. 20 minutes later, the "ink" was blew off by using nitrogen gas. Then the raised regions were brought into conformal contact with a substrate in order to print the ink onto the substrate surface. The material of interest was transferred from the PDMS stamp onto the substrate surface. The microscopy of the PDMS stamps were shown in figure 3.

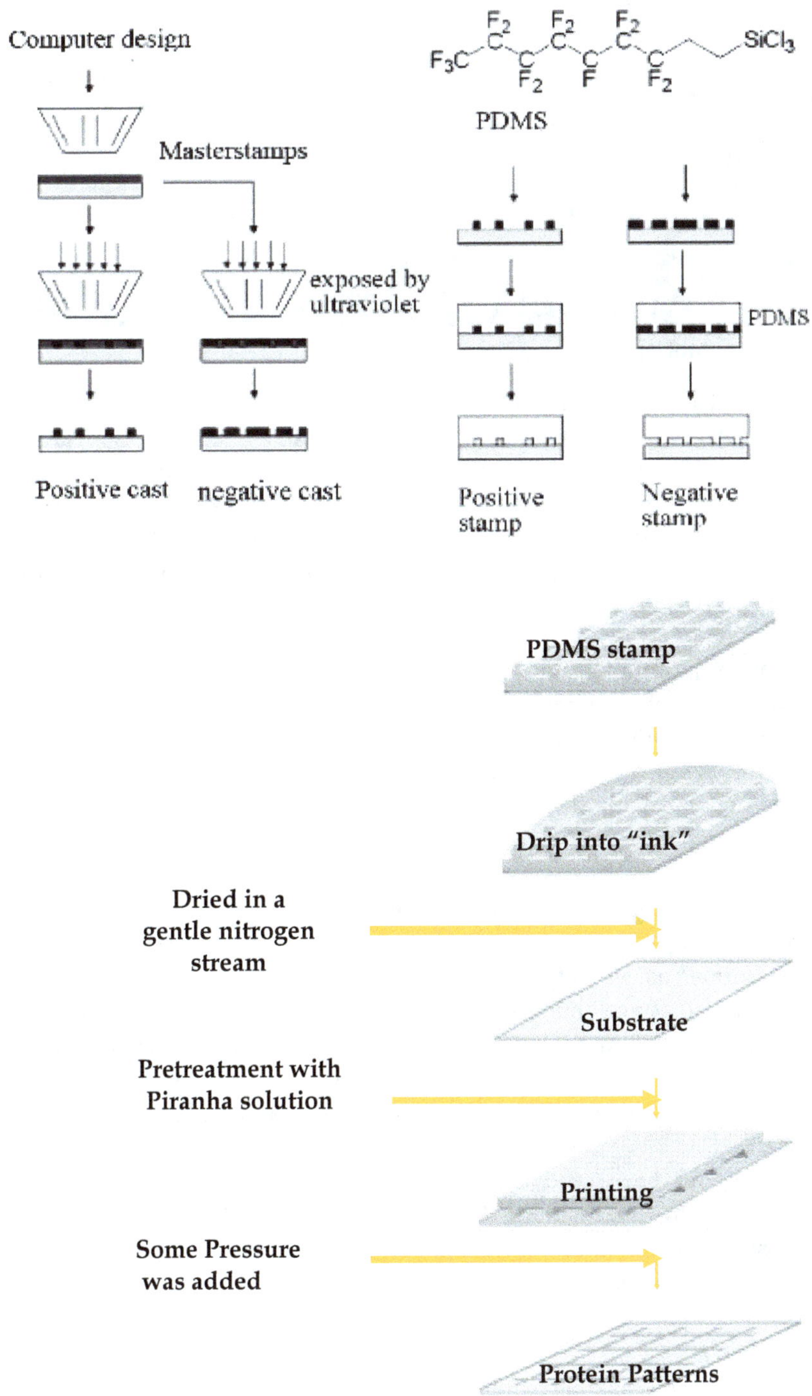

Fig. 2. Schematics of the processes of μCP

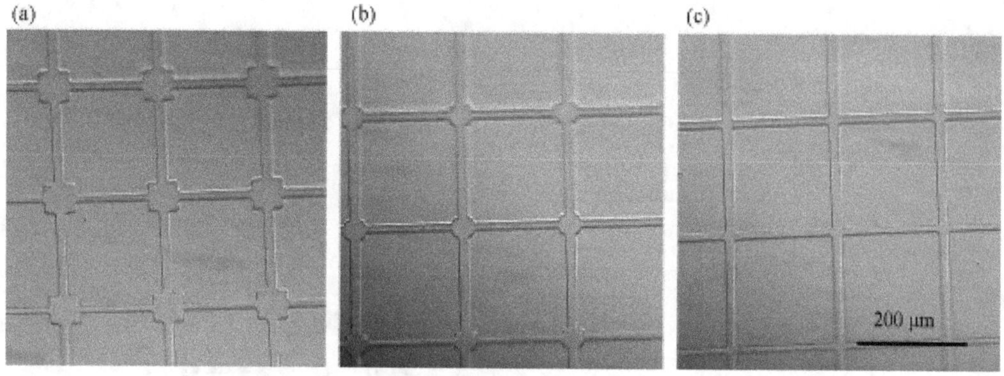

Fig. 3. Microscopy of the PDMS stamps

The bare areas of substrate surface that the PDMS stamp has not touched can be exposed to another coating material. μCP provides the patterning of self-assembled monolayers(SAMs) of alkanethiols on gold, and the resulting control over the adsorption of adhesive proteins facilitates the patterning of cells on substrates.

μCP enables easy stamp replication, fast printing using parallelization, and low-cost batch production. A conformal contact between the stamp and the surface of the substrate is the key to its success. The polymer stamps also minimize the problems of sample carry-over and cross contamination. Printing has the advantage of simplicity and convenience: Once the stamp is available, multiple copies of the pattern can be produced using straightforward experimental techniques. Printing is an additive process; the waste of material is minimized. Printing also has the potential to be used for patterning large areas.

However, μCP has some limitations that are mainly caused by the use of a soft polymer stamp. The swelling of a stamp during inking often results in an increase in the pattern size by diffusion of the excessive printed molecules on the substrate.

3.2 Microfluidic patterning using microchannels

The difference between μCP and microfluidic patterning is that PDMS stamps are soaked in the "ink" in the former usage, but the stamps contact the substrate forming microchannels delivering the materials for cell adhesion or cell suspension to the desired area in the latter usage because of the elastic nature and hydrophobicity of PDMS. The substrate tilted 45 degree, drop of liquids were injected to the PDMS microchannels by the pipette. Then the substrate was put on the test tube rack vertically for 25 to 30 min. Patterns were formed of after the liquid dried.

While this method has been used primarily for surface attachment of cells, it may be possible to adapt this method to three-dimensional tissue constructs. In many cases three dimensional tissue constructs promote cellular differentiation and more authentic cellular morphology and metabolism.

3.3 Microstencil on MEA

The former two methods enable patterning adhesion molecules and guiding cultured cells grow physically. But the cells' communication and interactions in co-cultures are difficult to be detected, which is important to research the function of the cells network.

MEA was a cell-based biosensor for extracellular electrophysiological investigations of neuronal networks. PDMS microstencil was designed to pattern adhesion molecules at the surface of MEA guiding cultured cells grow along the patterns. PDMS microstencil mold was fabricated by a complex photomask aligning method, shown in figure 4.

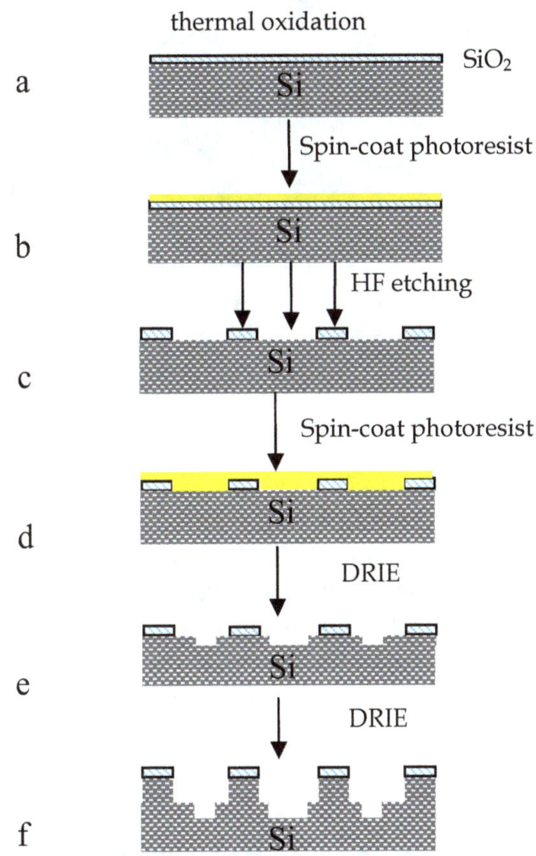

Fig. 4. Schematics of the fabrication of microstencil mold

525μm silicon was thermal oxidated with 4000 Å SiO2 (see Fig. 4a). The substrate was first coated with a thin photoresist (AZ AZ9912) for 30 s at 3000 rpm (see Fig. 4b). 8 × 8 SiO2 arrays were fabricated by wet etching as the RIE mask which was the same as the MEA structure (see Fig. 4c). Photoresist was spin-coated to the silicon with SiO2 mask again (see Fig. 4d) and selectively exposed to UV under a chromium photomask. The silicon was etched 30μm by deep reactive ion etching (DRIE) with photoresist in order to construct the microchannel , (see Fig. 5a). Then photoresist was removed by ultrasonication in acetone. The silicon was selectively etched 70μm by DRIE with SiO2 resist, forming the topographic PDMS microstencil mold (see Fig. 5b). The high depth-to-width ratio silicon-based mold was designed to penetrate through the PDMS membrane on the MEA to exposure the electrodes and form the microchannel between the electrodes so that the MEA could also work.

(a) (b)

Fig. 5. SEM images of PDMS microstencil mold

The mixture of PDMS prepolymer and curing agent was spin-coated on the mold for 40 s at 4000 rpm. The coated mold was cured for 2 hours at 110◦C in a convection oven. The fully-cured PDMS-coated mold was soaked in an acetone ultra-sonication bath until the PDMS layer released from the mold. The detached microstencil was rinsed with IPA and DI water. The upside and downside of PDMS microstencil with microholes and microchannels SEM images was shown in figure 6.

Fig. 6. SEM images of PDMS microstencil

However, PDMS microstencil is difficult to practise because it is hard to align with MEA and the silicon mold is easy to fracture when lifting off.

3.4 MEA with microchannels for patterning

Planar MEA are developed to study electrogenic tissues such as dissociated neuronal cultures (Hiroaki Oka, et al. 1999). They have been widely used with dissociated cultures for a variety of neuroscience investigation including learning and memory and cell-based biosensors for the detection of neurotoxins (Conrad D. James, et al. 2004). But the neurons grow disorderly and cannot form a network so that the function is not the same as the cells in vitro. Combining the patterning technology and MEA forming neuronal networks is the efficient method to research the neurobiology.

μCP, microfluidic patterning technique and microstencil are difficult to operate because the space is too small and the PDMS stamp or stencil can hardly align with MEA. The best and easiest way to forming neuronal networks is to fabricate the microchannels on MEA with polyimide (PI) guiding the cell growing.

MEA were fabricated using a conventional semiconductor process (Guangxin Xiang, et al. 2007). After cleaning the polished quartz glass wafer, the conductive layer of Au/Ti film (Au 3000 Å and Ti 700 Å) was sputtered. 8 × 8 electrode arrays were left with the photomask protection by standard photolithography. Then, a combination of SiO2/Si3N4/SiO2 (3000 Å /4000 Å /3000 Å) passivation layers was deposited onto the substrate using plasma enhanced chemical vapor deposition (PECVD), and the insulating layers on the electrodes and the bonding-pads were removed by inductively coupled plasma (ICP) (see Fig. 7a). Finally, Negative photosensitive polyimide (AP2210B, Fujifilm Electronic Materials Inc) was spin-coated to form microchannels having a thickness of 3~4 μm and photo-etched by the standard procedure to expose the microelectrodes and the terminals (see Fig. 7b).

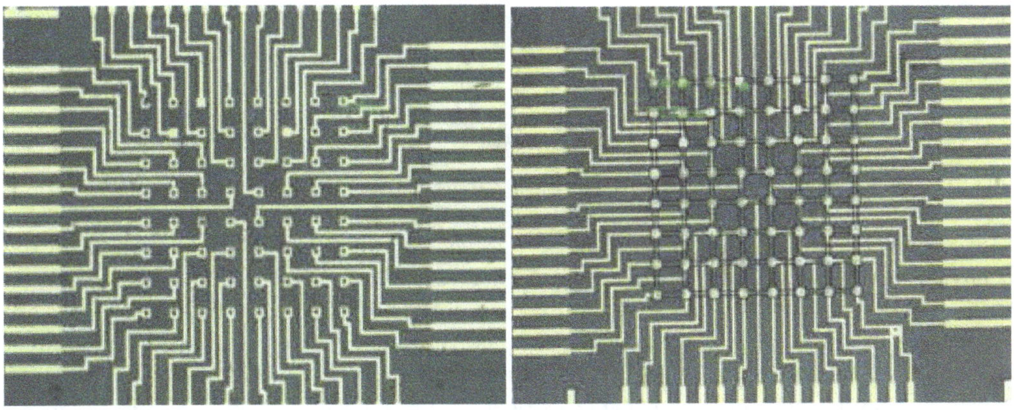

(a) (b)

Fig. 7. Microscopy of the MEA with PI microchannels

GABAergic neurons in the striatum and PC12 cells were cultured on MEA with PI microchannels which were coated with poly-l-lysine (PLL) to promote cell adhesion, (see Fig. 8a, 8b). PI microchannels could be seen between the electrodes and the neural cell can grow along the microchannels. However the nerve cell synapse could not formed along the microchannels. Because the depth of microchannels could not match the neurons and the

PLL could not be guarantee to coat the microchannels effectively after days. There is still a lot of work to do to construct the neuronal networks on MEA to study the cells function as in vitro.

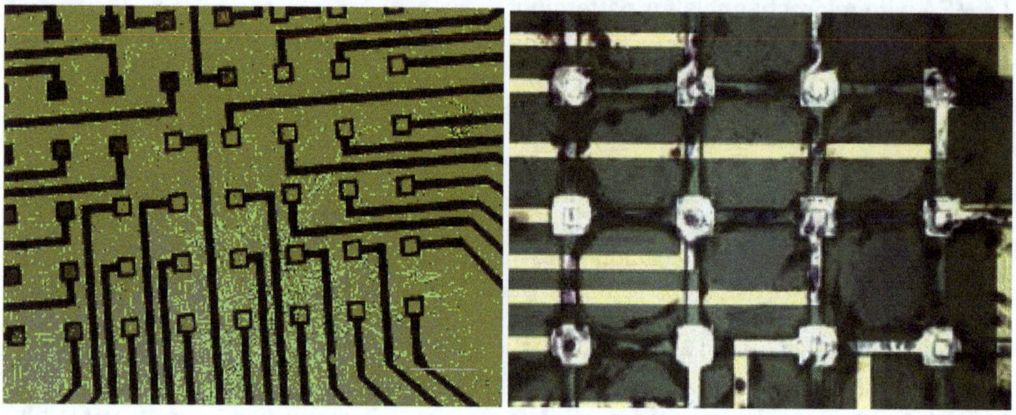

(a) (b)

Fig. 8. Microscopy of the cells cultured on MEA with PI microchannels (a) GABAergic neurons in the striatum cultured at 3 days (b) PC 12 cells cultured at 6 days

4. Construction of neural network by applying soft lithography

A man-made neural network on electrode can be applied to do basic research of neuroscience, be a kind of biosensor for drug discovery [9] and even be implanted into brain to establish artificial connections that could form the basis of a neural prosthesis[10]. These fields have caused much attention in the world [11]. In the past, there are three strategies to realize a simplified neural network in vitro, such as mechanical fixation-applying spatial restrictions[12,13], physical modification of surface roughness and surface topography[14,15], chemical polymer microfabrication on surface—using soft lithography.
Soft lithography is created by Whitesides in 1993[16]. It is used to create chemical structures on surfaces, including μCP, μFN and other downstream techniques. These microfabrication techniques that control both the size and shape of the cell anchored to a particular surface are extremely useful for understanding the influence of the cell-material interface on the behavior of cells [17, 18].The adhesion and survival of neural cells should be considered firstly for the patterned neural cell culture in vitro. So the selection of appropriate cell-attracting substances is an important step for pattern design in micro-contact printing. Ideal substances encourage good cell-substrate interactions, constantly stimulate the cells by substrate-bound chemical, biological, electrical or mechanical signals [19, 20] and even regulate neuritis growth on designed patterns. The most commonly used coating reagents to promote cell adhesion are extracellular matrix (ECM) proteins like laminin (LN), positively charged polymers such as poly-l-lysine (PLL) and synthetic amide-linkage-free compounds such as polyethylenimine (PEI). Therefore, the characteristics of three different substrates, PEI, PLL, LN were compared by the primary neuron culture in our previous work. The PEI characterized with strong positive surface charges was validated to fabricate more continuous and integrated micro-contact printing neural patterns under serum-free culture conditions than PLL, LN[21].

For the functional neural network construction on MEA, an inevitable question that should be addressed finally is how to realize accurate opposite between neurons and the electrode under neurons. We assume firstly microfluidic technique may have more advantages than μCP. In subsequent research, we achieved satisfied patterns by microfluidic technique for further research with the help of the progress on parameters of template. Specific neural network were constructed by applying advanced soft lithography above to do the primary cell culture, such as dopaminergic neurons in the substantial nigra and GABAergic neurons in the striatum. The conditions of neuronal adhesion on different patterns (grids and lines) were also observed using several techniques, including atomic force microscopy, immunohistochemistry, transmission electron microscope and scanning electron microscope.

4.1 Neural network with rat fetal hippocampal cells by μCP patterns

In previous study, we examined the ability of another positively charged polymer, polyethyleneimine (PEI), to promote neuronal adhesion, growth and the formation of a functional neuronal network in vitro. PEI, PLL and LN were used to produce grid-shape patterns on glass coverslips by μCP. Post-mitotic neurons from the rat fetal hippocampus were cultured on the different polymers and the viability and morphology of these neurons under serum-free culture conditions were observed

4.1.1 Cells adhesion

The number of cells that adhere to the different substrates after 24 h in culture is shown in Fig. 9. The adhesive effects were evaluated by calculating the ratio of cell numbers that adhere to the grid-like patterns divided by the total area of printed polymer. We found that the positively charged polymers (PLL and PEI) had a significantly higher level of cell attachment than LN ($p < 0.05$)[13].

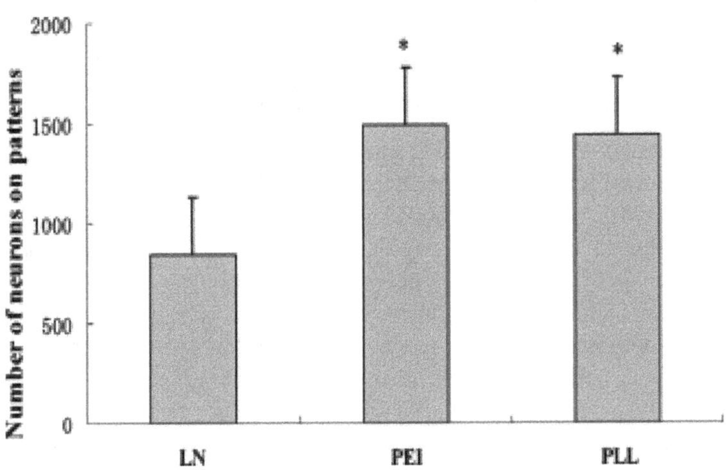

Fig. 9. The adhesive effects were test by analysis of the number of neurons on the area (mm2) of LN, PEI and PLL grid patterns after 24 h in culture. The asterisks indicated neurons on PEI and PLL patterns had significantly higher lever than on LN patterns, n = 12, $p < 0.05$.

4.1.2 Cells viability

In Fig.10, We show that neurons cultured on the PEI- and PLL-coated surfaces adhered to and extended neurites along the grid-shape patterns, whereas neurons cultured on the LN-coated coverslips clustered into clumps of cells. In addition, we found that the neurons on the PEI and PLL-coated grids survived for more than 2 weeks in serum-free conditions, whereas most neurons cultured on the LN-coated grids died after 1 week[13].

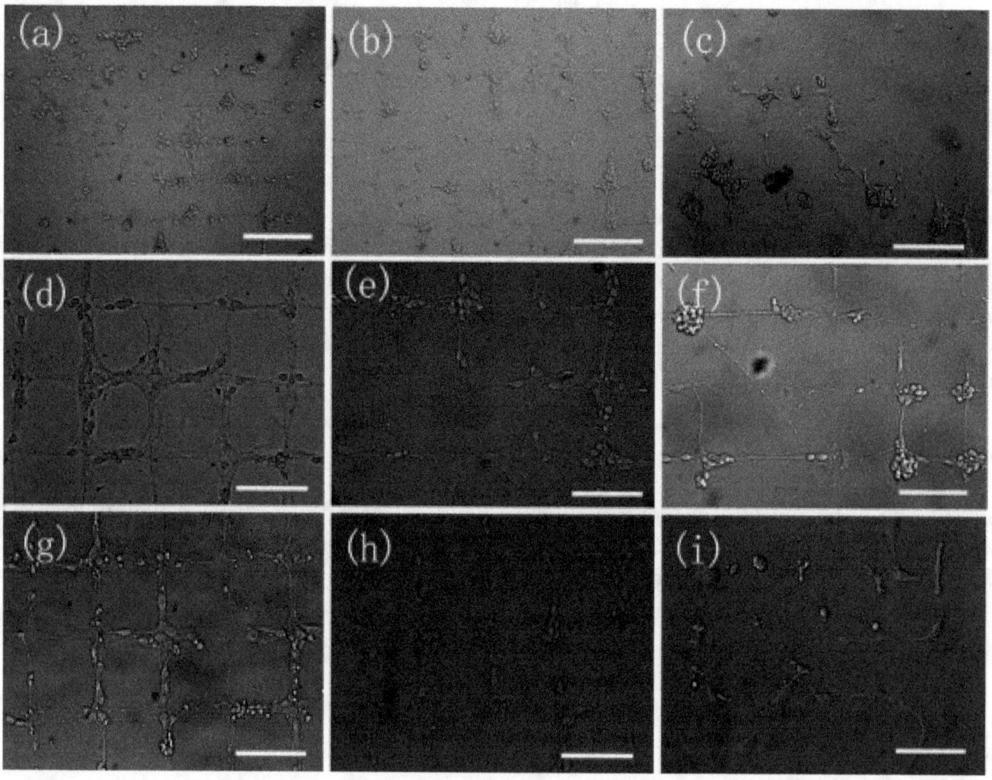

Fig. 10. Images obtained using the phase contrast microscope, showing cells cultured on PEI, PLL and LN polymeric films at different time points. (a, d, and g) Representative images of neurons cultured on PEI grid patterns at 24 h, 7 days, and 14 days, respectively, show that neurons adhere and grow accurately along the PEI grids at differently time points. (b, e, and h) Representative images of neurons cultured on PLL grid patterns at 24 h, 7 days, and 14 days, respectively. (c, f, and i) Representative images of neurons cultured on LN grid patterns at 24 h, 7 days, and 14 days, respectively. The images (c and f) show that the neurons on the LNgrid patterns often accumulate at the cross points of the grids. The image (i) shows that most neurons disappear after 14 days in culture and only small areas of the neural cells exist. (a–c) Bar = 100 μm; (d–i) bar = 50 μm.

4.2 Specific neural network with two relative neurons, such as dopaminergic neurons and GABAergic neurons

There is a closely relationship in the respect of function and structure between dopaminergic neurons in the substantial nigra and GABAergic neurons in the striatum. As well-known,

Parkinson's disease is due to the loss and injuries of dopaminergic neuron in substantia nigra which cause a decrease in nerve fibers projected to the new striatum. Finally, reasonable synaptic connections and neural network can not be established. Therefore, it is expected to establish cell models to investigate the relationship between these two kinds of coherent neurons and construct an artificial neural network in vitro by the application of soft lithography. In present work, specific neural network with dopaminergic neurons and GABAergic neurons co-culture was established by μCP PEI grid patterns. Meanwhile, PEI was validated again to fabricate more continuous and integrated neural patterns by using μCP and microfluidic technique both.

4.2.1 Neural network by μCP

Different kinds of neural network by μCP were established with neuron from the striatum, dopaminergic neurons from the substantial nigra and both of them co-culture. The conditions of neuronal adhesion on different pattern figures were observed using several techniques, including immunocytochemical staining, transmission electron microscope and scanning electron microscope. Using immunocytochemical staining, transmission electron microscope, we identified the types of neural cells and observed some neurosynapse-like structures near the neuronal soma on PEI-coated coverslips. These findings indicate that PEI is a suitable surface for establishing a functional neuronal network in vitro.

4.2.1.1 Investigation of neural cell types and neurite elongation along the grid-like patterns by immunocytochemical staining and SEM

PEI, PLL and LN were used to produce grid-shape patterns on glass coverslips by micro-contact printing. GABAergic neurons and medium spiny neuron from the rat striatum, dopaminergic neurons from the rat substantial nigra and both of them co-culture were researched separately on the different polymers coated surface. The viability and morphology of these neurons under serum-free culture conditions were observed using fluorescent microscopy in Fig. 11, Fig. 13, Fig. 14. After 7 days in culture, we found that the neural cell bodies on the PEI patterns were located mostly at the cross-points of the grid, whereas neurites extended along the line of the grid-like patterns. More continuous and integrated neural network was achieved finally. On the PLL-coated coverslips, the neural patterns appeared to be integrated. But several cells clustered at the cross-points of the grid disappeared gradually after the media was replaced. In contrast, cells cultured on the LN-coated grids were generally clustered into clumps and cannot form satisfied patterns. Different sizes of PEI pattern were produced by microcontact printing. In Fig. 12, compared with 50 μm, 100 μm, 200 μm pattern sizes, we found few difference early. After 7 days or 14 days culture, most neural cells on 200 μm size grew well and were seldom found to overlap each other, unlike those on 50 μm size clustered into clumps at the cross-points of grid and disappeared gradually. Identified with immunocytochemical staining, we found that neural cells from the rat substantial nigra were TH positive, synaptic vesicle protein were synaptophysin positive in Fig. 13 and cells from the rat striatum were GABA positive or DARPP-32 positive in Fig. 11, Fig. 12, Fig. 14. SEM show neurons outgrowth on PEI-coated patterns and validate the findings by immunocytochemical staining in Fig. 15, Fig. 16.

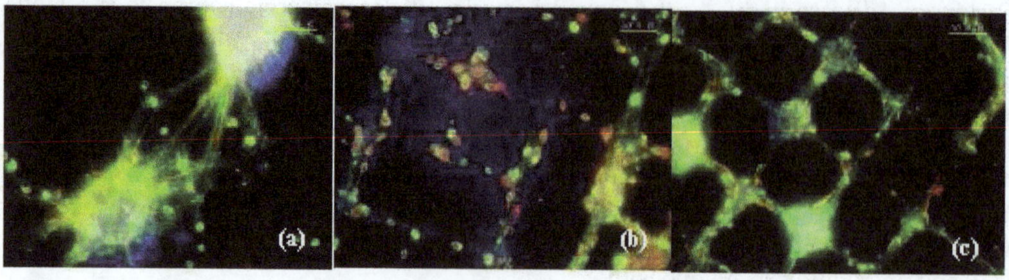

(a) on LN grid patterns (b) on PEI 100 μ m grid patterns (c) on PLL 100 μ m grid patterns bar=50 μ m

Fig. 11. Immunofluorescent image of anti-GABA (green fluorescence) +anti-MAP2 (red fluorescence) labelled striatal neurons cultured for 7 days on different substrates, the nuclei of neurons were stained with Hochest x400

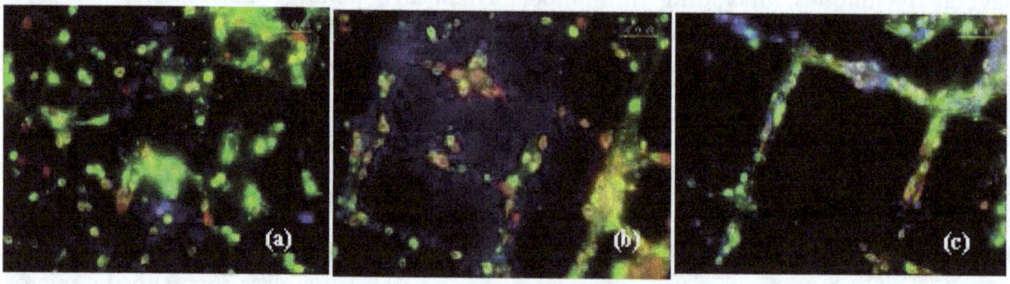

(a)on 50 μ m patterns (b) on 100 μ m patterns (c) on 200 μ m patterns bar=50 μ m

Fig. 12. Immunofluorescent image of anti-GABA(green fluorescence)+anti-MAP2(red fluorescence) labeled striatal neurons cultured for 7 days on different sizes of PEI patterns, the nuclei of neurons were stained with Hochest x400

(a) on LN grid patterns (b) on PEI 100 μ m grid patterns (c) on LN+PEI bar=20 μ m

Fig. 13. Immunofluorescent image of anti-TH(red fluorescence) labeled dopaminergic neurons from the substantial nigra and anti-synaptophysin(green fluorescence) labeled synaptic vesicle protein cultured for 7 days on different substrates. x200

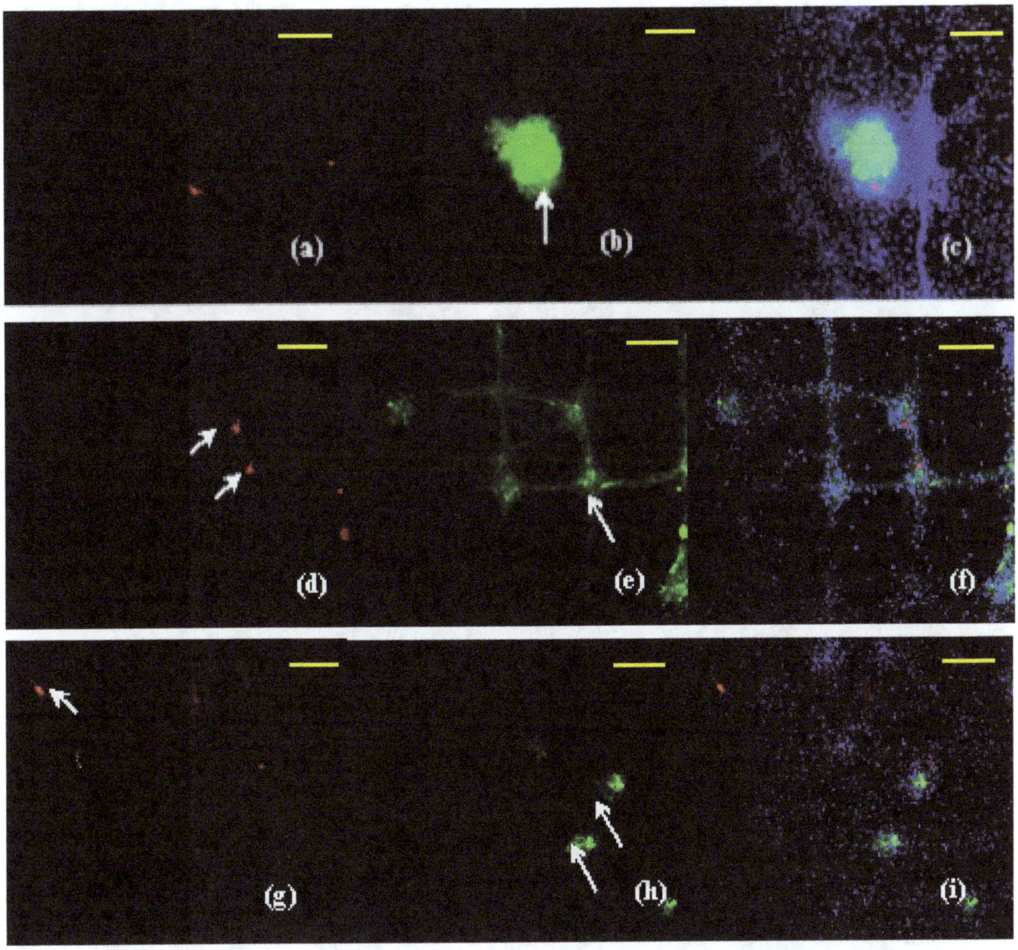

(a–c) co-culture neurons growing on LN grid patterns. (a) Immunostaining with anti-TH (red fluorescence) labelled dopaminergic neurons. (b) Immunostaining with anti-DARPP-32 (green fluorescence) labeled medium spiny neuron from the striatum(arrow). (c) Merged image of (a) and (b), showing TH positive neurons on the LN-coated patterns actually adhere to the cluster formed by medium spiny neuron. (d–f) co-culture neurons growing on PEI grid patterns. (d) Immunostaining with anti-TH (red fluorescence) labelled dopaminergic neurons(arrows). (e) Immunostaining with anti-DARPP-32 (green fluorescence) labeled medium spiny neuron from the striatum(arrow). (f) Merged image of (d) and (e) showing that DARPP32 positive neurons cultured on the PEI-coated patterns form a continuous and integrated neural network, and two TH positive neurons adhere to the cross-points of grid. (g–i) co-culture neurons growing on PLL grid patterns. (g) Immunostaining with anti-TH (red fluorescence) labeled dopaminergic neurons(arrow). (h) Immunostaining with anti-DARPP-32 (green fluorescence) labeled medium spiny neuron from the striatum(arrows). (i) Merged image of (g) and (h)showing only two DARPP32 positive neurons adhere to the cross-points of grid. TH positive neurons were not restricted by the grid pattern. bar=100 μm

Fig. 14. Immunofluorescent image of anti-TH labelled dopaminergic neurons from the substantial nigra and anti-DARPP32 labeled medium spiny neuron from the striatum co-culture for 7days on different substrates, the nuclei of neurons were stained with Hochest x200

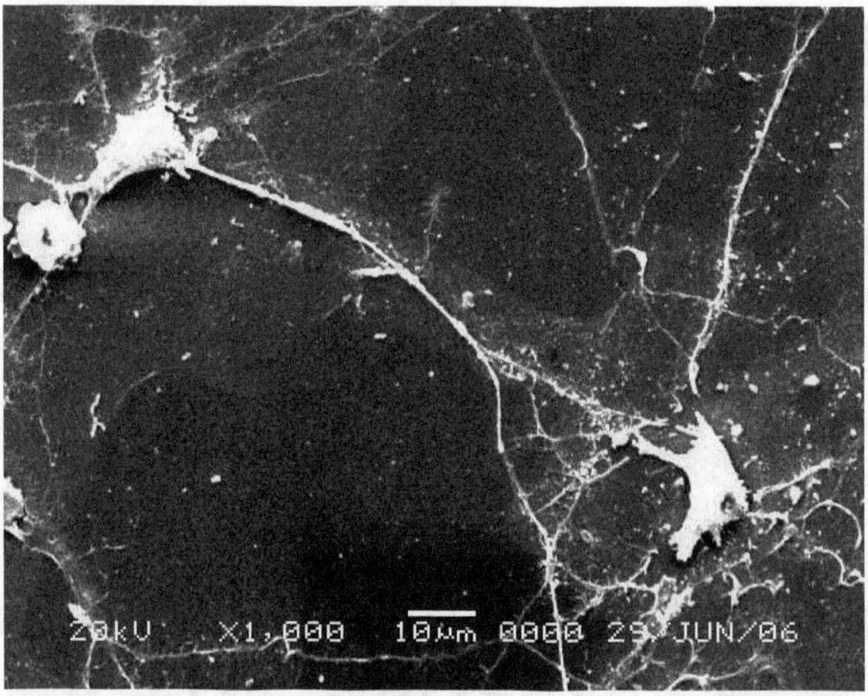

Fig. 15. SEM show neurons outgrowth on PEI-coated patterns. Somata of neuons located on the cross point and neurites extend along the lines

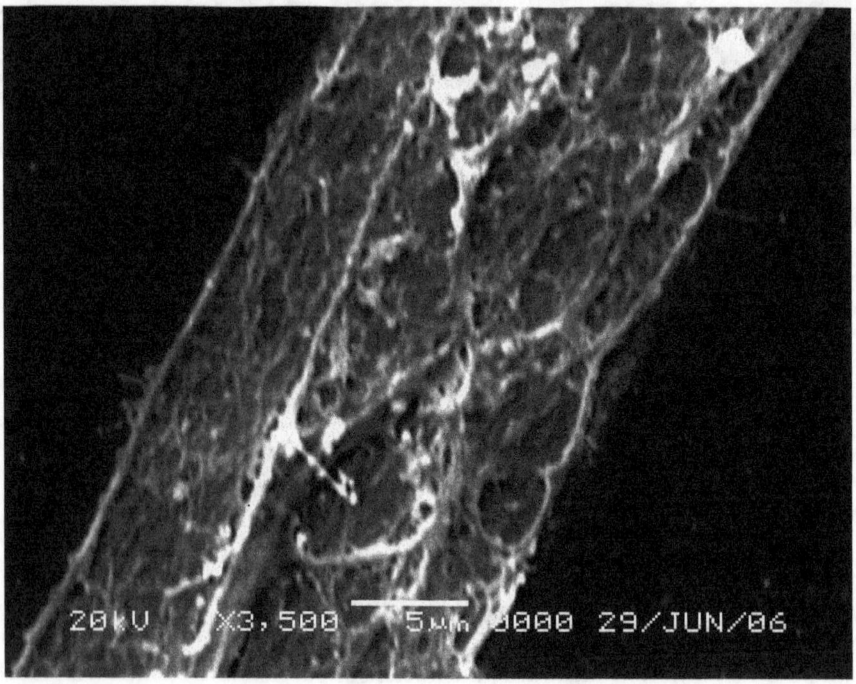

Fig. 16. SEM show that Line of grid were occupied by abound neurites.

4.2.1.2 Examination of synaptic formations by immunocytochemical staining and TEM

To further understand functional activities of neurons on the grid pattern, we observed the microstructure of the neuronal cell body, neuritis extension and processes of the synapse formation by using laser confocal microscope after 7 days in culture. It is found that several neural cell bodies aggregated on the cross point of the grid pattern (Fig.17.b), which neurites extended along the lines clearly by cytoskeletal proteins MAP2 and Synaptophysin double immunocytochemical staining of patterned neurons on PEI(Fig.17.a). Some processes can even gather into a bundle, span the distance between two cross-points and make connection with another neuron. Synaptophysin was a kind of granular protein scattered around the cell bodies and neurites, The neurites from neuron at the cross-point seemed to communicate with those at the lines which express a large number of synaptic vesicle protein(Fig.17.c).

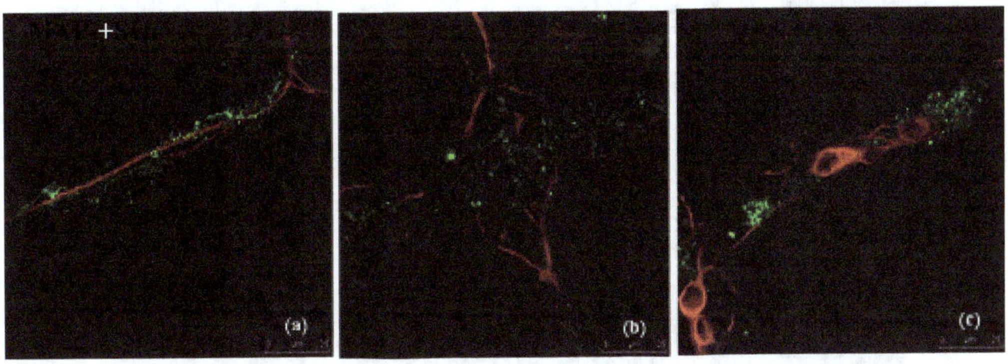

(a) neurites extended along the lines and synaptic vesicle protein around
(b) neural cell bodies at the cross-point and synaptic vesicle protein scattering around them
(c) synapse connection between two neurons at the grid pattern bar=25 μ m

Fig. 17. Immunofluorescent image of anti-MAP2(red fluorescence) labeled neurons and anti-synaptophysin(green fluorescence) labeled synaptic vesicle protein cultured for 7 days on PEI patterns x400

We chose PEI group which can construct more clearly neural network and continued to observe intercellular ultrastructure under transmission electron microscope (Figure18). Two periphery of neurites thicken show high electron density to form a synaptic contact (Red border). Width of synaptic cleft was measured to 30 ~ 50nm. Figure18(a) shows clear synaptic vesicles. Figure18(b) shows the synaptic cleft is relatively narrow and suspected to be electrical synapse structure due to double-membrane structure adjacent closely.

4.2.2 Neural network by microfluidic technique

In this experiment,, we made a big progress on microfluidic technique by re-designing the parameters and enhancing the photoresist coating thickness of the Cr template. After 7 days in culture, poly-l-lysine and laminin+polyethyleneimine were found to be formed more complete and clearer flow patterns by the application of microfluidic technique. On LN group, neurons were easy to cluster into clumps when channel width was 150μm,almost overshadowed flow pattern itself; On PEI group, even though the flow patterns are more complete, the neurites extend short and cannot constitute a connection between some cells

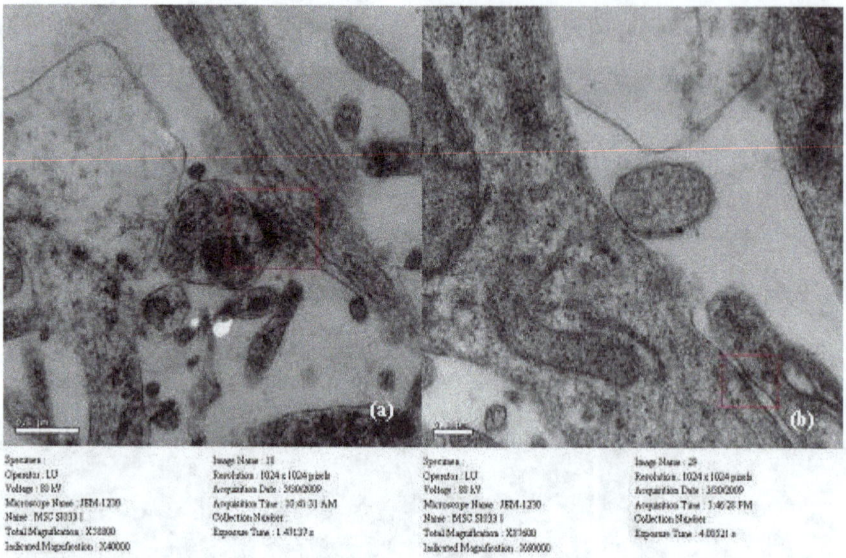

(a)chemical synapse structure x40000 bar=0.5 μ m (b)electrical synapse structure x60000 bar=0.2 μ m

Fig. 18. Under TEM, synapse like structure between neurons from the striatum and the substantial nigra on PEI patterns

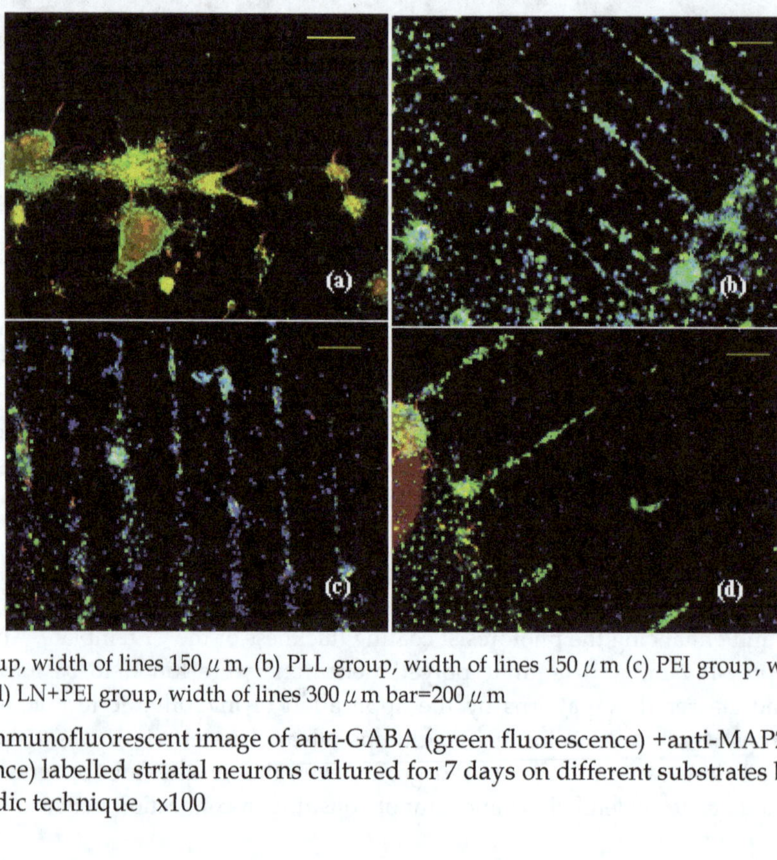

(a) LN group, width of lines 150 μ m, (b) PLL group, width of lines 150 μ m (c) PEI group, width of lines 200 μ m , (d) LN+PEI group, width of lines 300 μ m bar=200 μ m

Fig. 19. Immunofluorescent image of anti-GABA (green fluorescence) +anti-MAP2 (red fluorescence) labelled striatal neurons cultured for 7 days on different substrates by microfluidic technique x100 sufficiently. When the channel spacing of 300μm, LN + PEI group of cells in the liquid injection port at the distribution, flow pattern clear, but the group off between individual cells, so that flow interruption. Identified with immunocytochemical staining, we found that most of the MAP-2 positive neural cells from the striatum cultured on the flow pattern were also labeled with GABA, as shown in Figure 19.

Observing the fine structure of the intercellular on PLL group under scanning electron microscopic in Fig. 20, most of the neural cell bodies were found adhere to the flow pattern and majority of neurites are constrained within the width of the channel to grow following the orientation of flow channels. It is still visible that a few neuronal cell bodies deviate from the flow patterns slightly or adhere on the blank between two lines by local amplification. By observing the cross-linked region off low channel, we can see two endings of neurites seem to have varicose, swelling structure liking synapse.

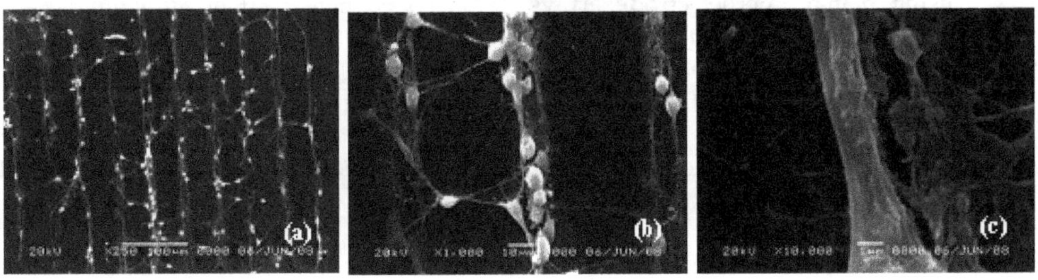

(a)x250 bar=100 μm (b)x1000 bar=10 μm (c)x10000 bar=1 μm

Fig. 20. Neurons from the striatum and the substantial nigra on PLL microfluidic patterns under SEM.

5. Conclusion

In summary, this study based on previous work improve the microfluidic technique, evaluate the influence of two different soft lithography, the micro-contact printing and microfluidic technique on various interface materials for the construction of neural network. Our future work can be divided into two levels. On the one hand continue to look for the intersection of microelectrode array and the micro-fabrication technology, trying to make cell grow in accordance with patterns of MEA electrode nodes. On the other hand, need to further improve the characteristics of electrode materials, enhance biocompatibility under the premise of improving signal-to-noise ratio and without adding resistance, do good to the survival of neurons and neurite extension.

The application of the micro-fabrication on microelectrode array may open up a broader platform of the technique for neurochip research and provide new ideas for the treatment of various injuries in the central nervous system. Ultimately, the combination both can achieve position fixing between neurons and electrodes precisely, to make the system as a real sensor, be able to accept electrical stimulation or chemical stimulation and record their signals, to analysis the transfer process of neural network information, and apply to drug screening of related diseases.

6. References

YN Xia, Whitesides G. M. (1998). Soft lithography. *Angewandte Chemie-International Edition*, Vol.37, No5, pp. 551-575.

Pilnam Kim, Kahp Yang Suh, et al. (2008). Soft Lithography for Microfluidics: a Review. *BIOCHIP* , Vol.2, No1, pp. 1-11,

Tai Hyun Park, Michael L. Shuler. (2003). Integration of Cell Culture and Microfabrication Technology, *Biotechnol. Prog,* Vol.19, pp. 243–253

Y. Nam, J. Chang, and B.C. Wheeler et al. (2004). Patterning to enhance activity of cultured neuronal networks. *Nanobiotechno,* Vol. 151, No. 3, pp. 109-115

Y.Nam, Katherine Musick, Bruce C. Wheeler. (2006)Application of a PDMS micromicrostencil as a replaceable insulator toward a single-use planar microelectrode array, *Biomed Microdevices,* Vol.8, No.4, pp. 375-381

Hiroaki Oka，et al.(1999). A new planar multielectrode array for extracellular recording: application to hippocampal acute slice, *J Neurosci methods,* Vol.93, pp. 61–67

Conrad D. James, et al.(2004). Extracellular recordings from patterned neuronal networks using planar microelectrode arrays, *IEEE Transactions on Biomedical Engineering,* Vol. 51, No. 9, pp. 1640 – 1648

Guangxin Xiang, et al.(2007). Microelectrode array-based system for neuropharmacological applications with cortical neurons cultured in vitro, *Biosensors and Bioelectronics* Vol.22, pp.2478–2484

Falconnet D ,Csucs G, Grandin HM ,et al. (2006). Surface engineering approaches to micropattern surfaces for cell-based assays.*Biomaterials,* Vol.27, No.16, pp. 3044 - 3063

Jackson A , Mavoori J , Fetz EE. (2006). Long-term motor cortex plasticity induced by an electronic neural implant. *Nature,* Vol.444, No.7115, pp. 56 -60

Lemon R. (2007). The Neurochip: Promoting Plasticity with a Neural Implant. *Current Biology,* Vol.17, No.2, pp. 54 -55

Maher M , Pine J , Wright J , et al. (1999). The neurochip: a new multielectrode device for stimulating and recording from cultured neurons. *J Neurosci Methods,* Vol.87, No.1, pp. 45 -56

Jenkner M , Muller B , Fromkerz P. (2001).Interfacing a silicon chip to pairs of snail neurons connected by electrical synapses. *Biol cyber,* Vol.84, No.4, pp. 239 -249

Fan YW , Cui FZ , Chen LN , et al. (2000). Improvement of neural cell adherence to silicon surface by hydroxyl ion implantation. *Surf Coat Tech,* Vol.131, No.1 23, pp. 355 - 359

Fan YW , Cui FZ , Chen LN , et al. (2002). Adhesion of neural cells on silicon wafer with nano-topographic surface. *Applied Surf Sci,* Vol.187, No.3 2 4, pp. 313 - 318

Kumar A , Whitesides GM. (1994). Patterned Condensation Figures as Optical Diffraction Grating. *Science,* Vol.263, No.5143, pp. 60 - 62

Chen, C S. Mrksich, M. Huang, S. Whitesides, G M, et al. (1997). Geometric Control of Cell Life and Death. *Science,* Vol.276, No.5317, pp. 1425–1428

Singhvi, R. Kumar, A.Lopez, G P, et al. (1994). Engineering cell shape and function. *Science,* Vol.264, No.5159, pp. 696–698

Bet, M R. Goissis, G. Vargas, S, et al. (2003). Cell Cell adhesion and cytotoxicity studies over polyanionic collagen surfaces with variable negative charge and wettability adhesion and cytotoxicity studies over polyanionic collagen surfaces with variable negative charge and wettability.*Biomaterials,* Vol.24, No.1, pp. 131–137

Folch, A. Jo, B H. Hurtado, O, et al. (2000). Microfabricated elastomeric stencils for micropatterning cell cultures. *J Biomed Mater Res,* Vol.52, No.2, pp. 346–353

Liu, B F. Ma, J. Xu, QY, et al. (2008). Development of an artificial neuronal network with post-mitotic rat fetal hippocampal cells by polyethylenimine. *Biosens Bioelectron* , Vol.23, No.8, pp. 1221 -1228

4

Photocontrolled Reversible Dimensional Changes of Microstructured Photochromic Polymers

Despina Fragouli, Roberto Cingolani and Athanassia Athanassiou
Center for Biomolecular Nanotechnologies @UNILE,
Istituto Italiano di Tecnologia
Italy

1. Introduction

Stimuli-responsive polymeric materials are able to change their chemistry and their conformation upon an external signal. The external signal may be derived from a change in temperature, chemical composition or applied mechanical force of the specific material, or can be triggered externally with exposure to an electric or magnetic field or to light irradiation. In this respect, a photochromic substance is a stimuli responsive material which is characterized by its ability to alternate between two different chemical forms having different absorption spectra, in response to light irradiation of appropriate wavelengths (Brown 1971). Due to this important property, a significant amount of effort has been devoted to the formation of polymeric materials functionalized with photochromic molecules for the creation of photosensitive "smart material" systems, that change reversibly their physical and chemical properties by the use of light. The corresponding reversible effects of the molecules such as dipole moment, surface energy, refractive index, and volume are preserved in the polymer matrix, and have numerous promising applications in devices for three-dimensional (3D) optical memories, (S. Kawata & Y. Kawata 2000), in actuators (Yu et al 2003, Athanassiou et al 2005), in holographic or diffractive optics, (Fu et al 2005, Tong et al 2005) or in microfluidics, (Caprioli et al 2007, Walsh et al 2010) etc. Concerning microfluidic devices using photochromic plastic films, the transportation of fluids happens without the need for their molecules to be charged, as done in other studies (Mitchel 2001). This is achieved by gradually modifying the surface tension, and thus the wettability, by irradiating with increasing time along the direction of the fluid movement (Ichimura et al 2000). The gradual wettability changes are exclusively based on the photochemical modification of the embedded photochromic molecules caused by the photoisomerization process. In addition, in the case of the diffraction gratings the development was generally done by interference of different polarized laser beams, or by electric-field application, and the modification of their diffraction efficiency is connected with the changes of the refractive index of the photochromic molecules during this procedure (Yamamoto et al 2001, Fu et al 2005).

Here we present how the volume changes induced to the photochromic polymers by the photoisomerization of their embedded photochromic molecules, can improve significantly the performance of these two different type of applications, namely the microfluidic devices and the diffraction gratings. In both cases the lithographic technique used for the microstructuring of the photochromic doped plastic films is the soft molding. Concerning the microfluidics applications, the presented microstructured photochromic plastic films exhibit a significant improvement on the reversible wetting characteristics compared to those on the flat surfaces. This improvement is due to the combination of the changes in the surface polarity and thus in the wetting properties with the modified surface conformation, both provoked by the light induced changes of the photochromic molecules. Moreover, regarding the optical gratings, we present a different approach in where the control of their diffraction efficiency relies on the dimensional variations of the gratings upon laser irradiation. Following this approach, the efficiency of the gratings is significantly improved with respect to previous works.

Such findings open the way for the production of optically switchable gratings based on reversible dimensional changes, and can be of great importance in all-optical signal processing systems. Moreover, the ability to control the wettability of surfaces by microstructuring and to tune it by using photochromic molecules, permits the application of these lithographically formed structures to all-optically controlled switches capable of operating with tunable speed, and to microfluidic actuation.

2. Materials and methods

2.1 Photochromic polymers and properties

Photochromic doped polymer films were prepared by incorporating photochromic molecules into polymer matrices. In particular, solutions of the polyethylmethacrylate-co-methylacrylate copolymer (PEMMA) (average molecular weight, M_w=100,000) (Aldrich) mixed with the photochromic molecule 1′,3′-dihydro-1′,3′,3′- trimethyl- 6-nitrospiro[2H-1-benzopyran-2,2′-(2H)-indole] or (6-NO_2 BIPS) (Aldrich) (weight ratio 90/10 respectively or 95/5) were prepared in toluene. Consequently a certain volume of this solution was spin-coated onto a glass substrate.

The photochromic dopant 6-NO_2 BIPS, belongs to the family of spiropyrans (SP) which have been extensively studied in the past decades (Görner 1998). Initially it has a 3D structure, and exists predominately in its non polar form (Figure 1). It is colorless, denoting that it is transparent in the visible range of the spectrum, but absorbs in the ultraviolet (UV). Upon irradiation with UV light, it is converted to its isomeric form, merocyanine (MC), through the photochemical cleavage of its carbon–oxygen (C_{spiro}-O) bond. The MC has a planar structure, it is colored and polar, and has a new absorption band in the visible range of the spectrum. MC can revert back to the SP form photochemically, using visible-light irradiation.

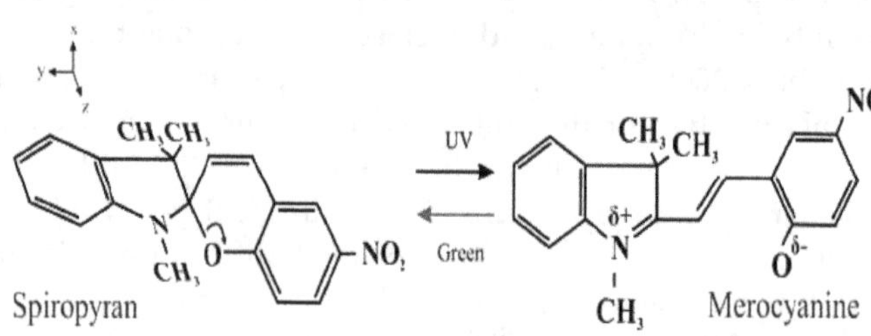

Fig. 1. The photochromic dopant spiropyran (SP) and the stable form of its isomer merocyanine (MC).

Briefly, UV and visible irradiation causes the reversible transformation of these chemical species, between two states (isomers) that have light absorption bands in characteristic spectral regions. This property is retained when the photochromic molecules are incorporated in polymer matrices, where they are homogeneously dispersed forming miscible systems. Specifically, the absorption properties of the photochromic polymer films prepared as described above change reversely upon UV-visible irradiation as shown at Figure 2. Initially the system is transparent at the visible range of the spectrum. Upon pulsed UV laser irradiation the SP is slowly converting to the MC isomer, fact indicated by the new absorption band in the visible region of the spectrum (ca. 565 nm). The intensity of the peak increases with the number of UV pulses until a plateau is reached, which suggests that the photoisomerization is completed and that the system has reached the equilibrium. The subsequent irradiation with green laser light, causes the decrease of the intensity of the previously formed MC peak, while after a certain number of pulses the spectrum reaches its initial form, indicating that MC reverts fully to the SP isomer. These data confirm that under the irradiation conditions mentioned in the figure caption of Figure 2, the reversible properties of SP are retained in the host polymer matrix. Depending on the irradiation conditions and the weight percentage of the photochromic molecules in the polymer matrix (usually ≤10%), it has been shown that typically about 4-10 irradiation cycles can be performed, while further irradiation causes the degradative photooxidation of the photochromic molecules, restricting thus the lifetime of the system (Athanassiou et al 2006c). Additionally, the degradative phenomena start to be evident usually after the third cycle. In order to exclude this parameter from the following study, results derived by the first three irradiation cycles are presented.

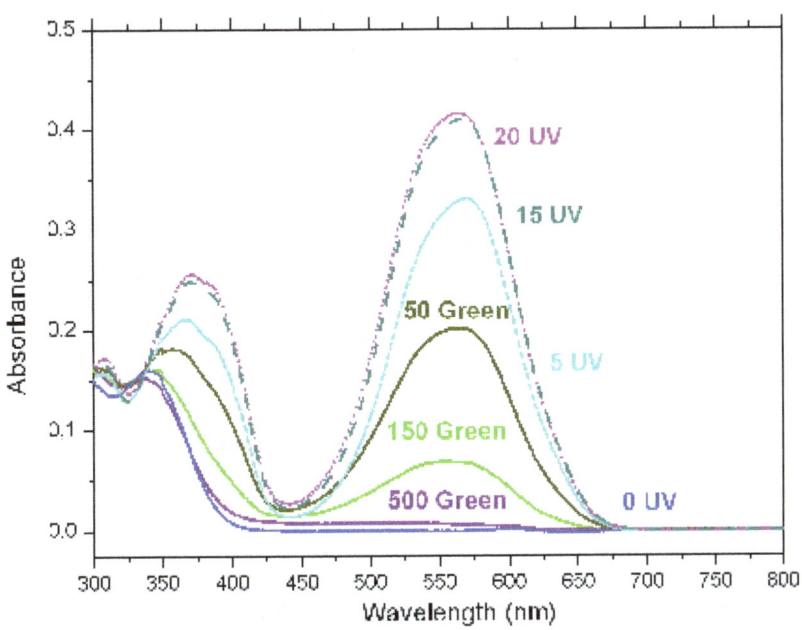

Fig. 2. Absorption spectra of the PEMMA/SP 10% wt upon UV and visible irradiation. For the specific study, the irradiation conditions used are: λ_{UV}=355 nm, fluence F_{UV}=20 mJ cm^{-2}, λ_{vis}=532 nm, F_{vis}=35 mJ cm^{-2}.

Except of the alteration in the optical properties, there are also other physical and chemical properties that change reversibly upon UV-visible irradiation even if the SP molecules are incorporated in the polymer matrix, such as dipole moment, surface energy, refractive index, and volume. Concerning the volume changes upon UV irradiation, occurs the formation of aggregates between different MC stereoisomers with zwitterionic character causing density fluctuations in the polymer matrix, reducing thus the MC partial molar volume. Consequently, takes place the short scale motion of the polymer chains in order to diminish the density fluctuations in the samples, and this leads to the macroscopic reduction of the dimensions of the matrix. This effect is reversible, since upon green irradiation, MC molecules return to the SP form, which does not form aggregates (Athanassiou et al. 2005)

2.2 Substrates microstructuring: soft molding

For the microstructuring of the photochromic plastic films it is used the soft molding lithography (SM). It is actually based on the conformal contact between the material to be patterned and an elastomeric replica of a master structure, and it combines soft and nanoimprint lithography, using elastomeric elements and exploiting the glass transition of organic compounds. Particularly, an elastomeric mold is placed onto a polymeric film applying the pressure of its own weight, and consequently is heated up above the films' glass transition temperature, T_g. The subsequent cooling down, below T_g, freezes the pattern into the polymer, and the replica is peeled off. The micropatterns formation is based on the capillarity effect that drives the polymer to penetrate into the recessed features of the elastomeric replica. The SM presents various advantages compared to the nanoimprint lithography. Specifically, since penetration of the polymer into recessed features of the replica is driven by capillarity effects, SM is only slightly affected by problems caused by difficult polymer transport. Moreover, it does not need any pressing setup to ensure contact between the mold and the polymer. Finally, SM does not present pattern shrinkage and distortion due to the solvents employed by other soft lithography techniques. (Pisignano et al 2004)

In Figure 3 it is represented the process followed for the SM. Initially, the original master structures are fabricated onto glass or Si by both photo- and electron-beam lithography. The realized masters are used as templates on which elastomeric replicas were realized using polydimethylsiloxane (PDMS, Sylgard 184, Dow Corning, Midland, MI) according to a standard replica molding procedure, and placed onto polymer films under their own weight. (Pisignano et al 2004)

In order to form the microstructured photochromic polymer substrates for the wettability study, elastomeric molds of PDMS having periods α= 1.3, 28.0, and 180.0 μm were placed onto flat spin-cast films. Then the system was heated at 50 °C, a temperature higher than the T_g of the PEMMA films (T_g = 48 °C). After the thermal cycle, the replica was easily peeled off from the photochromic polymer substrates, on which the patterns are transferred. For the preparation of the gratings on the films for the diffraction efficiency study, the substrates were placed on a hot plate and heated until they reach a temperature $T_{grating}$ of 65 °C. Then an elastomeric mold with α=4 μm was placed on the substrate for 10 min, resulting in the formation of the gratings. In both cases the SM procedure was carried out in nitrogen atmosphere to avoid the deterioration of the photochromic molecules upon heating.

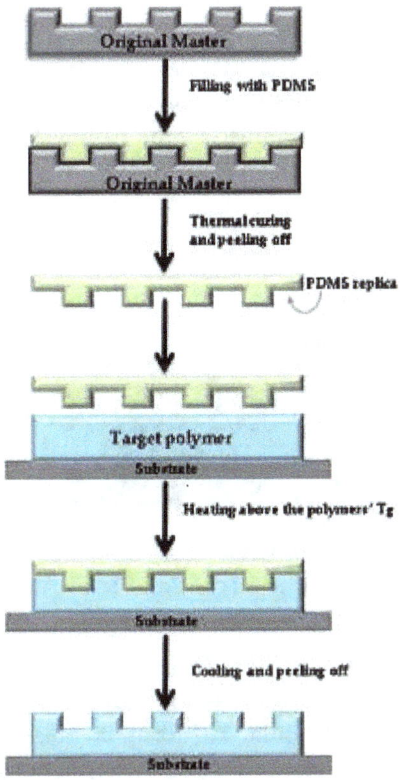

Fig. 3. Schematic diagram of the process of master replication and soft molding.

3. Light induced wettability changes of patterned substrates

3.1 Introduction

As abovementioned, the modification of the wetting characteristics of photochromic surfaces depends mainly on the photochemical processes which modify the surface tension and are caused by the UV-visible irradiation cycle. However, studies on patterned surfaces have demonstrated that the surface roughness affects significantly the wettability properties (Patankar 2003). Here it is studied the combined effect of the two aforesaid factors on the wettability properties of patterned photochromic polymeric surfaces. The wettability changes induced by photomechanical and photochemical changes are reversible upon UV-visible light irradiation, resulting in reversible changes of the wetting properties of the surfaces.

In particular, it is shown that the hydrophilicity of the photochromic polymeric surfaces is increased upon UV laser irradiation due to the polarity change caused by the photoisomerization, while the process is reversed upon green laser irradiation. The microstructuring of the surfaces enhances significantly the hydrophobicity of the system due to the increased surface roughness, and the light-induced wettability variations of the structured surfaces are enhanced by a factor of 3 compared to those on the flat surfaces. (Athanassiou et al 2006a, 2006b) In addition, by changing the topological parameters of the introduced pattern (e.g. by decreasing the period), are achieved higher differences in the surface wetting properties (Lygeraki et al 2008).

3.2 Results and discussion

Figure 4, shows the apparent water contact angle (WCA) of a drop on a flat photochromic polymer surface before and after pulsed UV and green laser irradiation. The WCA on the flat substrate, θ_Y, is formed when the liquid is in contact with a solid surface in static equilibrium with its vapor, and is determined by the Young's equation:

$$\gamma_{LV} \cos\theta_Y = \gamma_{SV} - \gamma_{SL} \quad (1)$$

where γ_{LV}, γ_{SV} and γ_{SL} represent the interfacial tensions at the boundaries between the liquid (L), vapor (V), and solid (S).

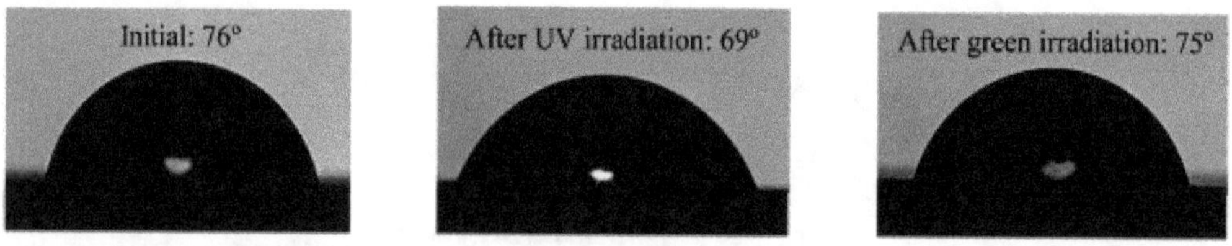

Fig. 4. WCA of a water drop of volume 3 μl on flat photochromic polymer surface before and after UV-green laser irradiation. (5% wt of SP in PEMMA, λ_{UV}=308 nm, F_{UV}=40 mJ cm^{-2}, λ_{green}=532 nm, F_{green}=45 mJ cm^{-2}) (Athanassiou et al 2006a)

As shown, before any irradiation the surface is hydrophilic with a WCA of 76°. After irradiation with enough UV laser pulses so as to reach the complete photoisomerization of all the non polar SP molecules to the polar MC form (50 pulses), the surface becomes more hydrophilic, with a WCA of 69°. The subsequent green irradiation with 500 laser pulses causes the reversible phenomenon, which is the conversion of the MC isomer to the SP form, and the increase of the WCA until it reaches its initial value. The maximum WCA difference measured on numerous flat surfaces upon UV irradiation was 7°±1°.

Fig. 5. Atomic force microscopy images of typical patterned surfaces of photochromic polymers. (Athanassiou et al 2006a)

In order to examine the effect of the photomechanical changes upon the same irradiation conditions, the aforementioned surfaces were microstructured using the SM technique. The replica used had a period α=1.3 μm, and by following the already described steps there were formed on the photochromic polymeric surface patterns of the same period (Figure 5). The WCA on the specific surface was found to be greatly affected by the patterning, showing an increased hydrophobicity with a value almost 30° greater than that of the flat surfaces (Figure 6). Moreover, always in comparison with the flat surfaces, the light induced WCA changes due to the photoisomerization effect are enhanced by a factor of 3, since the WCA change before and after UV irradiation is ca. 20° (WCA change on the flat surface, ca. 7°). This guides to the conclusion that the microstructuring affects significantly the reversible photoinduced wettability changes of the surfaces.

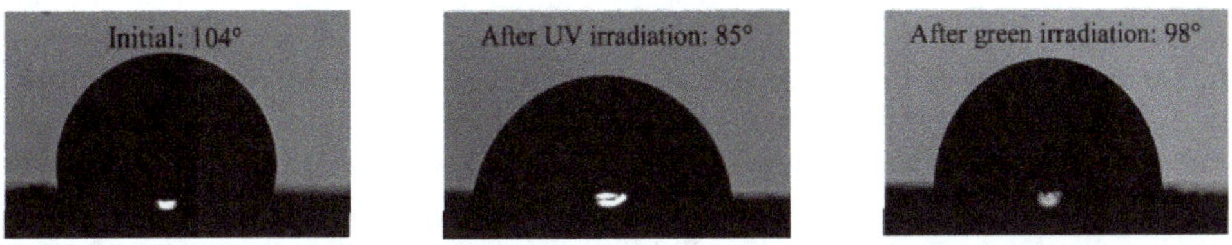

Fig. 6. WCA images obtained on patterned surfaces before and after laser irradiation. (5% wt of SP in PEMMA, λ_{UV}=308 nm, F_{UV}=40 mJ cm^{-2}, λ_{green}=532 nm, F_{green}=45 mJ cm^{-2}) (Athanassiou et al 2006a)

In order to explain the effect of roughness on the wetting characteristics of a surface, there are proposed two theories. The first is referred to as the Cassie-Baxter model (Cassie and Baxter 1944) (Figure 7a), and describes the wettability of rough surfaces, where only partial wetting may occur due to the trapping of air underneath the drop at the recessed regions of the surfaces. Since the drop is situated partially on air, the surface exhibits an enhanced hydrophobic behavior. The second one is the Wenzel model (Wenzel 1936), and it proposes that roughness increases the liquid-solid interfacial area, and thus hydrophilic surfaces (θ<90°) become more hydrophilic, and hydrophobic (θ>90°) more hydrophobic (Figure 7b).

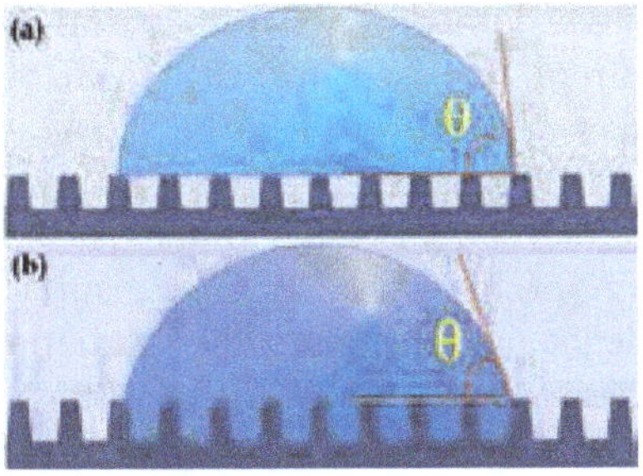

Fig. 7. Representation of a drop on a patterned surface, according to Cassie-Baxter model (a) and to Wenzel model (b).

In the presented cases the WCA of the drop on the flat surface is 76° thus hydrophilic (<90°) and according to the Wenzel model the patterning should increase its hydrophilicity. However, the experimental results presented above show that the WCA after the patterning is significantly increased, reaching a maximum value of 104°, indicating that the surface became hydrophobic (>90°). Thus, the presented surfaces follow the Cassie-Baxter model, where the relation with the WCA of the flat surface (θ_Y) is given by the following equation:

$$\cos\theta_r = -1 + f_s(1 + \cos\theta_Y) \tag{2}$$

with f_s the solid fraction of the surface in contact with the liquid. According to this equation, the WCA decreases when the f_s is increasing. Using the WCA (θ_Y) measured on the flat surface and the one of the patterned in all three cases, before, after UV, and after green irradiation, the f_s is calculated. The results show that the fraction of the patterned surface in contact with the liquid is increased after UV irradiation (f_s=0.8) compared to the no irradiated sample (f_s=0.6). AFM microscopy studies of the topological changes of the gratings after UV irradiation, showed that the average volume decrease of each nanoimprinted stripe is ca. 30 nm. However, since the f_s is higher, and the stripes of the pattern are narrower, it is concluded that the water drop penetrates deeper into the channels of the UV irradiated pattern (partial wetting-Figure 8b), decreasing thus the WCA value. After irradiation with green laser pulses the MC molecules return to their SP isomers, the stripes recover their previous volume, and thus the drop returns to its previous condition, demonstrating once more that the wetting behavior is greatly influenced by both, the changes in the surface polarity and the volume of the stripes (Athanassiou et al 2006a).

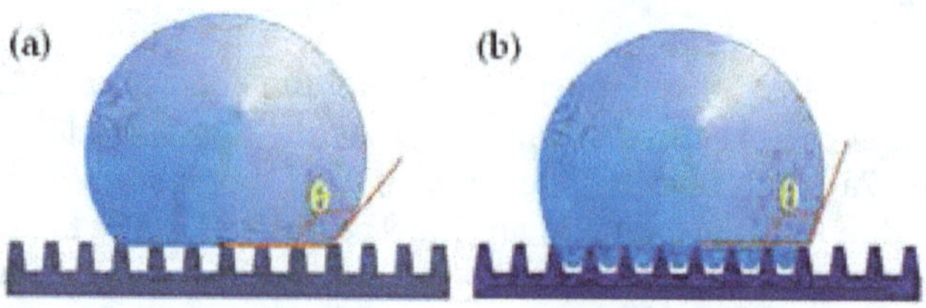

Fig. 8. Representation of a drop on a patterned surface, according to Cassie-Baxter model, before (a) and after (b) UV irradiation.

A further demonstration of the abovementioned statement is the study of the wetting properties of patterned surfaces with different topological parameters. In particular, the periods of the elastomeric replicas used for the formation of the patterns on the photochromic polymer surfaces (10% wt of SP in PEMMA) were 1.3, 28.0, and 180.0 μm. Figure 9a demonstrates the WCA changes upon three UV-green irradiation cycles for each of the patterned substrates and the comparison with the flat surface. For comparison reasons it is presented also the WCA of the pure polymeric surface, without any addition of photochromic molecules, which is not affected by the irradiation as expected. As shown, the patterning of the surfaces increases the initial WCA in all three cases. However, the smaller the period of the pattern the more hydrophobic is its behavior compared to the flat surface. After UV irradiation, all substrates become more hydrophilic, but again the WCA difference from the initial one is greater at the pattern with the smaller period, and becomes smaller as the period is increasing (Figure 9b).
The observed difference between the various patterns cannot be attributed to the surface chemistry changes upon UV-green irradiation, since this is the same in all samples. Additionally, if this was happening, it should be observed the inverse phenomenon, thus the greater change for the pattern with the greater period, since in these cases the drops are in contact with a greater percentage of the solid surface, as seen from the calculated value of f_s at Table 1. Thus, the reported differences can be attributed to the volume shrinkage of the patterns upon UV irradiation. This reduction of the patterns volume affects much more the contact angle of the samples with patterns of smaller period, since the drop lies on a greater number of patterned features. Indeed it is calculated that the radius of the drop is 20 times

larger than the size of each feature projected on the plane of the interface in the case of the patterns with a 180 μm period, about 140 times greater in the case of the patterns with a 28 μm period, and finally more than 2950 times greater in the case of the patterns with a 1.3 μm period. Accordingly, as shown at the table, for the pattern with period of 180 μm there is a slight change of the *f* upon UV-green irradiation. This is not the case for the smaller pattern where the change upon UV-green irradiation is almost 15%, while for the pattern of period 28 μm there is an intermediate change of ca. 6%.

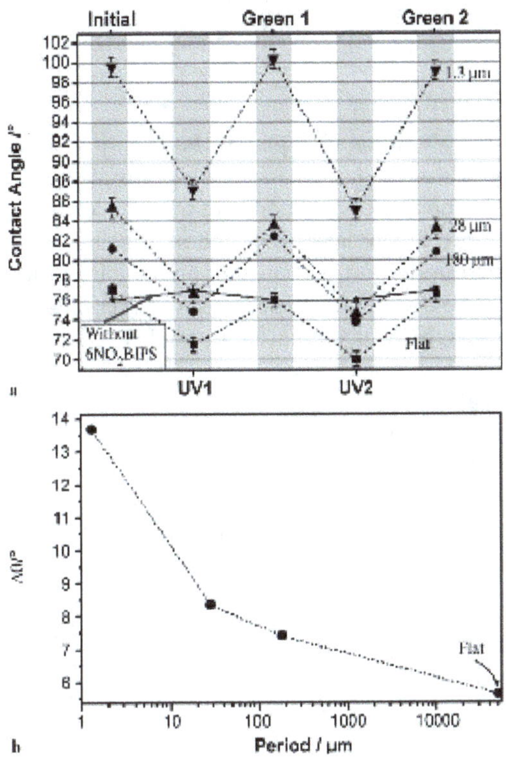

Fig. 9. (a) Mean WCA values of patterned surfaces of 10% wt SP in PEMMA for each pattern, for the flat surface and for the PEMMA surface. The media is taken after studying 10 samples of each case. (b) WCA difference (Δθ) from the initial one after UV irradiation. (λ_{UV}=308 nm, F_{UV}=20 mJ cm^{-2}, λ_{green}=532 nm, F_{green}=25 mJ cm^{-2}) (Lygeraki et al 2008)

Period	1.3μm	28μm	180μm
Initial	0.68	0.88	0.94
UV1	0.80	0.94	0.96
Green1	0.66	0.89	0.91
UV2	0.81	0.94	0.95
Green2	0.68	0.91	0.94

Table 1. Factor *f* calculated by the Cassie–Baxter model under different irradiation conditions.

In conclusion, at this section it is presented the possibility to create both hydrophobic and hydrophilic surfaces starting from the same photochromic polymeric sample by changing the topological parameters of its surface features using soft molding lithography. Due to the photochromic transformations taking place upon alternating UV and green irradiation, these surfaces can reversibly change their wettability. By careful control of the surface topology these changes can be fully controlled and tuned, in such a way that the surfaces can be wetted in a reversible manner.

4. Reversible diffraction efficiency changes of photochromic polymer gratings

4.1 Introduction

The photoinduced reversible volume changes of the lithographically patterned features of such materials are also used for realizing diffraction gratings, operating in the Raman-Nath regime, with controlled and reversible diffraction efficiency (DE). In particular, the alternating pulsed UV and green laser irradiation on the gratings causes the doped photochromic molecules to undergo transformations, which induce reversible dimensional changes to the samples. These volume changes cause reversible changes to the diffraction efficiency of the gratings, which is increased upon UV irradiation and decreased after irradiation with green laser light for various irradiation cycles. The experimental results are confirmed by a theoretical diffraction model. It is proved that the diffraction efficiency changes are attributed exclusively to the reversible dimensional changes of the imprinted structures (Fragouli et al 2008), and not to the refractive index changes as is the case in the majority of previous works. Specifically, most of the gratings with similar thickness as the here presented, are produced after irradiation with interfering beams that cause variation of the refractive index inside the samples mainly consisting of azobenzene polymers and liquid crystals. In these studies the switching procedure in the DE is due to the formation and deletion of the gratings (Yamamoto et al 2001, Fu et al 2005). Moreover, although the thickness of the produced gratings is very small, they exhibit higher DE than most of the volume gratings prepared in this regime (Tong et al 2005, Yamamoto et al 2001, Fu et al. 2005).

4.2 Results and discussion

A grating of period 4 μm and thickness of ca. 240 nm was formed on the photochromic polymeric surface by the SM technique as described at section 2. The experimental setup used for the study of the diffraction efficiency (DE) of the formed grating, is demonstrated in Figure 10. As shown, a continuous wave diode laser operating at λ=822nm, with an incident angle adjusted to have the maximum intensity of the first-order diffracted line (I_1), was used as a reading beam. This wavelength was chosen in order not to be absorbed by the photochromic sample. The zero-order transmitted (I_0) and the I_1 lines, are measured by two photodiodes, and were used for the calculation of the DE (equation 3).

$$DE = I_1/I_0 \tag{3}$$

The DE was calculated for three irradiation cycles, and Figure 11 shows the DE relative change in each case. Initially, the DE of the grating was ca 2.2%, and upon UV laser irradiation it was increasing until it reached a final value. After two minutes in the dark in order to leave time to the system to relax, subsequent irradiation with green laser pulses caused the DE to recover close to its initial value. It is worth noticing that the relative changes of the DE during the first irradiation cycle exhibit big variations between the various examined samples, in contrary with the following irradiation cycles, where the changes are similar for all the examined gratings. This is mainly attributed to internal stresses of the polymer matrix, produced during the preparation of the gratings, that are released in a random way upon irradiation (Liang et al 2007). Nevertheless, at the second cycle during UV irradiation the DE increases with increasing number of pulses, until it stabilizes to an average value of approximately 7.4±3.0% with respect to its initial value. After green irradiation the DE is slowly reaching its initial value with increasing number of pulses. This behavior is repeated also at the third cycle.

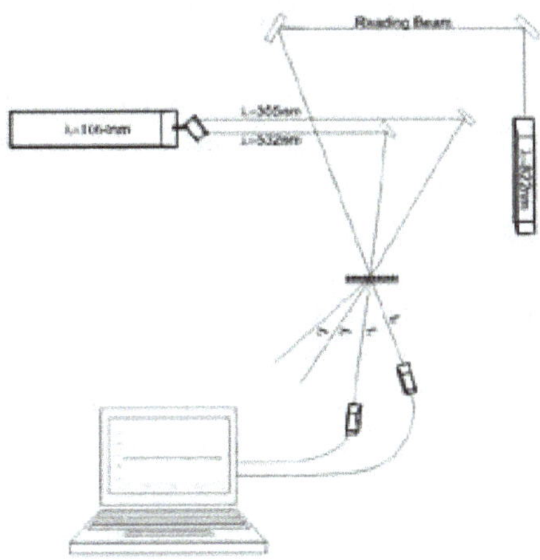

Fig. 10. Experimental setup for the measurement of the diffraction efficiency of the photochromic gratings (10% wt. SP in PEMMA) (λ_{UV}=355nm, F_{UV}=20 mJ cm^{-2}, λ_{green}=532 nm, F_{green}=35 mJ cm^{-2}, $\lambda_{reading\ beam}$=822 nm.)

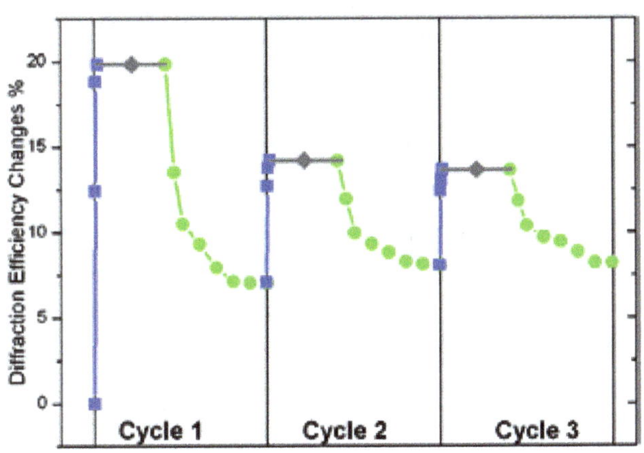

Fig. 11. Diffraction efficiency changes of the grating upon UV-green irradiation. (Fragouli et al 2008)

In order to examine the effect of the refractive index (n) change of the photochromic polymer sample upon UV-green irradiation on the observed change to the DE, ellipsometric measurements on a similar sample were conducted. The results show that before any irradiation for λ=822 nm, the refractive index of the sample is n=1.509. After UV irradiation, the n is higher by Δn=0.029. This difference is very small compared to the periodic refractive index variations in the grating between the photochromic polymer (1.509) and the air (1), which is actually what causes diffraction to occur. Thus it is believed that this change plays a negligible role in the measured DE relative changes. Furthermore, the small thickness of the gratings (ca. 240 nm) reduces the importance of the Δn even more.

Thus, the reversible DE changes can be attributed exclusively to the light-induced macroscopic deformations of the gratings. Specifically, Figure 12 illustrates the reversible macroscopic changes of the grating before and after UV-green irradiation as taken by AFM microscopy. As shown, the width of the stripes of the grating ($\alpha-\beta$) is decreased by ca. 13% after UV irradiation while the distance between the two stripes (β) is increased. A small decrease is also observed in the period of the grating (α) (α and β before UV, 3.971 μm and 2.366 μm respectively; α and β after UV, 3.842 μm and 2.449 μm respectively). After the subsequent irradiation with green light the values recover very close to the initial ones. It is worth noticing that, as shown at Figure 12, there is a dip separating each stripe in two equal parts. As already mentioned at section 2.2, the SM technique which is followed for realizing the grating relies mainly on the capillarity that allows the viscous polymer to spontaneously fill the vertical channels that are made of the recessed features of the elastomeric mold, since the wetting lowers the overall free energy. There is always the possibility that the photochromic polymer may not fill completely such regions, thus may be mostly accumulated in the regions that are adjacent to the protruding areas of the mold, forming thus dips in the central part of the growing capillarity features. This behavior is common to different imprint lithography methods (Zankovych et al 2001, Hong and Lee 2003, Pisignano et al 2004). However, this dip is useful for the AFM morphological analysis of the patterned surfaces, since it makes the volume changes upon UV-green irradiation cycles much clearer. Moreover, it is too narrow to give any contribution to the diffracted light from the grating.

In order to compare the experimental result with the existing theory, the basic equation that describes the intensity distribution of monochromatic light passing through a grating, was used (equation 5). (Born and Wolf 1999)

$$\frac{I}{I_0}=\left[\frac{\sin(\pi\beta p/\lambda)}{\pi\beta p/\lambda}\right]^2\left[\frac{\sin(N\pi\alpha p/\lambda)}{\sin(\pi\alpha p/\lambda)}\right]^2 \tag{5}$$

I and I_0 are the intensities of the light after the grating at various orders of diffraction and at zero order respectively, β is the distance between two successive stripes, α is the period of the grating, N is the number of stripes, λ is the wavelength of the reading beam, and $p = \sin\theta - \sin\theta_0 = m\lambda/\alpha$ (m=0, ±1, ±2 etc) where θ_0 is the angle of incidence and θ the angle of diffraction. The number of the stripes covered by the reading beam was calculated by dividing the diameter of the spot of the beam by the period of the grating in each case. The angle of incidence of the reading beam was θ_0=20°. In each case, by the AFM images it was measured the different value of α and β before and after UV-green irradiation. Taking into account the measured parameters by the experiment and using the equation 5, it was calculated the ratio I_1/I_0, and consequently the DE for the gratings before and after the UV-green irradiation. The theoretical calculations confirm that there is an increase of the DE after the photoisomerization process, by 8.7%. The agreement between the calculated and experimental (ca. 7.4±3.0%) values is notable, taking into account the experimental error due to factors such as the exact value of the laser beam spot size, which leads to an approximate value of the observed stripes, the imperfections of the surface introduced during the grating formation, and so on. The theoretical calculations presented, demonstrate that the decrease of the dimensions of the stripes of the gratings and of the period, are the main parameters that define the change in the DE upon irradiation (Fragouli et al 2008).

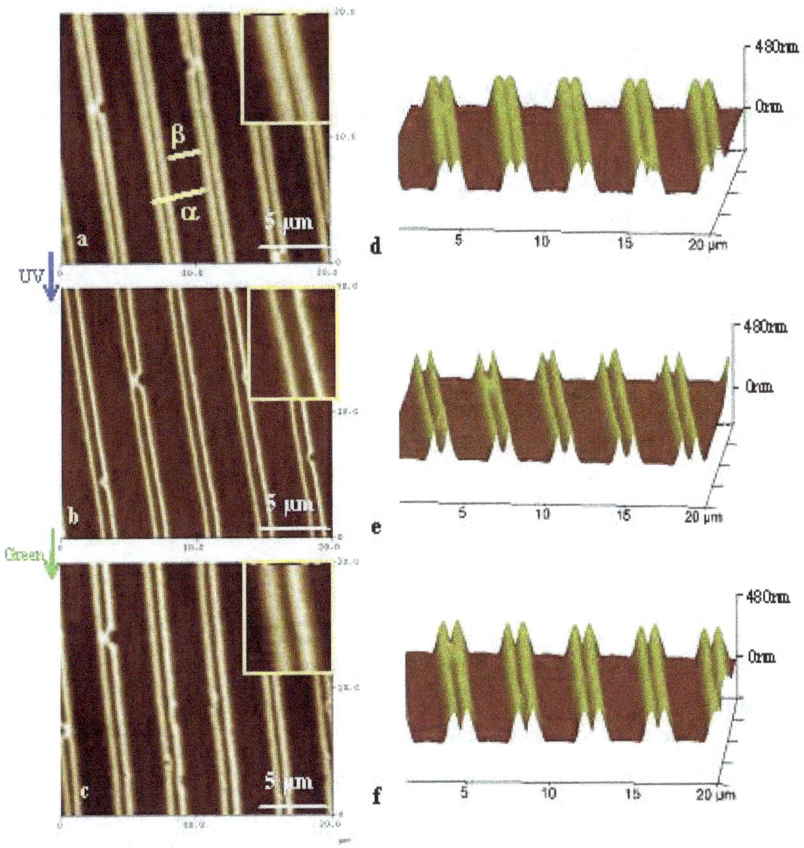

Fig. 12. AFM images of a grating before any irradiation (a) after irradiation with 20 UV pulses (b), and after exposure to 600 green pulses (c). The insets demonstrate a single stripe in magnification. 3D images (d–f) of the grating shown in (a–c), respectively. (α is the period and β is the distance between the stripes, measured at the full width at half-maximum of each feature. (Fragouli et al 2008)

5. Conclusions

In conclusion, it is demonstrated how soft molding lithography can be employed for the preparation of microstructured photochromic polymeric films, which undergo light controlled photomechanical changes responsible for the control of some functional characteristics of the patterned surfaces, namely the wetting properties and the diffraction efficiency. In particular, the light-induced isomerization of the embedded photochromic molecules in the flat surfaces is exclusively responsible for the reversible changes in their wetting properties. When the surface is microstructured by realizing patterns with the SM technique, these wetting properties are greatly enhanced. Moreover the control of the characteristics of the patterns (eg. the period), makes possible to control the light induced alterations in the wetting properties of the structured surface, demonstrating that they are influenced by both the changes in the surface polarity and the volume changes of the patterned structures. Finally, last but not least, it is demonstrated the possibility of fully manipulate the diffraction efficiency of thin photochromic polymer gratings. It is shown, that the produced gratings change their diffraction efficiency in a reversible way upon UV-green laser irradiation. This effect, which is verified also by a theoretical diffraction model, is attributed to the reversible dimensional changes of the imprinted structures, and not to the refractive index changes as is the case in the majority of previous work. Such findings open a way for the production of optically switchable gratings based on reversible dimensional changes. Moreover, the ability to control the wettability of surfaces by microstructuring and to tune it by using photochromic molecules opens the way to the application of these optimized patterns to various microfluidic devices.

6. Acknowledgment

The authors would like to thank Dr. D. Pisignano, and Dr. E. Mele of the Center for Biomolecular Nanotechnologies @UNILE, Istituto Italiano di Tecnologia Via Barsanti, 73010 Arnesano (LE) and Dr. L. Persano of the National Nanotechnology Laboratory (NNL) of CNR - Istituto di Nanoscienze, via per Arnesano, 73100 Lecce, Italy for the formation of the patterned structures presented in this work.

7. References

Athanassiou A.; Kalyva M.; Lakiotaki K.; Georgiou S. & Fotakis C. (2005). All-optical reversible actuation of photochromic-polymer microsystems. *Advanced Materials*, Vol.17, No.8, (April 2005), pp. (988-992)

Athanassiou A.; Lygeraki M. I.; Pisignano D.; Lakiotaki K.; Varda M.; Mele E.; Fotakis, C.; Cingolani R. & Anastasiadis S. H. (2006) Photocontrolled Variations in the Wetting Capability of Photochromic Polymers Enhanced by Surface Nanostructuring. *Langmuir*, Vol.22, pp. (2329-2333)

Athanassiou A.; Varda M.; Mele E.; Lygeraki M. I.; Pisignano D.; Farsari M.; Fotakis, C.; Cingolani R. & Anastasiadis S. H. (2006) Combination of microstructuring and laser-light irradiation for the reversible wettability of photosensitised polymer surfaces. *Applied Physics A*, Vol.83, pp. (351-356)

Athanassiou, A.; Sahinidou, D. ; Arima, V.; Georgiou, S.; Cingolani, R. & Fotakis, C. (2006) Influence of laser wavelength and pulse duration on the degradation of polymeric films embedding photochromic molecules. *Journal of Photochemistry and Photobiology A*, Vol.183, pp.(182-189)

Born M. & Wolf E. (1999) Principles of Optics: Electromagnetic theory of propagation, interference and diffraction of light, 7th Edition, Cambridge University Press.

Brown G. H. (1971) Photochromism, Willey-Interscience, New York

Caprioli L.; Mele E.; Angilè F. E.; Girardo S.; Athanassiou A.; Camposeo A.; Cingolani R. & Pisignano D. (2007) Photochromic spiropyran monolithic polymers: Molecular photo-controllable electroosmotic pumps for micro-fluidic devices. *Applied Physics Letters*, Vol.91, 113113

Cassie, A. B. D. & Baxter, S. (1944) Wettability of porous surfaces. *Transaction of the Faraday Society*, Vol.40, pp.(546-551).

Fragouli, D.; Persano, L.; Paladini, G.; Pisignano, D.; Carzino, R.; Pignatelli, F.; Cingolani, R. & Athanassiou A. (2008) Reversible Diffraction Efficiency of Photochromic Polymer Gratings Related to Photoinduced Dimensional Changes. *Advanced Functional Materials*, Vol.18, pp.(1617–1623).

Fu, S.; Liu, Y.; Dong, L.; Lu, Z.; Hu, W. & Xie, M. (2005). Photo-dynamics of polarization holographic recording in spirooxazine-doped polymer films. *Materials Letters*, Vol.59, pp. (1449-1452)

Fu, S.; Liu, Y.; Lu, Z.; Dong, L.; Lu, Z.; Hu, W. & Xie, M. (2005) Real-time holographic gratings recorded by He–Ne laser in polymer films containing spirooxazine compounds pre-irradiated by UV light. *Optical Materials*, Vol.27, pp.(1567-1570).

Görner H. (1997). Photoprocesses in spiropyrans and their merocyanine isomers: Efects of temperature and viscosity. *Chemical Physics*, Vol.222, pp. (315-329)

Hong, P. S. & Lee, H. H. (2003) Pattern uniformity control in room-temperature imprint lithography. *Applied Physics Letters*, Vol.83, 2441.

K. Ichimura, S.-K. Oh, M. Nakagawa, (2000) Light-Driven Motion of Liquids on a Photoresponsive Surface. Science, Vol.288, pp. (1624-1626).

Kawata, S. & Kawata, Y. (2000). Three -dimensional optical data storage using photochromic materials. *Chemical Reviews*, Vol.100, No.5, pp. (1777-1788)

Liang, G.-D.; Xu, J.-T.; Fan, Z.-Q.; Mai, S.-M. & Ryan, A. J. (2007) Effect of Substrate and Molecular Weight on the Stability of Thin Films of Semicrystalline Block Copolymers. *Langmuir*, Vol.23, pp. (3673-3679).

Lygeraki M. I.; Tsiranidou E.; Anastasiadis S. H.; Fotakis, C.; Pisignano D.; Cingolani R. & Athanassiou A. (2008) Controlling the reversible wetting capability of smart photochromic-polymer surfaces by micro patterning. *Applied Physics A*, Vol.91, pp. (397 - 401)

Mitchell, P. (2001) (2001) Microfluidics—downsizing large-scale biology. *Nature Biotechnology*, Vol.19, pp. (717-721).

Patankar, N. A. (2003) On the Modeling of Hydrophobic Contact Angles on Rough Surfaces. *Langmuir*, Vol.19 No.4, pp (1249–1253).

Pisignano D.; Persano L.; Cingolani R.; Gigli G.; Babudri F.; Farinola G. M. & Naso F. (2004) Soft molding lithography of conjugated polymers. *Applied Physics Letters*, Vol.84, No.8, (February 2004), pp. (1365-1367)

Tong, X.; Wang, G.; Yavrian, A.; Galstian, T. & Zhao Y. (2005) Dual-mode switching of diffraction gratings based on azobenzene-polymer-stabilized liquid crystals. *Advanced Materials*, Vol. 17, No.3, pp. (370-374).

Walsha Z.; Scarmagnani S.; Benito-López F.; Abele S.; Nie F.-Q.; Slater C.; Byrne R., Diamond D.; Paull B. & Macka M. (2010) Photochromic spiropyran monolithic polymers: Molecular photo-controllable electroosmotic pumps for micro-fluidic devices. (2010) *Sensors and Actuators B*, Vol.148, pp. (569–576)

Wenzel, R. N. (1936) Resistance of solid surfaces to wetting by water. *Industrial and Engineering Chemistry*, Vol.28, No.8, pp.(988-994).

Yamamoto, T.; Ohashi, A.; Yoneyama, S.; Hasegawa, M.; Tsutsumi, O.; Kanazawa, A.; Shiono, T. & Ikeda, T. (2001) Phase-Type Gratings Formed by Photochemical Phase Transition of Polymer Azobenzene Liquid Crystal. 2. Rapid Switching of Diffraction Beams in Thin Films. *Journal of Physical Chemistry B*, Vol.105, pp. (2308-2313).

Yu Y.; Nakano M. & Ikeda T. (2003). Directed bending of a polymer film by light. *Nature*, Vol.425, (September 2003), pp. 145

Zankovych, S.; Hoffmann, T.; Seekamp, J.; Bruch, J.-U. & Sotomayor Torres C. M. (2001) *Nanotechnology*, Vol.12, 91. Nanoimprint lithography: challenges and prospects.

5

Fabrication of Surfaces with Bimodal Roughness through Polyelectrolyte/ Colloid Assembly

Christine C. Dupont-Gillain*, Cristèle J. Nonckreman, Yasmine Adriaensen and Paul G. Rouxhet

Institute of Condensed Matter and Nanosciences - Bio & Soft Matter, Université Catholique de Louvain
Belgium

1. Introduction

From bioengineering to optics and electronics, a great deal of work has been conducted on the development of new materials with structured surfaces. A large range of methods has been used, such as plasma etching, electron beam and colloidal lithography (Denis et al., 2002), electrical deposition (Yang et al., 2009), phase separation (Dekeyser et al., 2004) and polyelectrolyte assembly in order to produce structured surfaces (Agheli et al., 2006). The combination of different methods is also more and more explored. Schaak et al. (2004) described a simple approach to achieve colloidal assembly on a patterned template obtained by lithography.

Densely packed layers of colloidal particles can be produced by lifting a substrate vertically from a suspension (Fustin et al., 2004; Li J. et al., 2007) or by spin coating (Yang et al., 2009). Colloidal lithography utilizes the ability of particles to adhere on oppositely charged surfaces (Johnson & Lenhoff, 1996; Hanarp et al., 2001, 2003)), possibly using surface modification by inorganic or organic polyelectrolytes. The surface coverage is influenced by several factors: ionic strength, particle size and time. The adhesion of microbial cells on various substrates was also achieved by surface treatments with inorganic or organic polycations (Changui et al., 1987; Van haecht et al., 1985) or with positively charged colloidal particles (Boonaert et al., 2002). A review on colloidal lithography and biological applications was published recently (Wood, 2007).

Adsorption of polyelectrolytes is influenced by ionic strength, pH and the polyelectrolyte characteristics (molecular mass, charge density) (Lindquist & Stratton, 1976; Davies et al., 1989; Choi & Rubner, 2005). At low ionic strength, highly charged polyelectrolytes adopt extended conformations and are fairly rigid due to the strong repulsion between charged units. The maximum adsorbed amount and the adsorbed layer thickness do not vary markedly according to molecular weight. As the salt concentration is increased and the electrostatic intrachain repulsion is decreased, the polyelectrolyte becomes more coiled. In this case, the maximum amount adsorbed (expressed in mass) increases as a function of molecular weight (Roberts, 1996; Lafuma, 1996; Claesson et al., 2005; Boonaert et al., 1999).

Build up of polyelectrolyte films may be achieved using layer-by-layer assembly through alternating adsorption of oppositely charged polyelectrolytes (Decher & Hong, 1991).

Polyethyleneimine (PEI), polyallylamine (PAH), poly-l-lysine (PLL) and polydiallyldimethylammonium chloride (PDDA) are common polycations used for multiple layer formation with polyanions such as polystyrene sulfonate (PSS) (Bertrand et al., 2000). Layer-by-layer assembly of polyelectrolytes has been combined with the use of colloidal particles. For instance, a film was made on silicon wafers precoated with thermally evaporated titanium, by adsorption of PDDA, followed by adsorption of PSS, followed by treatment with an aluminium chloride hydroxide solution (Hanarp et al., 2001, 2003). In another study, chemically patterned surfaces made by self-assembled monolayers (SAMs) were covered with a polyelectrolyte multilayer film, before adhesion of SiO_2 silica particles or functionalized polystyrene latex particles (Chen et al., 2000). While a polyelectrolyte layer may provide a strong bond to colloidal particles, the drying process applied after particle adhesion may be crucial to obtain a regular and homogeneous monolayer (Hanarp et al., 2003).

Structured hydrophobic surfaces have gained increasing interest because the roughness amplifies the hydrophobicity (Wenzel, 1936). This is exemplified by the Lotus effect, in which a dual size roughness seems to be important (Barthlott & Neinhuis, 1997; Patankar, 2004). Raspberry-like surface morphologies were created in different ways : styrene polymerization (Perro et al., 2006; Reculusa et al., 2002) or gold sputtering (Xiu et al., 2006) on silica particles , controlled aggregation of different surface-functionalized silica particles (Ming et al., 2005) or direct electrochemical synthesis of gold microaggregates (Li Z. et al., 2007) and immobilization on a specific substrate.

In this paper, we prepare surfaces covered with a homogeneous monolayer of colloidal particles, using adhesion of negatively charged polystyrene latex beads on a polycation-precoated glass substrate. The method is then extended to prepare surfaces presenting a bimodal roughness, by using latex particles of different sizes. The influence of substrate surface roughness on the behavior of mammalian cells has been of great concern in the last years (Nonckreman et al., 2010). Therefore, the stability of fabricated samples is tested in phosphate buffer saline (PBS), which simulates the pH and ionic strength of biological fluids. Note that here, the term "colloid" is used with the restrictive sense of lyophobic colloidal particle, and thus distinguished from polyelectrolytes.

2. Materials and methods

2.1 Materials

Glass substrates were microscope coverslips (12 mm diameter, Menzel-Gläzer, Germany). Polyethyleneimine solution (PEI, M_w = 750 000), polyallylamine hydrochloride (PAH, M_w~70 000), poly-l-lysine hydrobromide (PLL, M_w = 70 000 to 150 000), poly(diallyldimethylammonium chloride) solution (lPDDA, M_w < 100 000; hPPDA, M_w = 400 000 to 500 000) were purchased from Sigma-Aldrich (St. Louis, MO, USA). Negatively charged polystyrene particles with two different sizes (sulfate latex, 65±6nm, 470±12nm) were purchased from Interfacial Dynamic Corporation (IDC, Tualatin, OR, USA). The composition of the phosphate buffer saline solution (PBS) was as follows: 137 mM NaCl, 6.4 mM KH_2PO_4, 2.68 mM KCl, 8 mM Na_2HPO_4; adjusted to pH 7.4 with 1M NaOH.

2.2 Sample preparation

Figure 1 presents the different steps of sample preparation, from the glass substrate to surfaces with bimodal roughness.

2.2.1 Glass substrate conditioning

The glass coverslips were cleaned by overnight immersion in sulfochromic solution and rinsed with water, prior to polycation adsorption (Figure 1a). The pH and ionic strength (I) of the polycation solution were adjusted with NaOH and HCl, and NaCl, respectively. The polycation solution (1 ml) was poured into the wells of a tissue culture plate (Falcon, Becton Dickinson, Belgium, Ref. 353226), where the glass coverslips had been placed earlier, and was left in contact with the substrate for at least 2 h. Unless stated otherwise, the polycation solution was 10^{-5} M at pH 11 and I 10^{-2} M. The samples were rinsed by 6 successive dilutions to avoid exposure to air. Each rinsing step was performed by adding 2 ml of deionized water (produced by a Milli-Q plus system from Millipore, Molsheim, France), stirring gently, and removing 2 ml of liquid. They were then dried under a gentle nitrogen flow (Figure 1b).

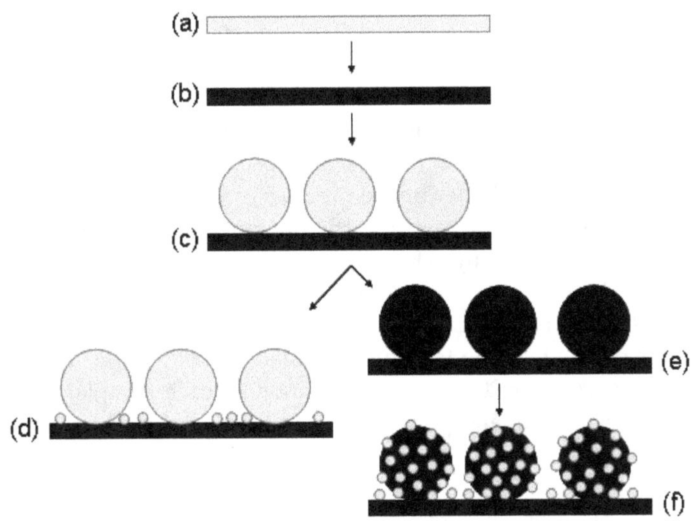

Fig. 1. Schematic representation of the steps used for sample fabrication. Glass substrate (a), polycation-conditioned glass substrate (b), substrate with a layer of adhering particles (c), surface with bimodal I roughness (d), substrate with a layer of adhering particles conditioned with the polycation (e), surface with bimodal II roughness (f).

2.2.2 Adhesion of a single colloid layer

Polycation-conditioned substrates were placed horizontally into wells of another culture plate. 1 ml of a suspension of the desired concentration, pH and ionic strength was poured into the wells and left in contact with the substrate for at least 3 h (Figure 1c). Unless stated otherwise, the procedure used for rinsing and drying the samples was performed as follows: rinsing 3 times with water and 3 times with isopropanol and drying overnight in air. The samples were rinsed by successive dilutions as detailed in the preceding paragraph. After the last rinsing, as much as possible of the solution was removed with a pipette.

2.2.3 Surfaces with bimodal I roughness

1 ml of a 0.1 % suspension of 470 nm particles was poured into the wells containing glass substrates conditioned with PAH (10^{-5} M, pH=11, I=10^{-2} M) (Figure 1b), placed in horizontal

position, and left in contact with the substrate for at least 3 h (Figure 1c). A suspension of 65 nm particles was then added to obtain a final concentration of 0.1 %, and the system was gently stirred and left resting for 1 more hour (Figure 1d). The samples were rinsed and dried using the same procedure as described for adhesion of a single colloid layer.

2.2.4 Surfaces with bimodal II roughness

Substrates with adhering 470 nm particles (Figure 1c), rinsed and dried, were again conditioned with PAH (10^{-5} M, pH=7, I=10^{-3} M) (Figure 1e), rinsed with water then with isopropanol and dried, as described for the adhesion of a single colloid layer. These samples were then placed horizontally in wells of another culture plate. 1ml of a 0.1% suspension of 65 nm particles was poured into wells and left in contact with the substrate for at least 2 h (Figure 1f). The samples were rinsed and dried using the same procedure as described for adhesion of a single colloid layer.

2.3 Surface characterization

The samples were examined by scanning electron microscopy (SEM) using a high resolution Digital Scanning microscope 982 Gemini (Leo Electron Microscopy, UK) operating at 1 or 2 kV, without any metal coating. SEM images of substrates covered with a monolayer of adhering particles were analyzed after image processing with Adobe Photoshop (version 8). In this way, the white rings defining latex particles were extracted and intensified in order to draw circles for digital processing. The images were binarized and the circles were filled before measurements (Visilog, version 6). A minimum of 500 particles was counted on more than 4 images with 10 000x magnification. The degree of particle coverage was measured as the area of particles divided by the total area. Manual counting was also performed on at least 3 images with 10 000x or 20 000x magnification of each sample. Regarding surfaces with bimodal roughness, no image processing was found suitable for extracting colloidal particles from the background. Manual counting of the 470 nm particles only was performed on at least 4 images.

The surface chemical analysis by X-ray photoelectron spectroscopy (XPS) was performed using a Kratos Axis Ultra Spectrometer (Kratos Analytical, UK) equipped with a monochromatized aluminium X-ray source (powered at 10 mA and 15 kV) and eight channeltrons detector. Charge stabilization was achieved using an electron source (filament current set at 1.9 A, bias -1.1 eV) mounted coaxial to the lens column and a charge balance plate (voltage set at -2.3 V). The rectangular analyzed area was about 0.7 mm × 0.3 mm. For recording individual peaks, the pass energy was set at 40 eV. In these conditions, the full width at half maximum (FWHM) of the Ag $3d_{5/2}$ peak was about 0.9 eV. The pressure in the analysis chamber was around 10^{-6} Pa. The following sequence of spectra was recorded: survey spectrum, C 1s, O 1s, N 1s, Si 2p, Cl 2p, B 1s, Na 1s, Zn 2p, Ti 2p, Al 2p, S 2p and C 1s again to check the stability of charge compensation and the absence of sample degradation as a function of time. The binding energy scale was set by fixing the C 1s component due to carbon only bound to carbon and hydrogen at 284.8 eV. The data analysis was performed with the CasaXPS program (Casa Software, UK). Molar concentration ratios were calculated using peak areas normalized on the basis of the acquisition parameters, sensitivity factors, and transmission function provided by the manufacturer. Angle-resolved XPS analyses were performed by using angles between the normal to the sample surface and the direction of photoelectron collection, φ, equal to 0°, 45° and 60°.

The streaming potential was determined at 20°C with an instrument purchased from the Department of Physical and Colloid Chemistry (Agricultural University, Wageningen, The Netherlands) (Elgersma et al., 1992) using glass microscope slides from Marienfeld (7.6 x 2.6 cm^2; Lauda-Königshofen, Germany). Two plates of the samples to characterize (clean glass, PAH-conditioned glass) were assembled to make a 100 µm-thick chamber. A solution of 0.01M KNO_3 was forced through these two plates. The difference of potential measured at the entrance and the exit of the chamber was used to compute to the zeta potential of the substrate (Rouxhet et al., 1993).

3. Results and discussion

3.1 Surface conditioning with PAH

The choice of pH = 11 for PAH adsorption was inspired by the following considerations. The pK_a of ethyl ammonium is 10.6. The apparent pK_a of the PAH used here is 8.7 in water and 9.3 in 0.5 M NaCl (Petrov et al., 2003; Choi and Rubner, 2005); however it is expected to be appreciably higher after adsorption by a negatively charged surface (Tagliazucchi et al., 2008). Conditions of low degree of protonation were chosen in order to allow adsorption of a thick layer. The adsorbed amount was indeed reported to be higher for polymers with low and intermediate cationicities, such as PAH or PLL compared to PEI, and to increase with the molecular mass (Roberts, 1996; Lafuma, 1996), owing to a more coiled conformation. Highly charged polycations, such as PEI, form flat adsorbed layers at low ionic strength (Claesson et al., 2005) and the adsorbed amount increases with pH (Meszaros, 2004).

The surface chemical composition of glass and glass conditioned with PAH (3 independent sets of results) is presented in Table 1. Non-conditioned glass showed the expected organic contamination and a low concentration of nitrogen. In addition, low concentrations of potassium (1-2 %), boron (2-3 %), sodium (1-1.5 %), and traces (< 1 %) of zinc, titanium and aluminium were found. As the analyzed depth decreased (increase of φ), the C/Si concentration ratio increased as expected, but N/Si did not vary significantly and N/C decreased. This indicates that nitrogen is associated with the glass matrix or the glass surface and not with the organic contaminants.

After PAH conditioning, polycation adsorption was evidenced by the increase of carbon and nitrogen concentrations and the decrease of silicon and oxygen concentrations. The N_{1s} peak showed components at 401.5 eV and 399.3 eV due to protonated and non protonated nitrogen, respectively. As the analyzed depth decreased (higher φ), the apparent concentration of carbon and nitrogen increased as expected, the N/C ratio remained constant, but the proportion of protonated nitrogen decreased appreciably.

The N/C ratio of about 0.14 must be compared with the value of 0.33 expected for pure PAH. This difference may not be due to an orientation of the polymer segments at the surface, given the structure of the repeat unit [$-CH_2-CH(-NH-CH_3)-$] and the analyzed depth. It is attributed to the simultaneous presence at the surface of PAH and organic contaminants (Caillou et al., 2008). As the N/C ratio does not vary with φ, the adsorbed organic layer appears to be a mere mixture, with no preferential accumulation of a component at the outermost surface. In an alternative way, the elemental composition of the adsorbed layer may be estimated by the difference of nitrogen concentration before and after PAH conditioning ratioed to the total carbon concentration, which provides a N/C ratio of 0.11. The adventitious contaminants observed on silica, which showed the same C1s peak shape as that observed here on glass, were modeled by the generic formula $C_{15}H_{28}O_4$

(Gerin et al., 1995). According to a N/C ratio of 0.11, the adsorbed layer obtained after PAH conditioning would be 33 wt% PAH (density 1.04 g/m³) and 67 wt% $C_{15}H_{28}O_4$ (density of 0.9 g/cm³).

Substrate	φ	Concentration (Mole fraction, in %) C	O	Si	N	S	Concentration ratios C/Si	N/C	N^+/N_{tot}
Glass coverslips									
sample 1	0°	12.6	57.2	22.8	0.6	-[a]	0.55	0.05	0.47
	45°	15.6	58.4	19.7	0.4	-[a]	0.79	0.03	0.37
	60°	20.5	56.5	18.0	0.4	-[a]	1.14	0.02	0.33
sample 2	0°	10.1	59.4	24.9	0.9	-[a]	0.41	0.09	-[b]
	60°	16.5	59.6	20.1	0.8	-[a]	0.82	0.05	-[b]
sample 3	0°	11.9	57.7	23.6	0.9	-[a]	0.50	0.07	0.46
PAH pre-coated glass									
sample 4	0°	24.1	46.6	18.6	3.5	-[a]	1.30	0.15	0.31
	45°	31.9	43.3	14.3	4.8	-[a]	2.23	0.15	0.18
	60°	42.5	36.9	11.3	5.6	-[a]	3.76	0.13	0.12
sample 5	0°	22.8	48.5	19.5	3.4	-[a]	1.17	0.15	-[b]
	60°	37.6	40.9	12.5	5.2	-[a]	3.01	0.14	-[b]
sample 6	0°	30.4	42.5	18.6	4.5	-[a]	1.63	0.15	0.49
PAH pre-coated glass with adherent colloidal particles									
sample 7 (470 nm)	0°	64.1	21.8	8.1	1.6	0.3	7.91	0.02	0.38
sample 8 (470 nm)	0°	41.1	38.0	13.9	2.7	0.2	2.96	0.06	-[b]
sample 9 (65 nm)	0°	52.2	30.4	11.3	2.7	0.4	4.62	0.05	-[b]

Table 1. Surface chemical composition determined by XPS (mole fraction with respect to all elements excluding hydrogen) at different photoelectron collection angles φ.

In order to evaluate the thickness of the organic adlayer, the experimental C/Si ratios were compared with ratios computed by considering a layer of constant thickness (t) on top of glass and using equation 1:

$$\frac{C}{Si}=\frac{i_{Si}}{i_N}\frac{\sigma_C}{\sigma_{Si}}\frac{\lambda_C^{ad}\,C_C^{ad}\left[1-\exp\left(-t/\lambda_C^{ad}\cos\varphi\right)\right]}{\left[\lambda_{Si}^{su}\,C_{Si}^{su}\exp\left(-t/\lambda_{Si}^{ad}\cos\varphi\right)\right]} \qquad (1)$$

where i_{Si} and i_N are the relative sensitivity factors provided by the manufacturer for Si (0.328) and C (0.278); σ_{Si} and σ_N are the photoionization cross sections for Si (0.817) and C

(1.000). The superscripts ad and su designate the adsorbed overlayer and the substrate, respectively. The concentrations C are: C_C^{ad} = 49.6 mmol/cm³ for contamination layer deposited on glass; C_C^{ad} = 46.9 mmol/cm³ for the layer made by a mixture of 67 % contaminents and 33 % PAH (N/C = 0.11); C_{Si}^{su} = 39 mmol/cm³ for Si in glass (considering that Si concentration in glass is about 90 % of that in pure SiO_2; density of 2.6 g/cm³). The electron inelastic mean free paths were calculated according to Tanuma et al. (1997) with an energy gap of 7 eV: λ_C^{ad} = 3.9 nm, λ_{Si}^{ad} = 3.6 nm, λ_C^{ad} = 4.4 nm. Figure 2 presents plots of C/Si molar ratios computed using different values of t, as a function of φ. The curves are only slightly different if a N/C ratio of 0.14 is considered. Comparison with the experimental data indicates that the organic layer adsorbed after conditioning is about 3 nm, which is about 3 times thicker than the contamination layer. This does not fit closely the apparent proportions of contaminants and PAH in a homogeneous adlayer. Such approach cannot be exploited further owing to the simplicity of the model (homogeneous, flat and compact adlayer) and limited data accuracy. However it indicates that the organic adlayer obtained after conditioning has a thickness of about 3 nm and that PAH coexists with contaminants. The latter may adsorb from the surrounding atmosphere between sulfochromic cleaning and PAH conditioning or after conditioning.

The degree of protonation of PAH in a solution at pH 11 is less than 10 %. The XPS spectra recorded after PAH conditioning (Table 1) gave values of 31 and 49 % for the degree of nitrogen protonation at a photoelectron collection angle φ = 0°. This is in agreement with expectations regarding the effect of local potential on protonation and with data indicating that the apparent pK_a of PAH may be appreciably increased on a negatively charged surface (Tagliazucchi et al., 2008).

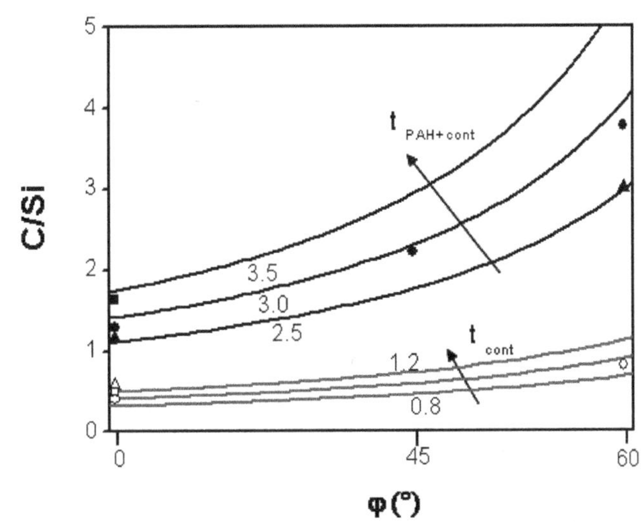

Fig. 2. Plot of computed C/Si molar concentration ratios vs the angle φ of photoelectron collection, for non-conditioned glass with a 0.8, 1.0, 1.2 nm thick (t_{cont}) contamination layer, and for PAH-conditioned glass with a 2.5, 3.0 and 3.5 nm thick ($t_{PAH+cont}$) adlayer made of PAH and contamination. Experimental data for glass (open symbols) and for PAH-conditioned glass (dark symbols). Circles: samples 1 and 4, triangles: samples 2 and 5, squares: samples 3 and 6 (see Table 1).

Figure 3 presents the variation of the zeta potential of glass and of PAH-conditioned glass as a function of pH (in 0.01 M KNO_3). As expected, the glass substrate was found to be negatively charged above pH 2. After polycation adsorption, the zeta potential became less negative, but did not reach positive values. It appears (Figure 3) that adsorbed PAH is shielding the negative charge of glass surface but is not reversing the surface charge. This is in contrast with data reported for substrates such as mica and titanium oxide (Adamczyk et al., 2006, 2007) and is attributed to the fact that here the samples were dried between adsorption and streaming potential measurement, which provoked irreversible shrinkage of the adsorbed layer. It may be related to the decrease of the degree of nitrogen protonation as φ increased (Table 1), revealing that it is much higher for segments in close contact with the glass surface. Note that drying after adsorption was a choice made in the context of a fabrication process.

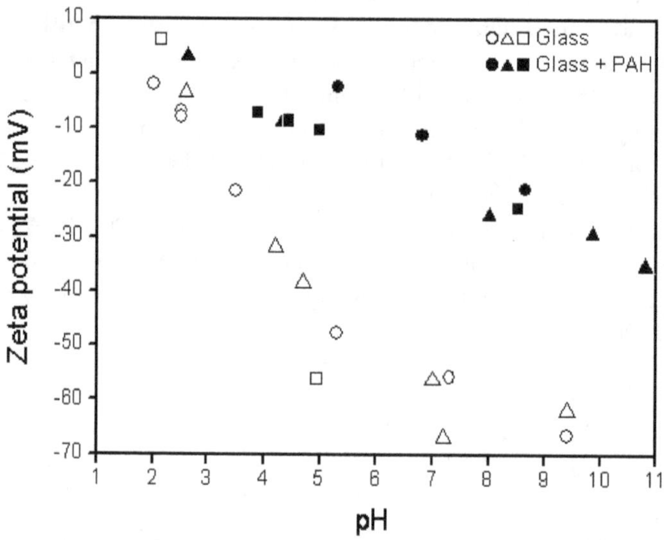

Fig. 3. Zeta potential (mV) of glass (open symbols) and PAH-conditioned glass (dark symbols) as a function of pH in 0.01 M KNO_3. Three experiments were performed on samples prepared independently.

3.2 Nanostructured surfaces with monomodal roughness

Figure 4a shows representative micrographs of samples prepared with substrates conditioned as described above (PAH 10^{-5} M, pH 11, I 10^{-2} M), incubated with 470 nm particles (0.1%, pH 7, I 10^{-3} M), rinsed 6 times with water and dried with a nitrogen flow. The adhering particles were under the form of aggregates, usually bidimensional along the surface plane, sometimes tridimensional (i.e. forming multilayers). Attempts were made to improve the homogeneity of the colloidal particle distribution, keeping the colloidal particle treatment unchanged but using polycations which differ according to functional groups, hydrophobicity and size, and changing polycation treatment conditions : (i) PAH, PLL, PEI, lPDDA and hPDDA; 10^{-7} M at pH 3 and 10^{-5} M at pH 11; I $10^{-3}$and 10^{-1} M; (ii) PAH and lPDDA; 10^{-7} and 10^{-5} M; pH 3, 7 and 11; I 10^{-3}, 10^{-2} and 10^{-1} M. In summary, polycation adsorption at pH 7 or 11 and low ionic strength (10^{-3}, 10^{-2} M) provided a higher degree of coverage by colloidal particles; PAH was as efficient as, or better than other polymers tested; a polycation concentration of 10^{-5} M was more efficient than 10^{-7} M. However in all cases, the adhering particles were under the form of aggregates as described above.

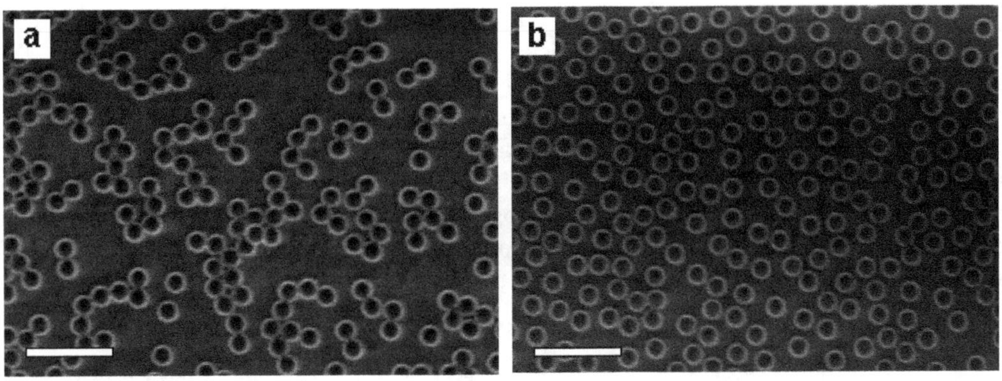

Fig. 4. SEM images obtained on PAH-conditioned (10^{-5}M, pH=11, I=10^{-2}M) glass samples, submitted to latex particle adhesion (470 nm in diameter, 0.1%, pH=7, I=10^{-3}M), rinsed with water and dried under a gentle nitrogen flow (a) or rinsed with water, then with isopropanol and air-dried (b). Scale bars: 2 μm.

A more homogeneous particle distribution was obtained when the sample was rinsed first with water then with isopropanol and left to dry in air overnight, as shown in Figure 4b. Both images of Figure 4 present the same degree of particle coverage (~ 40 %), but in image b, no aggregation is present and the particle distribution is more random. The influence of rinsing and drying procedures on the layer of latex particles may be due to their gathering together by the movement of the liquid film or by the nitrogen flow. In order to avoid particle aggregation at high coverage, Hanarp et al. (2001) attempted to retain particles in place during drying by adsorbing smaller silica particles between the latex particles or by heating the samples with adhering particles in boiling water in order to deform the particles and increase the contact area with the substrate. In the present work, rinsing with isopropanol was found to be efficient (Figure 4b) ; this is attributed to the reduction of capillary forces owing to the 3 times lower surface tension of isopropanol compared to water ((21.7 mN/m compared to 73.0 mN/m (Weast, 1972).

Particle adhesion was further studied using substrates conditioned with PAH as described above, rinsing 3 times with water and 3 times with isopropanol, followed by air drying. Figure 5 presents representative results obtained by incubating the conditioned substrate during 2 hours with the colloids at 0.1 % concentration, using 3 pH values and 2 ionic strengths. At pH 7 with a low ionic strength and at pH 11, substrates were covered by a homogeneous layer of particles. In contrast, a high degree of aggregation was observed at pH 3 whatever the ionic strength, and at pH 7 with a high ionic strength.

Figure 6 (a, b) presents micrographs obtained after adhesion of the two kinds of latex particles separately (470 nm or 65 nm; 0.1%, pH 7, I 10^{-3} M), with isopropanol rinsing and air drying. Two counting methods were used in order to assess the degree of substrate coverage. Binarization and automatic processing of SEM images allow counting to be made more rapidly and on a large number of images in comparison with manual counting.

However, it involves operations which could possibly create systematic variations. Binarization of SEM images gave degrees of coverage of 50.2 ± 5 % and 14.6 ± 4 % on surfaces layered with 470 and 65 nm particles, respectively. A consistent degree of substrate coverage was obtained using manual counting (470 nm : 45.2 ± 2 %; 65 nm : 10.7 ± 1 %). The differences between the two modes of counting were thus in the range of the standard deviation. Samples with adherent particles were also analyzed by XPS (Table 1). For samples 7 and 8 , 470 nm particles were used and the degree of coverage was 53 and 27 %, respectively; for sample 9, 65 nm particles were used and the degree of coverage was 15 %, as assessed by SEM image analysis. As compared with PAH conditioned substrates with no adhering latex, these three samples gave a higher concentration of carbon, a lower concentration of silicon, oxygen and nitrogen and indicated the presence of sulfur, characterized by a S 2p peak (showing the $2p_{3/2}$ and $2p_{1/2}$ components with S $2p_{3/2}$ observed at 169 eV), characteristic of sulfate and due to the latex surface.

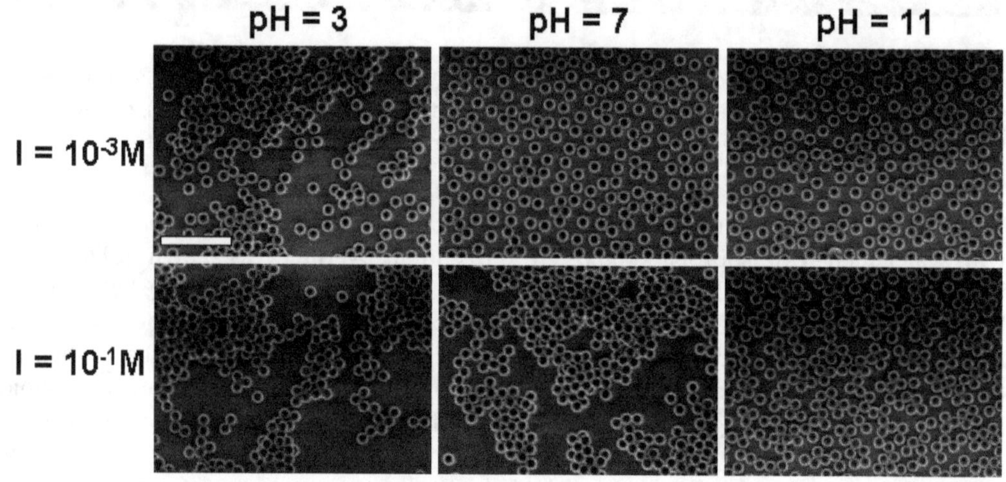

Fig. 5. SEM images obtained on PAH-conditioned (10^{-5}M, pH=11, I=10^{-2}M) glass samples, submitted to latex particle adhesion in different conditions of pH and ionic strength (470 nm in diameter particles, 0.1%), rinsed with water, then with isopropanol and air dried. Scale bars: 2 μm.

A variation of the duration of incubation of the PAH-conditioned substrate with the 470 nm latex (1 to 6 hours, 1 to 6 days) showed that an incubation time of at least 2 hours was required in order to obtain these degrees of coverage. However no increase of coverage was obtained when incubating for longer periods of times. Using a 1 % latex concentration instead of 0.1 % did not provide any rise of the degree of coverage (data not shown).

Obtaining a layer of adhering particles deserves discussion, considering not only interfacial interactions but also the amount of colloidal particles involved and mass transfer. Table 2 presents data computed for the experimental conditions of the treatments with 0.1 % colloidal suspensions. Similar data for PAH 10^{-5} M solution are given for comparison, considering the polycation as a non hydrated sphere of 3.0 nm radius; note that the hydrodynamic radius at pH 7.4 is about 5 times larger (Adamczyk et al., 2006).

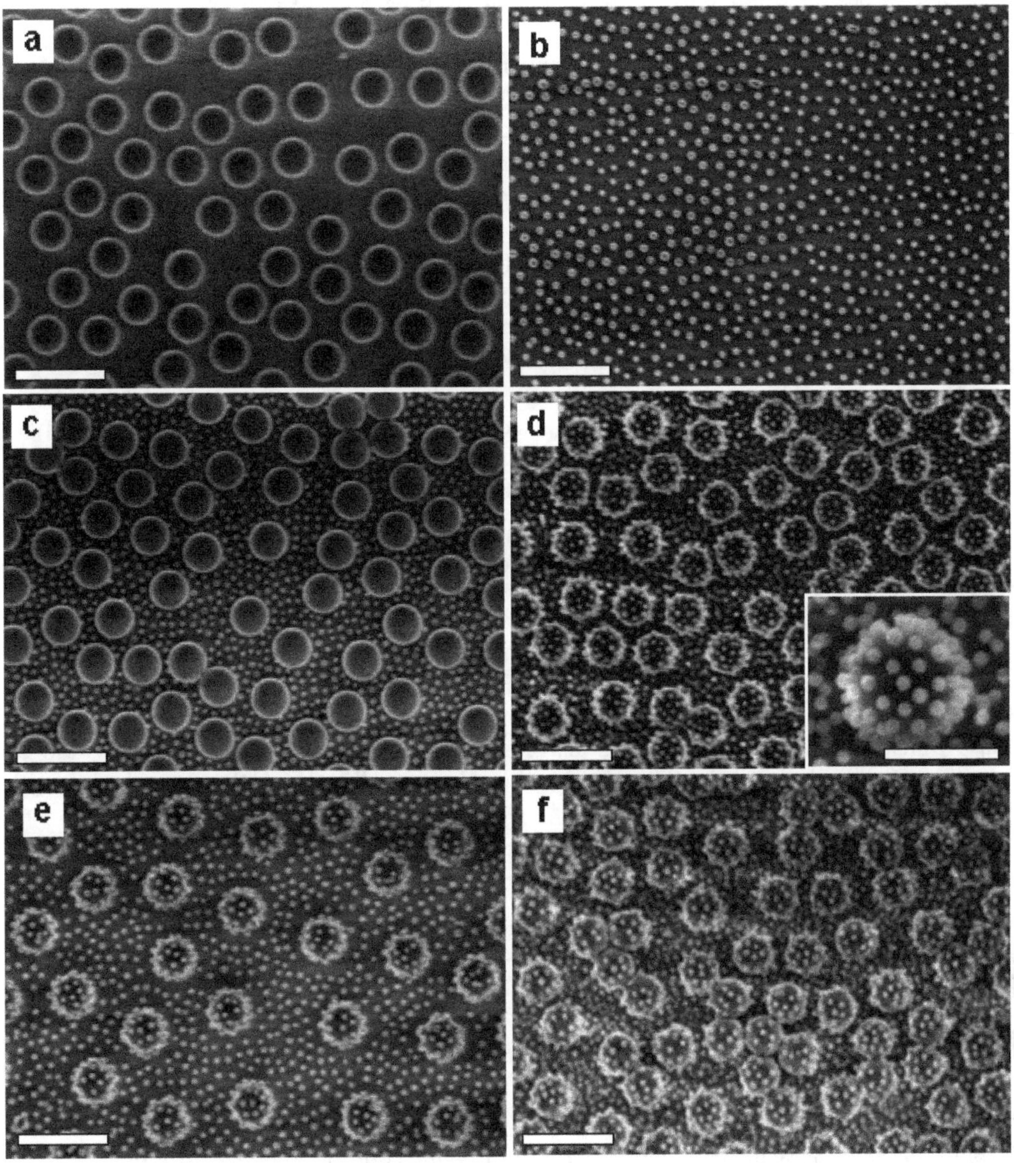

Fig. 6. SEM images obtained on PAH-conditioned (10^{-5}M, pH=11, I=10^{-2}M) glass samples, submitted to adhesion of latex particles (0.1%, pH=7, I=10^{-3}M, rinsing with water then with isopropanol) with 470 nm (a) and 65 nm diameter (b), and submitted to sequential adhesion of 470 nm and 65 nm particles without (c, bimodal I roughness) and with (d, e, f, bimodal II roughness) PAH adsorption after adhesion of the 470 nm particles. Image f was obtained on a sample prepared in the same experimental set as d, then submitted to immersion in PBS at 37°C for 24h. Images d and e were obtained on samples fabricated independently. The inset shows an enlargement of structures present on surfaces with bimodal II roughness. Scale bars: 1 μm. Scale bar inset: 500 nm.

- Area A occupied per particle in hexagonal close packing (3.46 a^2, a being the particle radius).
- Amount Q of particles required to cover the surface with a monolayer in hexagonal close packing.
- Concentration C of particles in the liquid phase, which, if expressed in ml^{-1}, is also the amount of particles involved in the preparation.
- Thickness e of the liquid layer containing the amount of particles required to make a monolayer.
- Diffusion coefficient D of the particles ($D = kT / 6\pi\eta a$), where k, T and η are the Boltzmann constant, the temperature and the viscosity, respectively.
- Time t_d required to produce a monolayer if the rate of adsorption is entirely controlled by diffusion through a concentration gradient created by adsorption itself (Crank, 1957): $t_d = \pi Q^2 / 4 C^2 D$.

It appears that the particles are in large excess of what is needed to make a monolayer. Mass transport by diffusion is fast for polycations and 65 nm latex particles. In contrast, a period of one day is expected to be needed for insuring the formation of a monolayer of 470 nm particles if diffusion is the rate controlling process. Thus sedimentation or convection should contribute to bringing the 470 nm particles to the surface. Considering a density of 1.05 g/cm^2, the rate of sedimentation v_s of 470 nm particles is about 6.0 nm/s according to Stokes law. Accordingly the time t_s required to produce a monolayer if the rate is entirely controlled by sedimentation is $t_s = e / v_s$ = 13 hours. Actually the amount of adhering 470 nm particles does not increase significantly if the incubation time is extended from 1 hour to 6 days, suggesting that convection due to manipulation insures mass transport. The data of Table 2 also help to figure out in what limits it would be possible to control the adsorption of polycations and small latex particles by playing with incubation time and concentration.

	PAH **10^{-5} M**	**Latex 65** **0.1 %**	**Latex 470** **0.1 %**
A (nm^2 $part^{-1}$)	32	$3.7\ 10^3$	$1.91\ 10^5$
Q (part. cm^{-2})	$3.1\ 10^{12}$	$2.7\ 10^{10}$	$5.2\ 10^8$
C (part. cm^{-3})	$6.0\ 10^{15}$	$7.0\ 10^{12}$	$1.8\ 10^{10}$
e (µm)	5.2	39	284
D ($cm^2 s^{-1}$)	$7.2\ 10^{-7}$	$6.7\ 10^{-8}$	$9.3\ 10^{-9}$
t_d	0.29 s	2.9 min	19 hours

Table 2. Particle characteristics relevant to experimental conditions. See text for details.

While mass transport is not a limiting factor for adhesion of 65 nm particles, the density of the monolayer remains fairly low (10 to 15 % degree of coverage). For this latex at the ionic strength of 10^{-2} M, the product κa, where κ is the inverse of the Debye length, is about 10. For this κa value, maximum coverages of about 35 % were reported for positively charged latex particles on mica (Johnson and Lenhoff, 1996) and for negatively charged latex particles on titanium oxide modified by successive adsorption of PDDA, polystyrene sulfonate and aluminium chloride hydroxide (Hanarp et al., 2001), in agreement with predictions based on repulsion between the particles considered as hard spheres. It is thus reasonable to consider that the coverage by the 65 nm latex is controlled by double layer repulsion between particles. For this latex, a surface coverage of 10 to 15 % corresponds to an apparent area of 37 10^3 to 24 10^3 nm^2/ particle, corresponding to an occupational diameter of 207 to 166 nm. The difference (142 to 101 nm) with the real particle size is larger that the latter, which demonstrates that the degree of coverage is limited by particle-particle repulsion.

If the particles were in hexagonal close packing, the area occupied per particle would be as presented in Table 2. A surface coverage of 45 % for the 470 nm latex thus corresponds to an apparent area per particle of 4.2 10^5 nm^2 and an occupational diameter of 700 nm. The difference of 230 nm between the occupational diameter and particle size is lower than the real particle diameter. Accordingly the degree of coverage is limited by the space available, considering a random distribution of particles. Consequently, the degree of coverage would be limited by the space available for 470 nm particles and by electrostatic repulsion for the 65 nm latex. Hanarp (2003) used the ionic strength (10^{-5} to 10^{-2} M), to control the density of 110 nm polystyrene particles adhering on titanium oxide conditioned with a solution of aluminium chloride hydroxide. Our data indicate that this method would not work for particles above 200 nm diameter.

Figure 5 shows aggregation of 470 nm particles when using a 10^{-1} M ionic strength at pH 3 or 7, or when using a 10^{-3} M ionic strength at pH 3. The surface charge of colloids used here is not affected by pH as confirmed by Schulz et al. (1994a, 1994b), who used latex particles of 131 nm from the same provider and with the same surface specifications as the particles used here. If the aggregates were formed in the suspension, the adhering amount would be expected to increase with time in the range of a few hours or when increasing the concentration from 0.1 to 1 %, in contrast with observations. This suggests that aggregation may take place at the sol - substrate interface owing to low particle - particle repulsion, in addition to the tendency to gather together induced by subsequent solvent evaporation. The aggregation may also be favored by partial desorption of the polycation, as it is enhanced at a pH 3 and 7, at which the polycation is more highly charged (Choi and Rubner, 2005).

3.3 Nanostructured surfaces with bimodal roughness

Surfaces with bimodal I roughness (Figure 6c) were created by adding 65 nm particles subsequently to the 470 nm particles (cf Figure 1c,d). The degree of substrate coverage with 470 nm particles was of 46.9 ± 1 % on this sample. This approach is similar to that of Takeshita et al. (2004) using poly(ethylene terephthalate) conditioned with PAH and adhesion of 350 nm carboxylated polystyrene latex particles followed by 100 nm particles. However, the layer obtained in that case was not regular, despite sonication of the solution to prevent latex aggregation. We attribute this to the fact that the samples were rinsed with water before drying.

Surfaces with bimodal II roughness (Figure 6d) were prepared by adsorbing PAH after the formation of the first adhering layer (cf Figure 1b,c,e,f). In this way, 65 nm particles were adhering not only to the glass substrate but also to the 470 nm particles, providing raspberry-like structures. This procedure was repeated five times independently and 3 to 15 samples were prepared in each experiment, demonstrating the repeatability as well as the reproducibility of the method. Figure 6 (d, e) presents the extreme results obtained, corresponding to degrees of coverage by 470 nm particles of 50 ± 3 % and 22 ± 1 %, respectively.

Owing to the possible interest of nanostructured surfaces in the field of biointerfaces, samples with adhering particles were incubated in phosphate buffer saline (the main constituent of culture media) for 24 h at 37°C, followed by rinsing with water, rinsing with isopropanol and drying. The result obtained with a bimodal II roughness is presented in Figure 6f. The observed morphology is similar to that of the sample not exposed to buffer (Figure 6d), demonstrating the robustness of the protocol and of the nanostructured surface obtained.

Recently published works aimed at creating this kind of roughness using different approaches. Suspensions of raspberry-like particles were prepared by styrene polymerization on silica particles (Perro et al., 2006; Reculusa et al., 2002). In another study, silica particles having different sizes and bearing functional groups were firstly synthesized independently and then mixed to react together. The obtained aggregates were then grafted on a specific substrate to obtain a dual-size roughness surface (Xiu et al., 2006). In another approach, a layer of silica particles in hexagonal close packing was created on a substrate, and gold nanoparticles were formed on the top of the silica spheres by sputtering (Ming et al., 2005). The method used in the present work has several advantages, such as the use of components which are commercially available (polycation, latex particles), and a simple procedure (sequential steps of polycation adsorption and colloid adhesion) which does not require sophisticated devices or complex reactions. The protocol could be extended to other particle sizes for obtaining a broader panel of roughness, the density of the two types of particles could be tuned, and architectures could be elaborated with more than two particle sizes. The bimodal surfaces of type II roughness mimic the particular topography observed on Lotus leaf. A superhydrophobic surface is thus expected to be obtained after treatment with compounds conferring a low surface energy (Bravo et al., 2007).

4. Conclusion

Nanostructured surfaces were fabricated through assembly of PAH and polystyrene latex particles. Thereby, different types of roughness were created, with a single layer of particles (65 or 470 nm), a layer of two types of particles (65 and 470 nm) - bimodal roughness of type I - or a layer of raspberry-like relief features (65 nm on 470 nm) – bimodal roughness of type II. The best conditions for glass conditioning by PAH were a high pH (11) and a low ionic strength (10^{-2}M). A neutral or alkaline pH with a low ionic strength was satisfactory for the adhesion of the colloidal particles on PAH-conditioned glass. Terminating the rinsing procedure with isopropanol before air drying was needed to avoid particle aggregation due to capillary forces. Incubation in PBS, a buffer solution mimicking the electrolyte composition of biological fluids, did not alter the structures obtained.

After PAH conditioning, the thickness of the adsorbed layer detected on glass was in the range of 2.5 to 3.5 nm ; this layer contained an appreciable amount of adventitious organic contaminants. The degree of ionization of PAH at the outermost part of the adsorbed layer was low and did not provoke a surface charge reversal. This indicates that the colloid particles were not attracted by long distance forces but rather suggests a change of PAH protonation with a redistribution of counterions when colloidal particles approached the PAH-modified glass.

The control of the degree of coverage by adhering particles for performing colloidal lithography may be pursued by playing with interfacial interactions, and thus with pH and ionic strength. On the other hand it may be pursued by playing with mass transfer:

substrate orientation (upward, downward, vertical), convection, particle concentration and contact time. The selection of the best approach depends primarily on the particle size, which is critical in the range of 100 nm, and secondarily on particle density and on the desired degree of coverage.

5. Acknowledgements

The support of the Foundation for Training in Industrial and Agricultural Research (FRIA), of the Belgian National Foundation for Scientific Research (FNRS), of the Région Wallonne and of the Federal Office for Scientific, Technical and Cultural Affairs (Interuniversity Poles of Attraction Program) is gratefully acknowledged.

6. References

Adamczyk, Z., Zembala, M., & Michna, A. (2006). Polyelectrolyte adsorption layers studied by streaming potential and particle deposition. *J. Colloid Interface Sci.*, 303, 353-364

Adamczyk, Z., Zembala, M., Kolasinska, M., & Warszynski, P. (2007) Characterization of polyelectrolyte multilayers on mica and oxidized titanium by steaming potential and wetting angle measurements. *Colloid Surf. A: Physicochem. Eng. Asp.*, 302, 455-460

Agheli, H., Malmstrom, J., Hanarp, P., & Sutherland, D.S. (2006). Nanostructured biointerfaces. *Mater. Sci. Eng. C: Biomimetic Supramol. Syst.*, 26, 911-917

Barthlott, W., & Neinhuis C. (1997). Purity of the sacred lotus, or escape from contamination in biological surfaces. *Planta*, 202, 1-8

Bertrand, P., Jonas, A., Laschewsky, A., & Legras R. (2000). Ultrathin polymer coatings by complexation of polyelectrolytes at interfaces: Suitable materials, structure and properties. *Macromol. Rapid Commun.*, 21, 319-348.

Boonaert, C.J.P., Dupont-Gillain, C.C., Dengis, P.B., Dufrêne Y.F., & Rouxhet, P.G. (1999). Cell separation, flocculation, In: *Encyclopedia of bioprocess technology: Fermentation, biocatalysis, and bioseparation,* M.C. Flickinger, S.W. Drew (Eds.), pp. 531-548, John Wiley & Sons, Inc., New York

Boonaert, C.J.P., Dufrêne, Y.F., & Rouxhet, P.G. (2002). Adhesion (primary) of microorganisms onto surfaces, In: *Encyclopedia of environmental microbiology,* G. Bitton (Ed.), pp. 113-132, John Wiley & Sons Inc., New York

Bravo, J., Zhai, L., Wu, Z., Cohen, R.E., & Rubner M.F. (2007). Transparent superhydrophobic films based on silica nanoparticles. *Langmuir*, 23, 7293-7298

Caillou, S. , Gerin, P.A., Nonckreman, C.J., Fleith, S., Dupont-Gillain, C.C., Landoulsi, J., Pancera, S.M., Genet , M.J., & Rouxhet, P.G. (2008). Enzymes at solid surfaces: nature of the interfaces and physico-chemical processes. *Electrochim. Acta*, 54, 116-122

Changui, C., Doren, A., Stone, W.E.E., Mozes, N, & Rouxhet, P.G. (1987). Surface properties of polycarbonate and promotion of yeast cell adhesion. *Journal de Chimie Physique et de Physico-Chimie Biologique*, 84, 275-281

Chen, K.M., Jiang, X., Kimerling, L.C., & Hammond, P.T. (2000). Selective self-organization of colloids on patterned polyelectrolyte templates. *Langmuir*, 16, 7825-7834

Choi, J., & Rubner, M.F. (2005). Influence of the degree of ionization on weak polyelectrolyte multilayer assembly. *Macromolecules*, 38, 116-124

Claesson, P.M., Poptoshev, E., Blomberg, E., & Dedinaite, A. (2005). Polyelectrolyte-mediated surface interactions. *Adv. Colloid Interface Sci.*, 114-115, 173-187

Crank, J. (1957). *The mathematics of diffusion*, Oxford University Press, p. 30-31

Davies, R.J., Dix, L.R., & Toprakcioglu C. (1989). Adsorption of poly-L-lysine to mica powder. *J. Colloid Interface Sci.*, 129, 145-152

Decher, G., & Hong, J.D. (1991). Build up of ultrathin multilayer films by a self-assembly process: Consecutive adsorption of anionic and cationic bipolar amphiphiles on charged surfaces. *Macromol. Chem. Macromol. Symp.*, 46, 321-327

Dekeyser, C.M., Biltresse, S., Marchand-Brynaert, J., Rouxhet, P.G., & Dupont-Gillain, C.C. (2004). Submicrometer-scale heterogeneous surfaces by PS-PMMA demixing. *Polymer,* 45, 2211-2219

Denis, F.A., Hanarp, P., Sutherland, D.S., & Dufrene, Y.F. (2002). Fabrication of nanostructured polymer surfaces using colloidal lithography and spin-coating. *Nano Lett.*, 2, 1419-1425

Elgersma, A.V., Zsom, R.L.J., Lyklema, J., & Norde, W. (1992). Kinetics of single and competitive protein adsorption studied by reflectometry and streaming potential measurements. *Colloid Surf.*, 65, 17-28

Fustin, C.A., Glasser, G., Spiess, H.W., & Jonas, U. (2004). Parameters influencing the templated growth of colloidal crystals on chemically patterned surfaces. *Langmuir,* 20, 9114-9123

Gerin, P.A., Dengis, P.B., & Rouxhet, P.G. (1995). Performance of XPS analysis of model biochemical coumpounds. *Journal de Chimie Physique*, 92, 1043-1065

Hanarp, P., Sutherland, D.S., Gold, J., & Kasemo B. (2001). Influence of polydispersity on adsorption of nanoparticles. *J. Colloid Interface Sci.,* 241, 26-31

Hanarp, P., Sutherland, D.S., Gold, J., & Kasemo, B. (2003). Control of nanoparticle film structure for colloidal lithography. *Colloid Surf. A: Physicochem. Eng. Asp.*, 214, 23-36.

Johnson, C.A., & Lenhoff, A.M. (1996). Adsorption of Charged Latex Particles on Mica Studied by Atomic Force Microscopy. *J. Colloid Interface Sci.*, 179, 587-599

Lafuma, F. (1996). Mechanisms of flocculation and stabilisation of suspensions by organic polymers, In: *Paper Chemistry,* J.C. Roberts (Ed.), Chapman & Hall, London

Li, J., Luan, S., Huang, W., & Han, Y. (2007). Colloidal crystal heterostructures by a two-step vertical deposition method. *Colloid Surf. A: Physicochem. Eng. Asp.* 295, 107-112

Li, Z., Ravaine, V., Ravaine, S., Garrigue, P., & Kuhn, A. (2007). Rapsberry-like gold microspheres: preparation and electrochemical characterization. *Adv. Functional Materials,* 17, 618-622

Lindquist, G.M., & Stratton, R.A. (1976). The role of polyelectrolyte charge density and molecular weight on the adsorption and flocculation of colloidal silica with polyethylenimine. *J. Colloid Interface Sci.*, 55, 45-59

Meszaros, R., Varga, I., & Gilanyi, T. (2004). Adsorption of poly(ethyleneimine) on silica surfaces: Effect of pH on the reversibility of adsorption. *Langmuir,* 20, 5026-5029

Ming, W., Wu, D., van Benthem, R., & de With, G. (2005). Superhydrophobic films from raspberry-like particles. *Nano Lett.*, 5, 2298-2301

Nonckreman, C.J., Fleith, S., Rouxhet, P.G., & Dupont-Gillain, C.C. (2010). Competitive adsorption of fibrinogen and albumin and blood palatelet adhesion on surfaces modified with nanoparticles and/or PEO. *Colloids and Surfaces B: Biointerfaces, 77*, 139-149

Patankar, N.A. (2004). Mimicking the Lotus Effect: Influence of double roughness structures and slender pillars. *Langmuir*, 20, 8209-8213

Perro, A., Reculusa, S., Bourgeat-Lami, E., Duguet, E. , & Ravaine, S. (2006). Synthesis of hybrid colloidal particles: From snowman-like to raspberry-like morphologies. *Colloid Surf. A: Physicochem. Eng. Asp.*, 284-285, 78-83

Petrov, A.I., Antipov, A.A., & Sukhorukov, G.B. (2003). Base-acid equilibria in polyelectrolyte systems: From weak polyelectrolytes to interpolyelectrolytes complexes ans multilayered polyelectrolyte shells. *Macromolecules*, 36, 10079-10086.

Reculusa, S., Poncet-Legrand, C., Ravaine, S., Mingotaud, C., Duguet, E., & Bourgeat-Lami, E. (2002). Syntheses of raspberrylike silica/polystyrene materials. *Chem. Mater.*, 14, 2354-2359

Roberts, J.C. (1996). The surface chemistry of paper and the paper-making system, In: *The Chemistry of Paper*, J.C. Roberts (Ed.), The Royal Society of Chemistry, Letchworth.

Rouxhet, P.G., Doren, A., Dewez, J.L., & Heuschling, O. (1993). Chemical composition and physico-chemical properties of polymer surfaces. *Prog. Org. Coat.*, 22, 327-344

Schaak, R.E., Cable, R.E., Leonard, B.M., & Norris, B.C. (2004). Colloidal crystal microarrays and two-dimensional superstructures: a versatile approach for patterned surface assembly. *Langmuir*, 20, 7293-7297

Schulz, S.F., & Sticher, H. (1994a). Surface charge densities and electrophoretic mobilities of aqueous colloidal suspensions of latex spheres with different ionizable groups. *Prog. Colloid Polym. Sci.*, 97, 85-88

Schulz, S.F., Gisler, T., Borkovec, M., & Sticher, H. (1994b). Surface charge on functionalized latex spheres in aqueous colloidal suspensions. *J. Colloid Interface Sci.*, 164, 88-98

Tagliazucchi, M., Calvo, E.J., & Szleifer, I. (2008). Redox and acid-base coupling in ultrathin polyelectrolyte films. *Langmuir*, 24, 2869-2877

Takeshita, N., Paradis, L.A., Oener, D., McCarthy, T.J., & Chen, W. (2004). Simultaneous tailoring of surface topography and chemical structure for controlled wettability. *Langmuir*, 20, 8131-8136

Tanuma, S., Powell, C.J., & Penn, D.R. (1997). Calculations of electron inelastic mean free paths (IMFPs). VI. Analysis of the Gries inelastic scattering model and predictive IMFP equation. *Surf. Interface Anal.*, 25, 25-35

Van Haecht, J.L., Bolipombo, M., & Rouxhet, P.G. (1985). Immobilization of Saccharomyces cerevisiae by adhesion: treatment of the cells by aluminum ions. *Biotechnol. Bioeng.*, 27, 217-224

Weast, R.C. (1972). *Handbook of Chemistry and Physics*, 52nd ed., The Chemical Rubber C.O., Cleveland.

Wenzel, R.N. (1936). Resistance of solid surfaces to wetting by water. *J. Ind. Eng. Chem.*, 28, 988-994

Wood, M.A. (2007). Colloidal lithography and current fabrication techniques producing in-plane nanotopography for biological applications. *J. R. Soc. Interface*, 4, 1-17

Xiu, Y., Zhu, L., Hess, D.W., & Wong, C.P. (2006). Biomimetic creation of hierarchical surface structures by combining colloidal self-assembly and Au sputter deposition. *Langmuir*, 22, 9676-9681

Yang S., Cai, W., Yang , J., & Zeng, H. (2009). General and simple route to micro/nanostructured hollow-sphere arrays based on electrophoresis of colloids induced by laser ablation in liquid. *Langmuir*, 25, 8287-9291

6

Approach to EUV Lithography Simulation

Atsushi Sekiguchi
Litho Tech Japan Corporation
Japan

1. Introduction

1.1 Simulation based on measured development rate measurements

EUV lithography is a reduced projection lithography technology based on 13.5 nm wavelength EUV (Extreme Ultraviolet). Development of EUV lithography is currently underway for the mass production of semiconductor devices for 90 nm design rule applications for ArF dry exposures and for 65 to 45 nm design rule applications for ArF immersion exposures [1-2]. EUV lithography is among the most promising next-generation lithography tools for the 32 nm technology node [3]. The evolving consensus is that EUV exposure technologies will be applied to mass production from the year 2011 [4]. Table 1 showed the relationship among technology node, exposure numerical aperture (NA), and process coefficient factor (k_1) [5]. Achieving the 32 nm node based on an ArF laser source exposure technology will require the development of an optical system with NA increased to 1.55 and k1 improved to 0.26. In contrast, an exposure technology based on an EUV light source will permit the use of an optical system with 0.25 NA for mass production of the 32 nm node with room to spare. The requirement for the k1 factor is an easy-to-meet value of 0.59. These factors underscore the promise and importance of EUV exposure technologies.

However, the development of EUV exposure equipment presents its own set of technology barriers, as does the development of ArF immersion exposure system. A wavelength of 13.5 nm requires a reflecting optical system with a combination of multiple multilayer reflecting mirrors [6], since no lens material can be used in the 13.5 nm wavelength range, if we rule out dioptric lenses. The development of EUV exposure equipment requires further examination of component technologies, including technologies related to light sources, illumination optical systems, projection optical systems, and masks. Although various exposure equipment manufacturers are actively promoting the development of EUV reduced projection exposure equipment [7-8], a resist material for EUV lithography must be developed before the first exposure system can be introduced. We have developed a new virtual lithography evaluation system with lithograph simulation that takes an approach completely different from conventional resist evaluation technologies (direct evaluation method), which require actual patterning to assess resists. The new evaluation system focuses on open-frame exposures using an EUV light source, measurements of development rates at various exposure doses, and lithography simulations based on development rate data. This chapter presents the results of our evaluations of EUV resists using this new system.

Wavelength (nm)	Tech. Node	65 nm	45 nm	32 nm	22 nm	16 nm
	NA	k1	k1	k1	K1	k1
193	0.93	0.31				
	1	0.4				
	1.2		0.28			
	1.35		0.31	0.22	0.15	
	1.35DP			0.20	0.18	
13.5	0.25			0.59	0.41	
	0.35				0.57	0.41
	0.45					0.53

Table 1. Relationship among technology node, numerical aperture (NA), and process factor (k_1)

1.2 System configuration

The virtual lithography evaluation system (VLES) proposed consists of an EUV open-frame exposure system, a resist development analyzer, and a lithography simulator. Fig. 1 is a schematic diagram of the VLES.

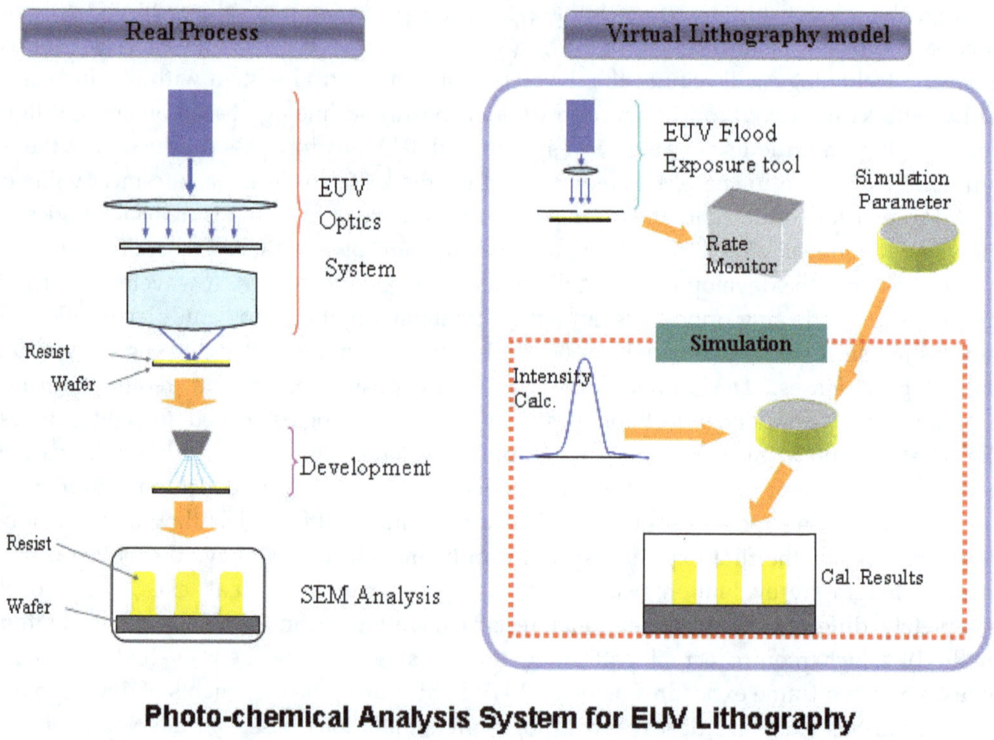

Fig. 1. Schematic diagram of the VLES

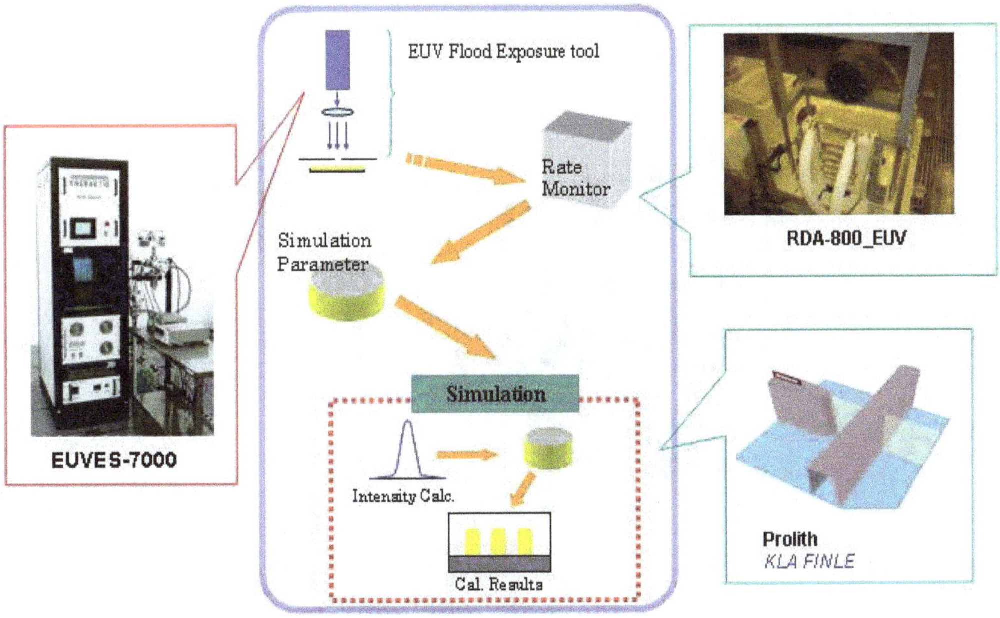

Fig. 2. Analyzers used in the VLES

Fig. 2 shows the analyzers comprising the VLES.

1.2.1 EUV open-frame exposure system (EUVES-7000)

This equipment uses an electrodeless Z-pinch discharge-excitation plasma light source [9] manufactured by Energetiq Technology Inc. It extracts 13.5 nm light using a Zr filter and multilayer reflecting mirrors. The exposure pattern is a 10 mm x 10 mm open frame; 12 exposures can be achieved per wafer at varying exposure doses. Fig. 3 gives an external view of this equipment and a picture of an exposure pattern (after exposure, PEB, and development).

The plasma emissions produced by the EQ-10M pass through the Zr filter to remove UV-region rays. Next, the Mo-Si multilayer reflector selectively reflects only 13.5 nm rays, which are shaped by the aperture into a 10 mm x 10 mm exposure region. The rotary Mo-Si multilayer reflector directs the light at a reflection angle of 45 degrees toward the exposure chamber at the upper section of the equipment during the exposure of a substrate. For power measurements, it rotates and directs the light to the power measurement diode chamber at the lower section of the equipment. Exposures are performed as the wafer rotates. A total of 12 exposures are possible per wafer at varying exposure doses.

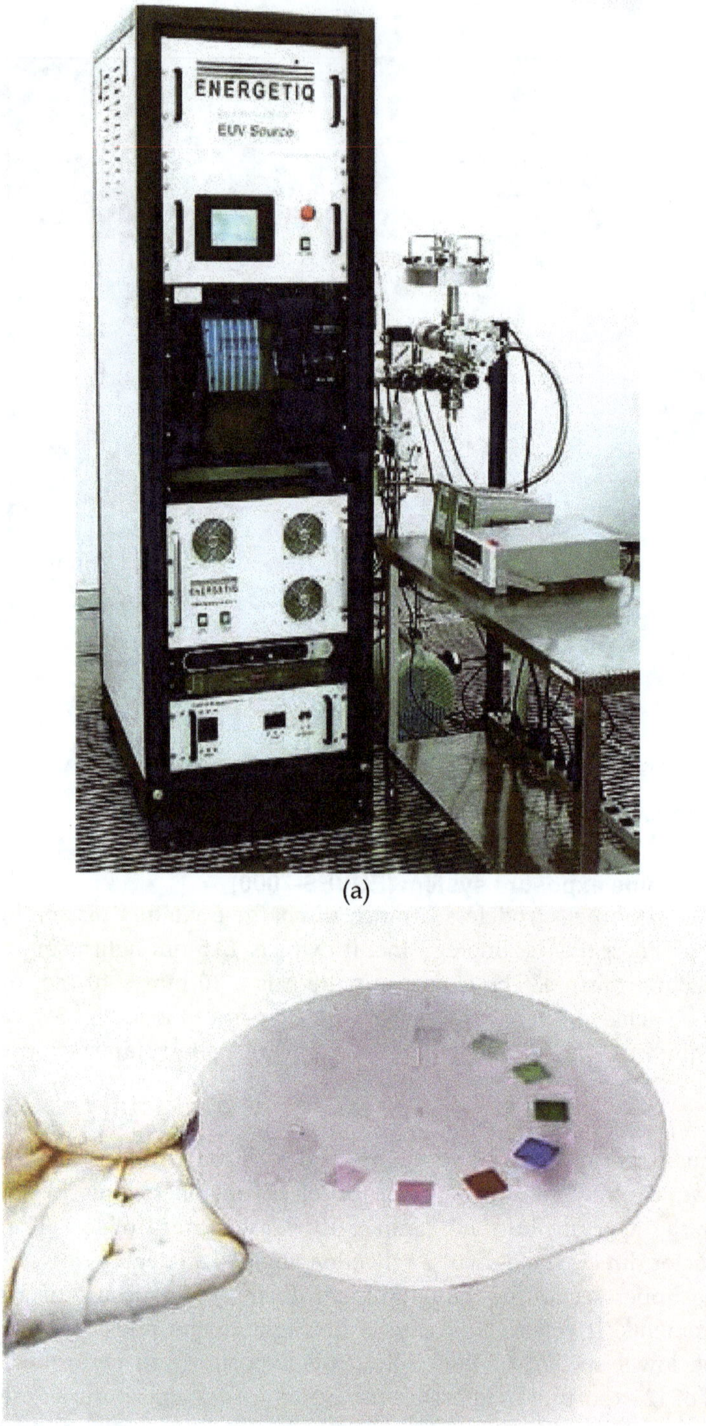

(a)

(b)

Fig. 3. (a) External view of EUVES-7000 and (b) exposure pattern

Fig. 4 is a picture of the beam line.

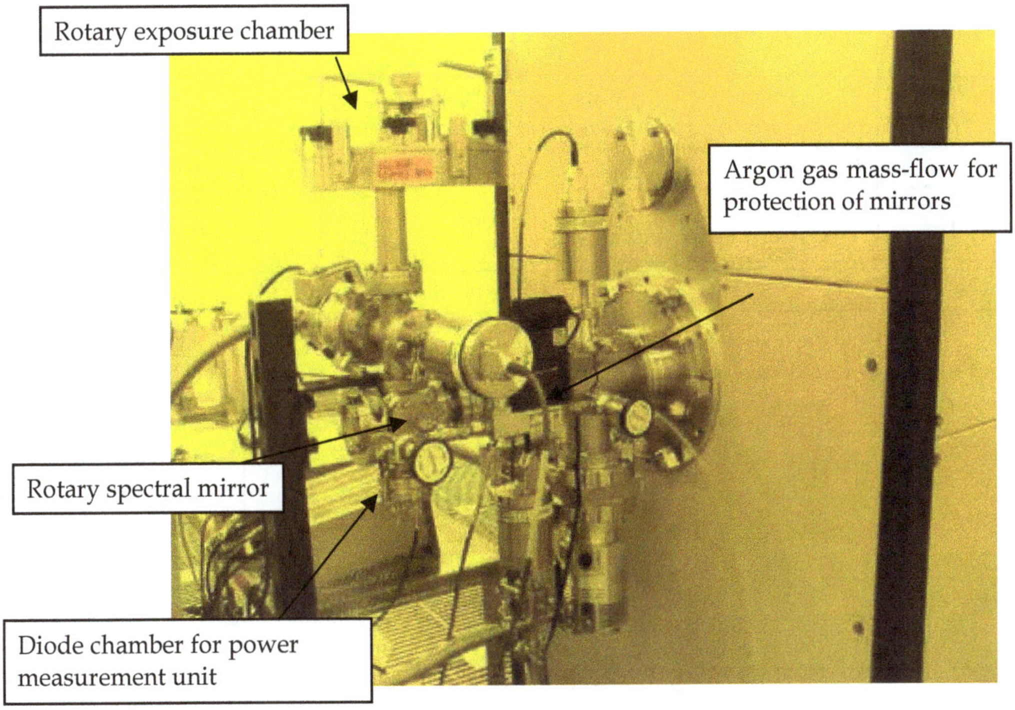

Fig. 4. Beam line for EUV exposure

1.2.2 Resist development analyzer (RDA-800EUV)

Following the exposure, a wafer is processed for PEB. Then, following measurement of film thickness, this resist development analyzer is used to measure the development rate of a resist corresponding to each exposure dose [10].

1.2.3 EUV lithography simulator (Prolith Ver. 9.3)

The obtained development rate data file is imported into the Prolith lithography simulator [11] (manufactured by KLA-Tencor) for EUV lithography simulation.

1.3 Experiment and results

We investigated the sensitivity of positive- and negative-type resists in EUV exposures with the system as described above, then performed simulations using the development rate data obtained.

Table 2 gives the conditions of the resists in our experiment.

The negative-type resists examined were the SAL-601 electron beam resist and SU-8 epoxy-resin-base chemically amplified resist. The positive-type resists used in our experiment were ZEP-520 non-chemically amplified electron beam resist, EUVR-1 and EUVR-2 acrylic-resin-base resists, and EUVR-3 low-molecular-weight resist.

Negative type						
Resist	Maker	Pre-bake		PEB		Thickness (nm)
		Temp. (deg.C)	Time (s)	Temp. (deg.C)	Time (s)	
SAL-601	Rohm & Hass	105	60	115	60	100
SU-8	Nippon Kayaku	90	90	95	100	100
Positive type						
Resist	Maker	Pre-bake		PEB		Thickness (nm)
		Temp. (deg.C)	Time (s)	Temp. (deg.C)	Time (s)	
ZEP-520A	Nippon Zeon	90	90	95	100	100
EUVR-1	TOK	120	90	120	90	100
EUVR-2	TOK	100	90	110	90	100
EUVR-3	TOK	110	90	100	90	100

Table 2. Conditions of resists in the experiment

Fig. 5.(a) shows discrimination curves for negative-type resists; Fig. 5.(b) shows discrimination curves for positive-type resists.

Posi-Type	Eth(60) mJ/cm2	γ60	tanθ
SAl-601	0.928	-1.445	-2.23
SU-8	0.478	-3.023	-3.73

Nega-Type	Eth(60) mJ/cm^2	γ60	tanθ
ZEP-520A	14.710	1.669	1.90
EUVR-1	2.562	2.325	5.06
EUVR-2	8.574	3.997	30.75
EUVR-3	8.497	1.528	14.53

Table 3. Development characteristics

Table 3. shows the results of development characteristic evaluations.
The results show EUVR-2 provides the highest contrast.

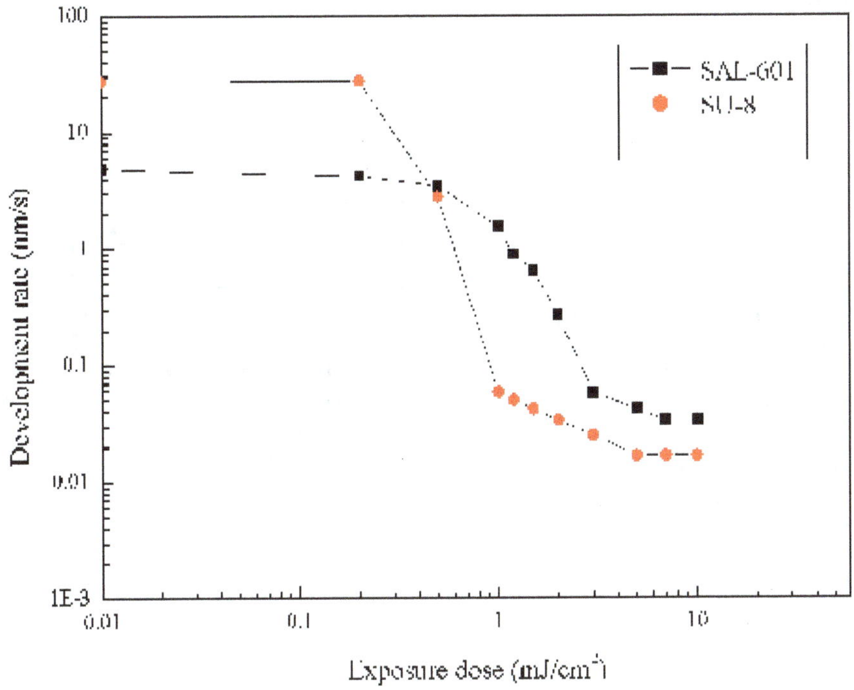

(a) Discrimination curves for negative-type resists

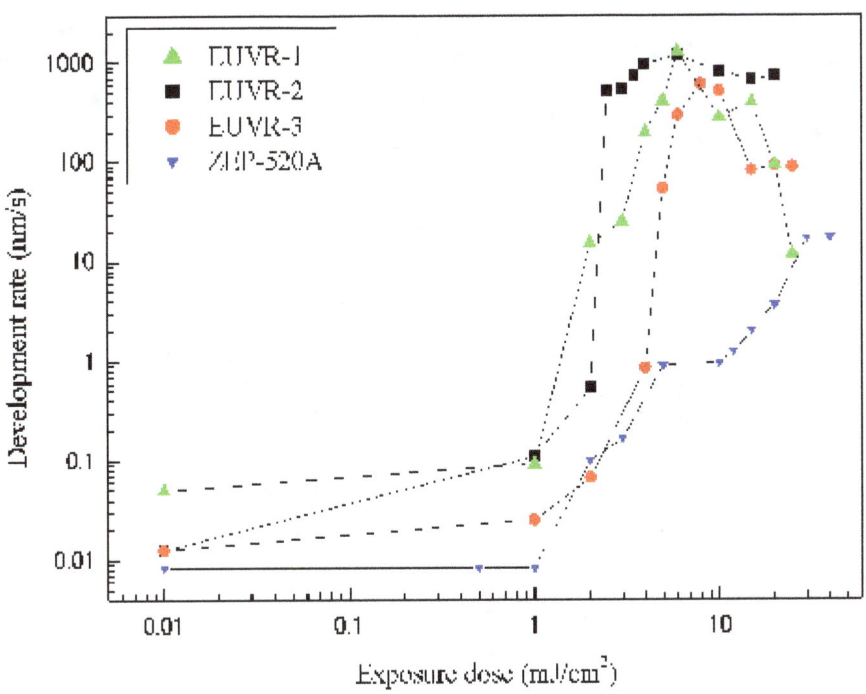

(b) Discrimination curves for positive-type resists

Fig. 5. Relationship between development rate and exposure dose

1.4 Simulation

We performed a simulation using the EUV-PM2 development data. Table 4 gives the simulation conditions.

Wavelength (nm)	13.5
NA	0.3
σ	0.8
Reduction	1/5

Table 4. Simulation conditions (with Nikon HiNA-3)

We examined L&S*[1] patterns and isolated patterns with pattern dimensions of 65, 55, 45, 32, and 22 nm. Defocus was examined using a 32 nm L&S pattern. Figures 6 through 8 show the simulation results. With L&S patterns, resolution can be maintained up to 32 nm. For isolated patterns, the results suggest that resolution on the order of 22 nm is within reach. In the defocus simulation, the simulation results support estimates of an attainable resolution range of -0.1 to +0.1 μm.

*[1]L&S: Line and space

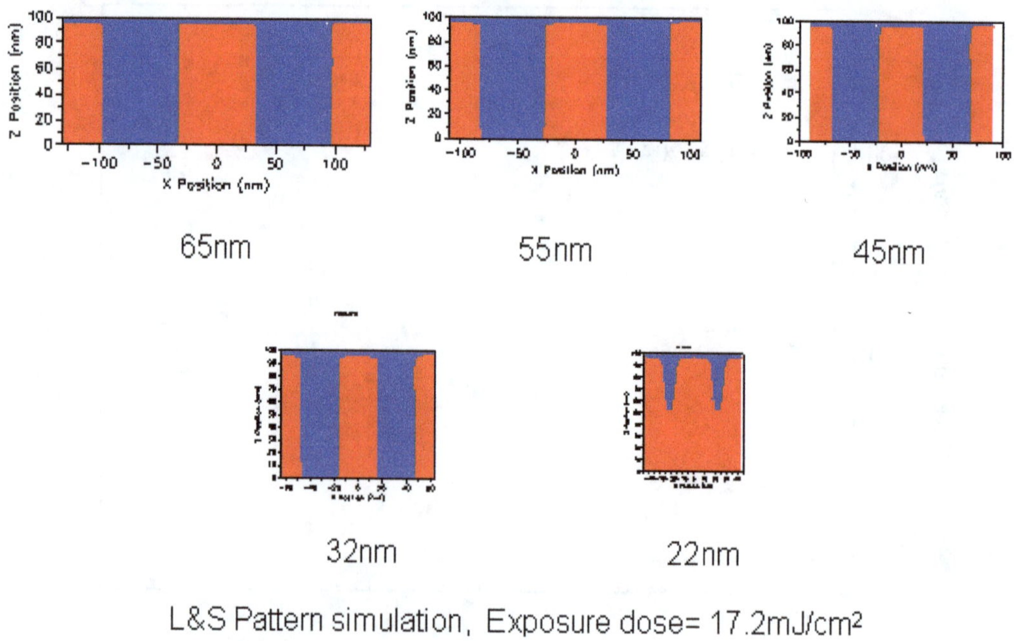

Fig. 6. Simulation results (65-22 nm Line and space patterns)

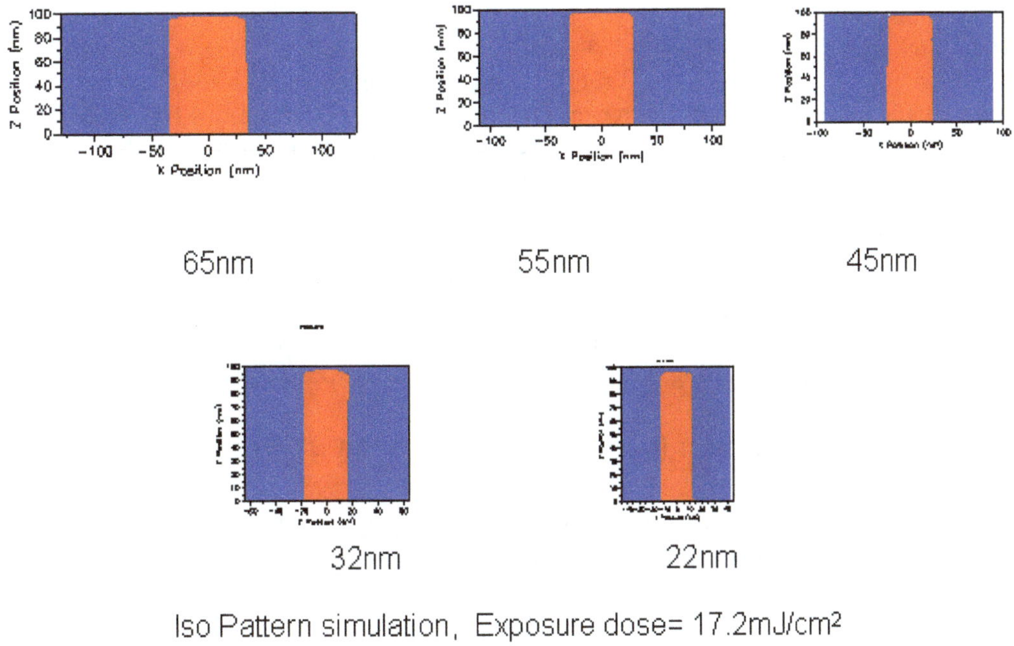

Fig. 7. Simulation results (65-22 nm Isolated patterns)

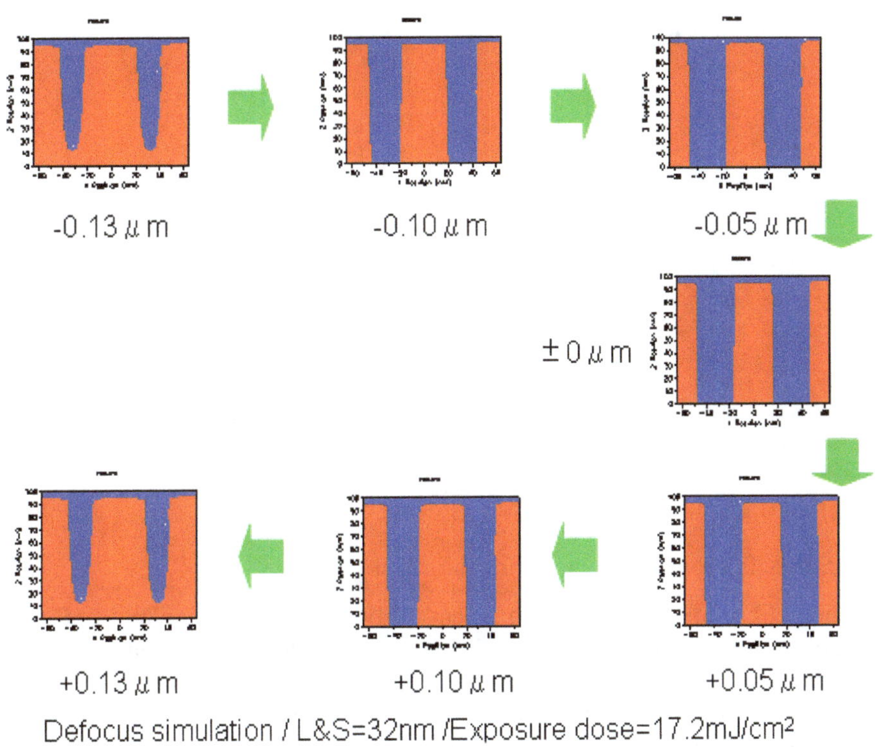

Fig. 8. Simulation results (32 nm L&S/ defocus -0.13∼+0.13μm)

1.5 Conclusion

The VLES consists of the EUVES-7000 EUV open-frame exposure system, RDA-800EUV development rate analyzer, and Prolith lithography simulator. We used the VLES to compare the sensitivity and development contrast of negative- and positive-type resists with EUV exposure. We also simulated EUV exposures using development rate data for the EUVR-2, which showed the highest development contrast of all resists tested. The results of the experiment suggest that it should be possible to obtain resolutions of 32 nm with L&S patterns and 22 nm with isolated patterns. We also calculated defocus characteristics with a 32 nm L&S pattern. Based on these calculations, we estimate a focus margin of approximately 0.2 μm in defocus width. We believe using the system as described in this paper will permit the development of photoresist materials for EUV and expedite process development without requiring the purchase of costly EUV exposure equipment.

2. Simulating EUV Resists (Comparison of KrF and EUV Exposures)

2.1 Introduction

According to ITRS Roadmap 2007 Update Version [12], EUVL is currently the most promising candidate for 22 nm half-pitch lithography. The component technologies required for EUVL mass production must be established before the start of mass production of DRAM half-pitch, currently scheduled for 2016. RLS specifications for realizing 22 nm half-pitch resolution were presented at the 7th EUVL Symposium [13] in Lake Tahoe, California, in October 2008.

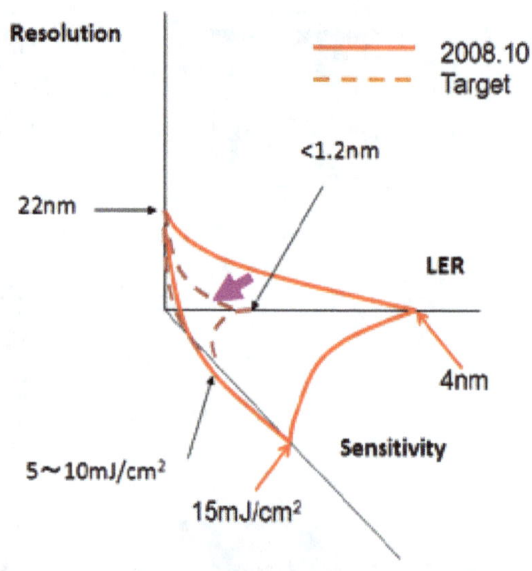

Fig. 9. RLS specifications targeting 22 nm half-pitch

A resolution of 22 nm half-pitch requires sensitivity of 5 to 10 mJ/cm^2 and LER of less than 1.2 nm. At an international conference, it has been pointed out that although resolutions have reached the target value, sensitivity lags, at 15 mJ/cm^2, while LER (Line Edge Roughness) is no less than 4 nm. These are the best values achieved to date. Lithography simulations should prove highly effective in advancing the state of current research, given the time required to perform experiments.

The conventional EUVL simulation method involves obtaining parameters by exposing the resist to EUV. However, EUV exposure equipment is costly, and the types of exposure equipment available are limited. For these reasons, we explored the possibility of performing EUVL simulations using parameters obtained with KrF exposures. The idea was that if we detected no significant differences between parameters obtained with KrF and EUV exposures, we could use the simpler KrF exposure method to obtain valid simulation parameters for EUVL. Using EUV resists, we obtained parameters by performing both KrF and EUV exposures, then compared the parameters and simulation results. This chapter discusses this comparison.

2.2 Simulation parameter measurement system

2.2.1 Exposure equipment for parameter measurement

Fig.10 shows the exposure equipment used in our parameter measurements. The exposure area is an open-frame pattern measuring 10 mm x 10 mm. We used a UVES-2000 for KrF exposures and an EUVES-7000 [14] for EUV exposures.

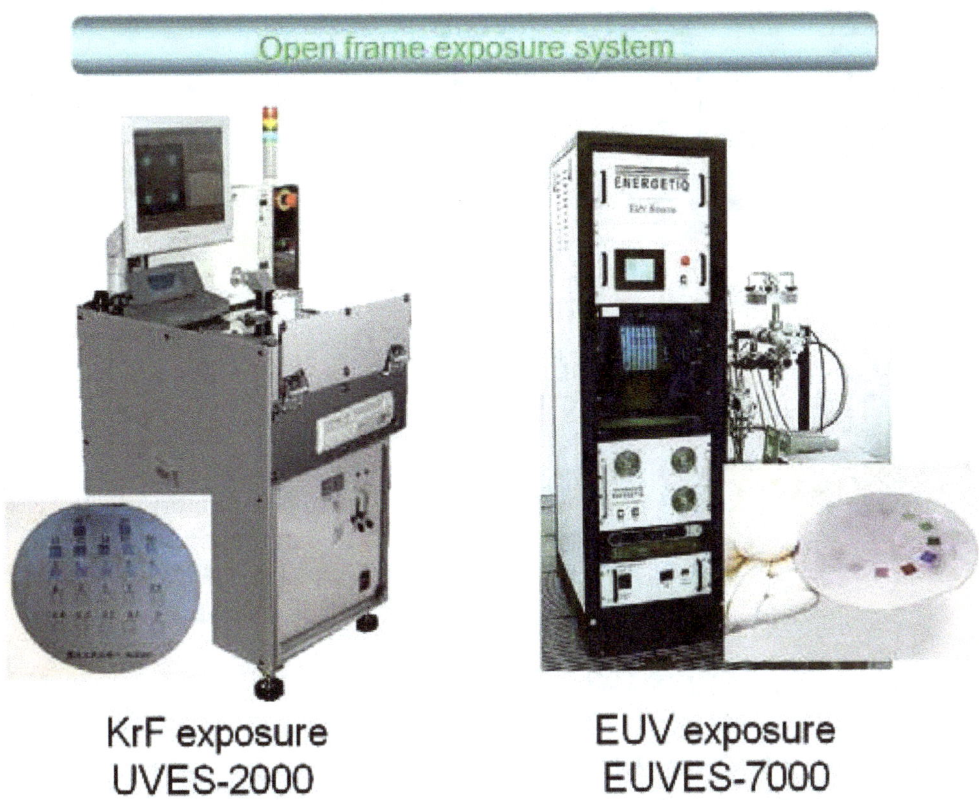

Fig. 10. Open frame exposure tool for KrF and EUV

These exposure tools permit resist exposures on Si wafers and the acquisition of development and PEB parameters.

2.2.2 Development parameter measurement system

We used a development analyzer to measure development parameters. When homogeneous light is irradiated onto a resist film during development, the light waves reflected from the resist surface and light waves reflected from the wafer surface interfere, generating unique waveforms. Analyzing the waveforms of the reflected light allows us to obtain resist development rates. By varying exposure values and measuring resist development rates at different exposures, we can calculate the development parameter, among the simulation parameters [15]. This measurement has been performed before using a monitor wavelength of 470 nm. However, thin films do not generate the interference needed, and a monitor wavelength of 470 nm limits us to resist film thicknesses exceeding 100 nm. Since the film thickness of EUV resists ranges from approximately 50 to 100 nm, we developed a measurement system for our experiments based on a monitor wavelength of 265 nm (Fig.11).

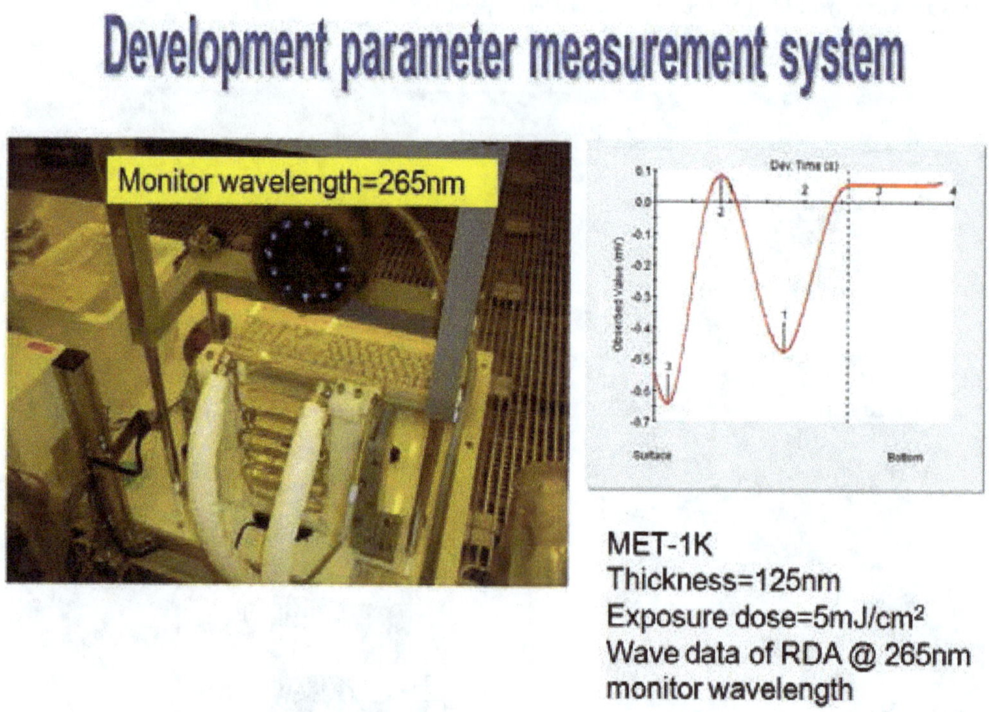

Fig. 11. Resist development analyzer RDA-800EUV

2.2.3 B parameter measurement system

We used the following equation to calculate the B parameter [16-17] of Dill based the resist transmission factor at the time overexposure completely breaks down the PAG.

$$B = -\frac{1}{d}\ln\left(T_{\infty}\right) \tag{1}$$

Here, *d* is resist film thickness and T_∞ the resist transmission factor at the time overexposure completely breaks down the PAG. We developed a system for measuring the resin transmission factor using EUV light. Incorporating a LPP light manufactured by Toyota Macs as its light source and using a solid Cu target, this system irradiates EUV light onto a Si/Mo multilayer reflecting mirror to measure reflection intensity, while mirror angles are varied. To calculate the spectral transmission factor, we used the difference in reflectance between the case in which resist is applied to the multilayer mirror and the case in which no resist is applied.

Fig.12 illustrates the measurement system and gives a chart of the results of spectral transmission factor calculations for the MET resist.

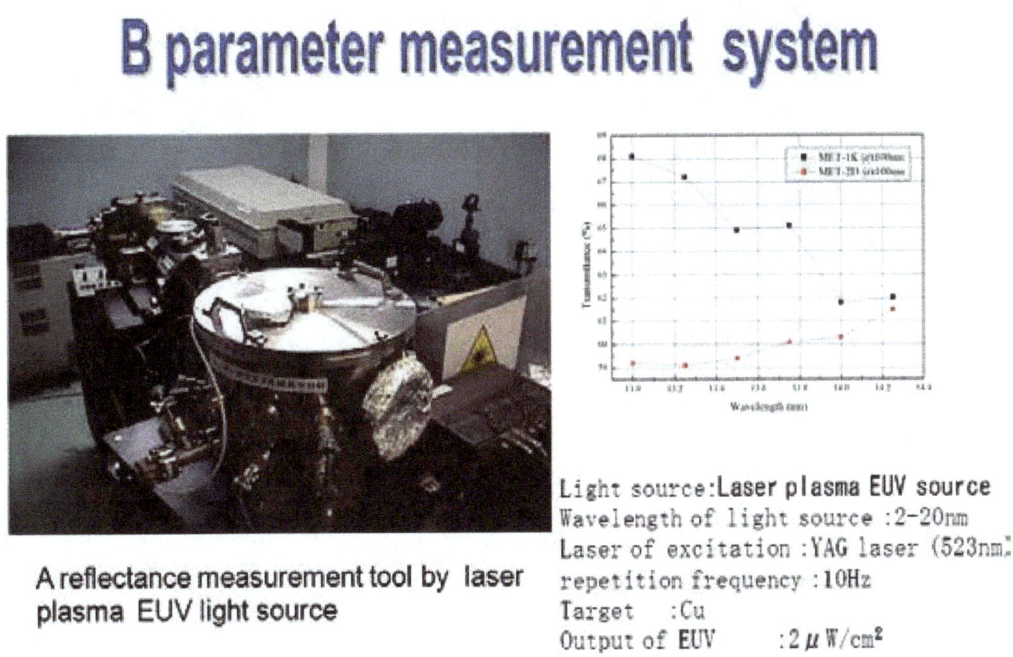

Fig. 12. B parameter measurement system using EUV exposure

2.2.4 De-protection reaction parameter and C parameter measurement system

Fig.13 gives an overview of the PEB parameter measurement system, which exposes resist on an Si wafer using KrF and EUV light. In the next step, we used an FT-IR system with a bake function to plot the de-protection reaction curve while performing PEB. We performed measurements at different PEB temperatures and measured the de-protection reaction parameter by fitting. During the course of fitting, we also obtained the C parameter for Dill. We modified the system [18] to allow irradiation of IR light for measurements on resist film at an angle of 45 degrees and to permit use with a resist film thickness of 50 nm. The resulting system was capable of handling extra-thin resist films ranging from 50 to 100 nm.

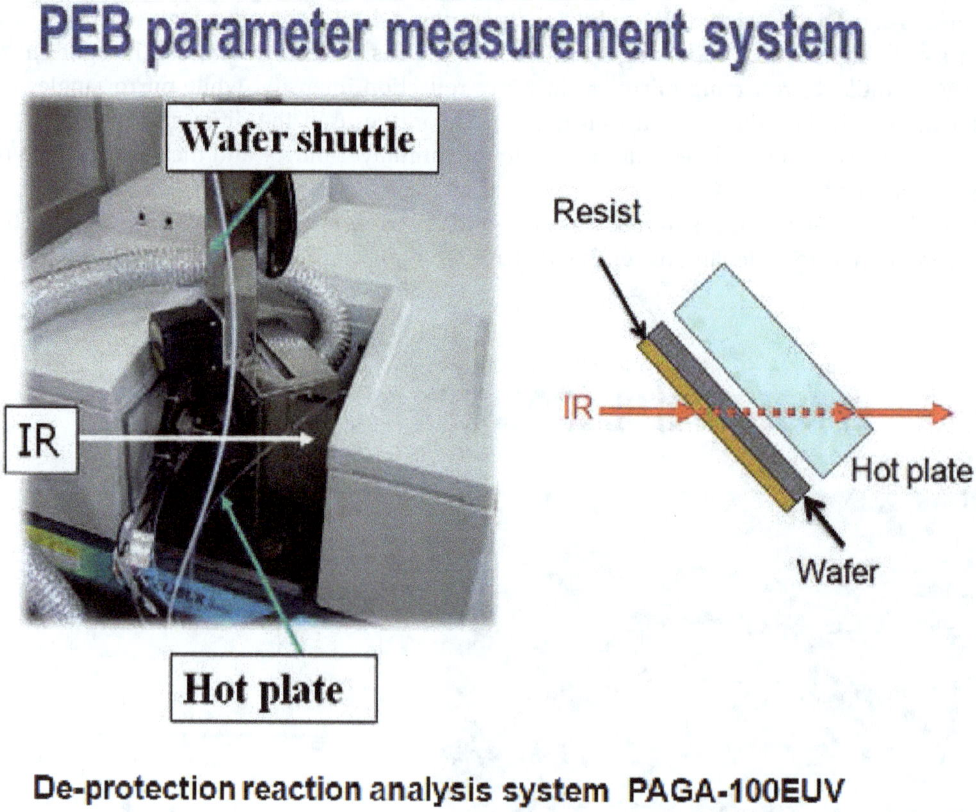

Fig. 13. PEB parameter measurement system

2.3 Parameter measurement results

We measured parameters using EUV chemically amplified resists MET-1K and MET-2D manufactured by Rohm and Haas.
Table 5 gives the process conditions.

	Vendor	Thickness (nm)	PAB		PEB	
			Temp. (deg.C)	time (s)	Temp. (deg.C)	time (s)
MET-1K	R&H	125	130	60	110	90
MET-2D	R&H	125	130	60	110	90

R&H: Rohm and Hass

Table 5. Measurement conditions

2.3.1 B parameter measurement results

Fig.14 shows B parameter measurement results. With KrF exposures, MET-1K and MET-2D yielded values of 0.726 and 0.788, respectively. With EUV exposures, MET-1K and MET-2D yielded 4.32 and 5.21, respectively, indicating greater absorption with EUV exposures.

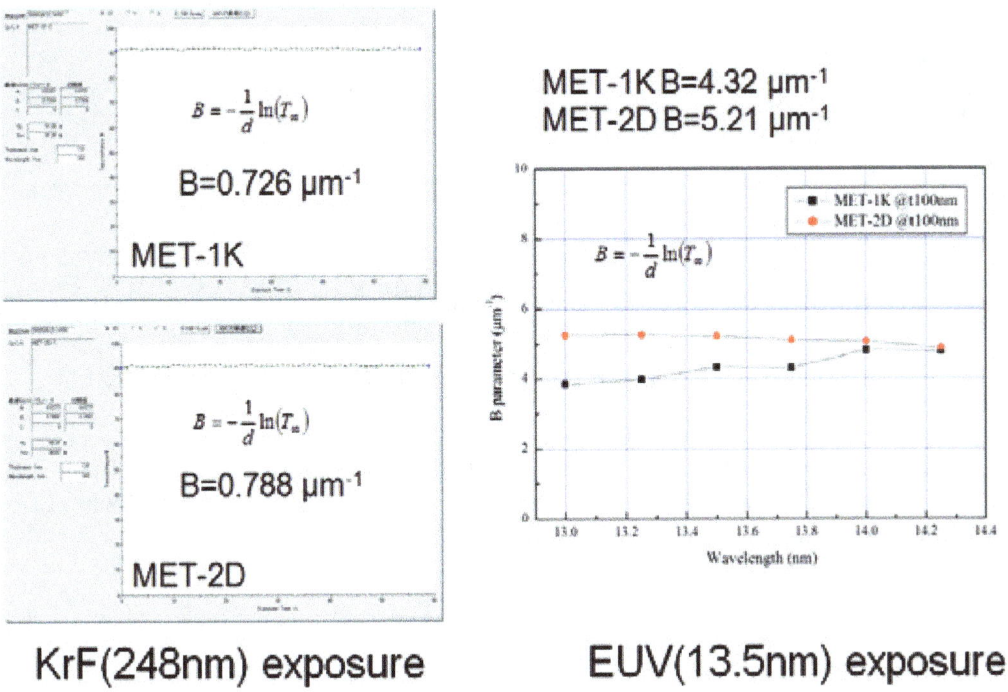

Fig. 14. B parameter measurement results

2.3.2 Development parameter measurement results

Fig.15 compares measurements of the discrimination curve (a logarithmic plot of development rates and exposure values) development parameter. We found no significant differences between development parameter values obtained with KrF and EUV exposures.

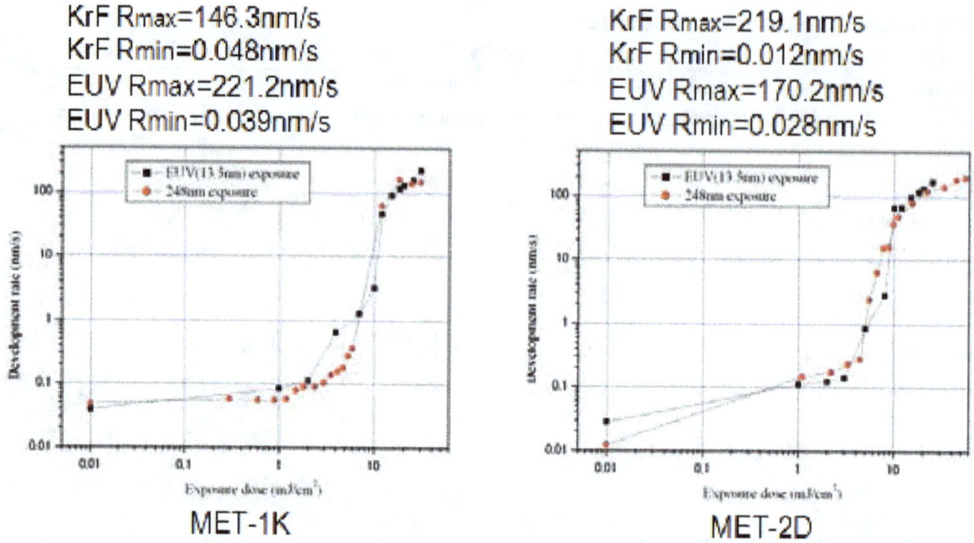

Fig. 15. Comparison of development parameter values

2.3.3 De-protection reaction observations and results of de-protection reaction parameter measurements

Fig.16 shows IR spectra obtained from the MET-1K before and after 16-mJ/cm^2 exposures. The figure indicates weakened (-CO) bonds due to de-protection reactions at 1,230 cm^{-1}. The extent of the peak decline with KrF exposure is roughly identical to that with EUV, indicating the absence of significant differences in de-protection reactions.

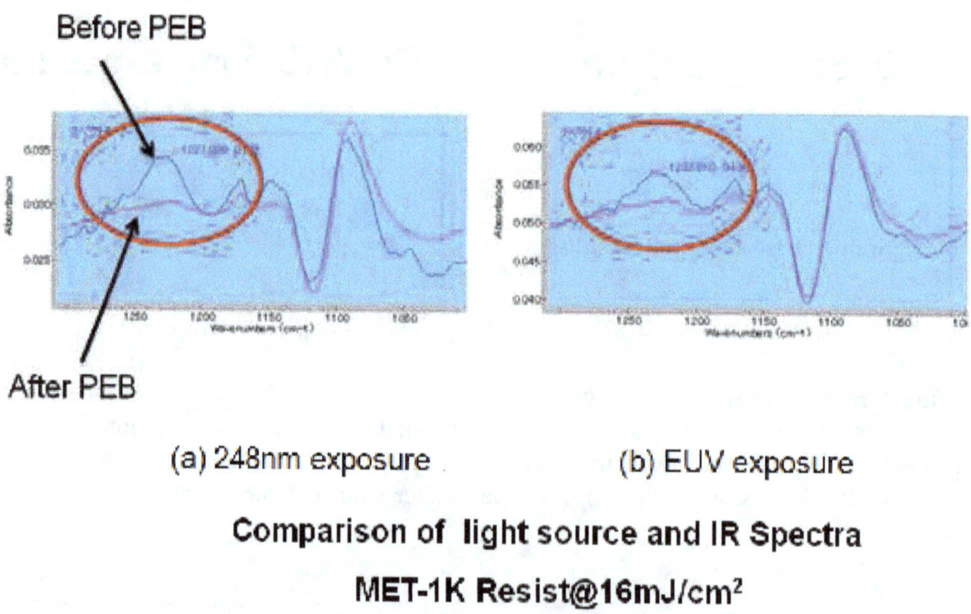

Fig. 16. Observations of de-protection reactions with KrF and EUV exposures

Table 6 is a list of simulation parameter measurement results.

	KrF(248nm) exposure		EUV(13.5nm) exposure	
	MET-1K	MET-2D	MET-1K	MET-2D
ABC-parameter				
A (1/μm)	0	0	0	0
B (1/μm)	0.726	0.788	4.32	5.21
C(cm^2/mJ)	0.017	0.021	0.086	0.090
Development parameters (Mack original model)				
Development R_{max} (nm/s)	146.3	219.1	221.2	170.2
Development R_{min} (nm/s)	0.048	0.012	0.009	0.028
Development M_{th}	0.773	0.783	0.624	0.518
Development n	15.21	17.30	12.65	18.96
PEB parameters				
PEB Amplification Ea (Kcal/mol)	12.21	11.30	3.82	2.23
PEB Amplification LN(Ar)(1/s)	14.89	13.65	5.53	3.55
PEB Diffusion Controlled Reaction Ea (Kcal/mol)	8.56	4.82	8.80	1.00
PEB Diffusion Controlled Reaction LN(Ar) (1/s)	10.14	5.54	11.50	1.78
PEB Diffusivity Ea (Kcal/mol)	5.00	5.00	5.00	5.00
PEB Diffusivity LN(Ar) (nm2/s)	23.22	21.24	30.00	30.00

Table 6. Simulation parameter measurement results

2.3.4 Examination of simulation

We performed EUVL simulations using the simulation parameters obtained. Table 7 gives the simulation conditions used.

Simulation condition

NA=0.30

Wavelength=13.5nm

Illumination= Annular (0.3/0.7)

Flare=1%

Line and space=28nm

Resist: MET-1K and MET-2D

Thickness:125nm

Table 7. Simulation conditions

For exposure equipment, our simulation assumed use of the Nikon EUV-1 installed at Selete [19]. Fig.17 shows the simulation results. The indicated exposure value is the exposure level (E_0) that achieved 1:1 resolution from a 28-nm L&S pattern. The development conditions called for 2.38% TMAH and development time of 60 seconds. The quencher diffusion length and PAG diffusion length were set to 20 nm and 10 nm, respectively.

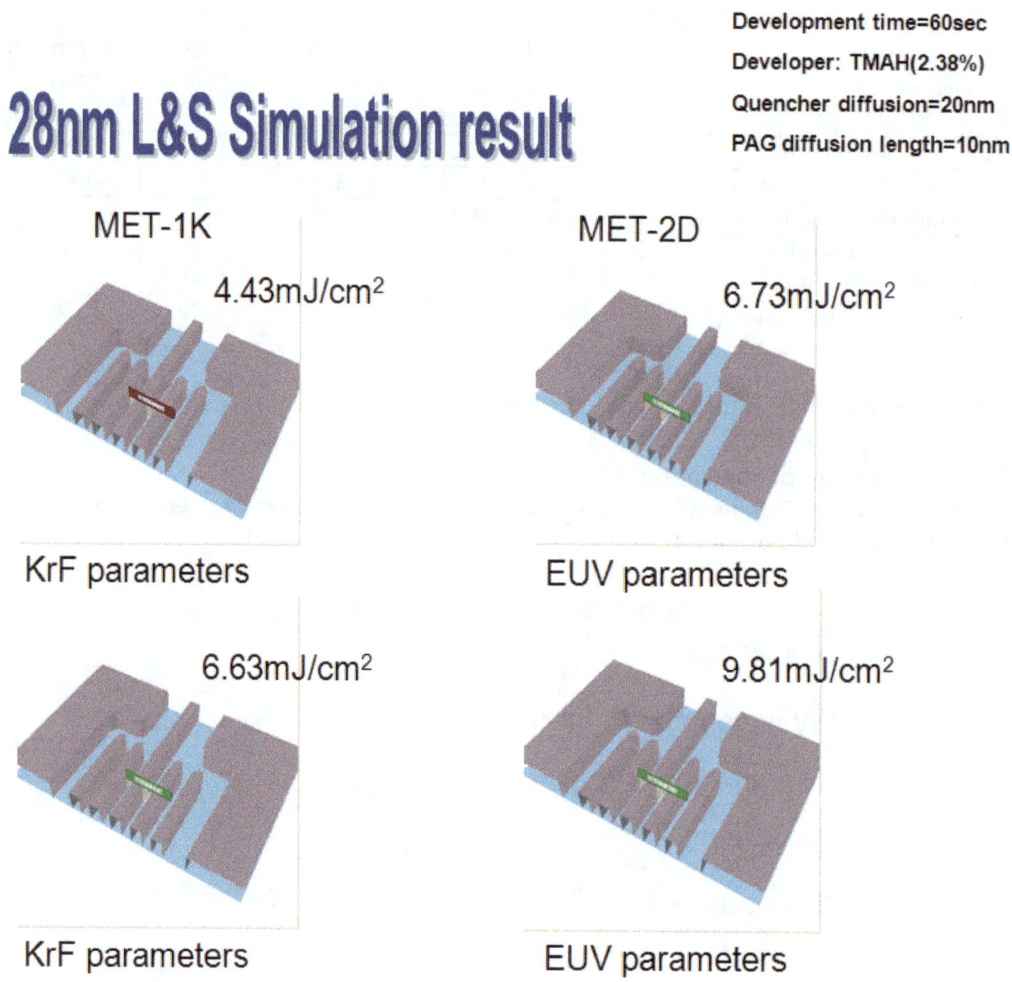

Fig. 17. Simulation results

We compared the results of EUVL simulations based on parameters obtained with KrF exposures to the results of EUVL simulations based on parameters obtained with EUV exposures. While the former simulation results indicated higher sensitivity (approximately 20% higher), we saw no major differences in shape.

2.4 Conclusion

We compared the results of EUVL simulations based on parameters obtained with KrF exposures to the results of EUVL simulations based on parameters obtained with EUV exposures. The former resulted in approximately 20% higher simulation sensitivity, but we saw no major differences in shape. Using parameters obtained with KrF exposure is a roundabout way to perform EUVL simulations. Since EUV exposures in many cases are not readily available, a valid option would appear to be to acquire simulation parameters through KrF exposures and to use these parameters as initial values in calculations for EUVL simulations.

3. References

B. J. Lin, *Proc. SPIE,* 4688, 11, 2002

IMEC 4th Immersion Workshop, Septemberin Belgium, 2005

V. N. Golovkina, P. F. Nealy, F.Cerrina, J. W. Taylor, H. H. Saolak, C. Davidand J. Gobrecht, *J. Vac. Sci., Technol.* B, 22(1), 99, 2004

Intel H.P.
http://www.intel.com/jp/technology/index.htm?iid=subhdr-JP+tech

ITRS 2005: International Technology Roadmap for Semiconductors 2005 Edition

H. B. Cao, W. Yueh, J. Roberts, B. Rice, R. Bristol and M. Chandhok. Proc. SPIE,5753, 459, 2005

NIKON H.P.

http://www.nikon.co.jp/news/2007/0711_nsr_01.htm

ASLMH.P.
URL: http://www.asml.co.jp

P. Blackborow, *Proc.SPIE,* 6151, 25, 2006

A. Sekiguchi, C.A. Mack, Y. Minami and T. Matsuzawa, *Proc. SPIE,* 2725, 49, 1996 Prolith-Version 9.3 User's manual

ITRS LOAD MAP 2007 Up-date in Web

7th EUVL symposium in Lake Tahoe, CA (2008.10)

A. Sekiguchi, Y. Kono, M. kadoi, Y. Minami,T. Kozawa, S.Tagawa, D. Gustafson and P. Blackborow, Proc. SPIE, 6519, 168 (2007).

A. Sekiguchi, Y. Minami, and Y. Sensu, The Electrochemical Society of Japan, Proc. of the 42nd Sysmp. on Semiconductors and Integrated Circuits Technology, Vol. 42, pp. 109-114 (1992).

Dill-B F. H. Dill, W. P. Hornberger, P. S. Hauge, and J. M. Shaw, IEEE Trans. Electron Dev., Vol. ED-22, No. 7, pp. 445-452 (1975).

C. A. Mack, T. Matsuzawa, A. Sekiguchi, and Y. Minami, Proc. SPIE, Vol. 2725, pp. 34-48 (1996).

A. Sekiguchi, and Y. Kono, Proc. SPIE, 6923, 92 (2008)

H. Oizumi, D. Kawamura, K. Kaneyama, S. Kobayashi and T. Itani, RE-03, 7th EUVL symposium (2008).

7

Physical Deposition Assisted Colloidal Lithography: A Technique to Ordered Micro/Nanostructured Arrays

Yue Li[1,3], Shuyan Gao[2], Guotao Duan[1], Naoto Koshizaki[3] and Weiping Cai[1]
[1]*Key Laboratory of Materials Physics, Institute of Solid State Physics, Chinese Academy of Sciences, Anhui,*
[2]*College of Chemistry and Environmental Science, Henan Normal University, Xinxiang*
[3]*Nanosystem Research Institute (NRI) National Institute of Advanced Industrial Science and Technology (AIST)*
[1,2]*China*
[3]*Japan*

1. Introduction

Ordered micro/nanostructured arrays have attracted much interest due to their important applications in microfluidic devices, optoelectronic devices, nanophotonics, field emitters, nanogenerators, sensors, nano-biotechnology, surface science, photocatalytic properties etc.[1-11] The traditional routes to create periodic micro/nanostructured arrays are generally divided into two step. Microsized structure arrays are first fabricated by traditional lithographic techniques (e.g. photo-lithography, electron-beam lithography, ion beam lithography, x-ray lithography)[12-15.] as well as soft lithography (e.g. the techniques of replica molding, microcontact printing, micromolding in capillaries)[16-19], the nanostructures are then modified on the microsized units in array,[20] thus hierarchical micro/nanostructured arrays are finally achieved. However, they cannot be afforded due to the high costs and time-consuming in the most laboratories. Recently, the monolayered colloidal crystals (or called colloidal monolayers), ordered monolayer colloidal sphere arrays with hexagonal close-packed lattice structures on a certain substrate by self-assembly,[21-35] can be used to prepare ordered structure arrays. [36-41] It has proved that it is a flexible approach to fabricate the periodic micro or nanostructure arrays (e.g. nanoparticle arrays,[42-49.] nanopore arrays,[50-59] hollow sphere arrays[60-65]) based on colloidal monolayer templates by the different routes, solution/sol-dipping route, electrochemical deposition etc. Their properties are morphology and arrangement parameter dependent. Besides these periodic structure arrays, the colloidal monolayer template also can be used to prepare hierarchical micro/nanostructured arrays. For example, the hierarchical micro/nanostructured polystyrene (PS) sphere/CNTs composite arrays were obtained by wet chemical self assembling;[66-68] hierarchical microsized PS sphere/silver nanoparticle composite arrays or microsized pore/silver nanoparticle arrays were made by thermal deposition of silver precursor;[69,70] gold hierarchical micro/nanostructured particle arrays were created by electrochemical deposition based two step replication of colloidal monolayer template.[71] However, these routes have been developed by basically chemical reaction. They have some disadvantages of impurities on surface of arrays due to incompletely decomposition of precursors, residua of surfactants in self-assembling or electrochemical deposition. Additionally, it is quite difficult to achieve very uniform morphology of hierarchical micro/nanostructure arrays on a large-scale. Another route of colloidal monolayer template combining with physical deposition is expected to resolve these problems. In this chapter, we focus on introducing the recent work to create micro/nanostructured arrays based on colloidal templates with physical deposition (pulsed laser deposition (PLD) and sputtering). The parameters of microstructure or nanostructure can be tuned by periodicities of colloidal templates or experimental conditions of physical deposition. The applications of nanorod arrays with controllable morphology and arrangement parameters in self-cleaning surfaces, enhanced catalytic properties, field emitters etc. are also presented in following section.

2. Pulsed laser deposition assisted colloidal lithography[73,74]

2.1 Method

A polystyrene (PS) colloidal monolayer was first fabricated on a substrate. The desired material was then deposited on this colloidal monolayer substrate by PLD at room temperature and oxygen was introduced into PLD chamber as the background gas. This periodic array has a special hierarchical micro/nanostructure array with a hexagonal-close-packed (hcp) arrangement, which originate from the pattern of colloidal monolayer. In this micro/nanostructure unit in array, the nanorod stands vertically on the microsized PS sphere tops, and nanobranches in each nanorod grow in a radiationlike manner, perpendicular to the PS sphere surface. The detailed experiments are described as following.

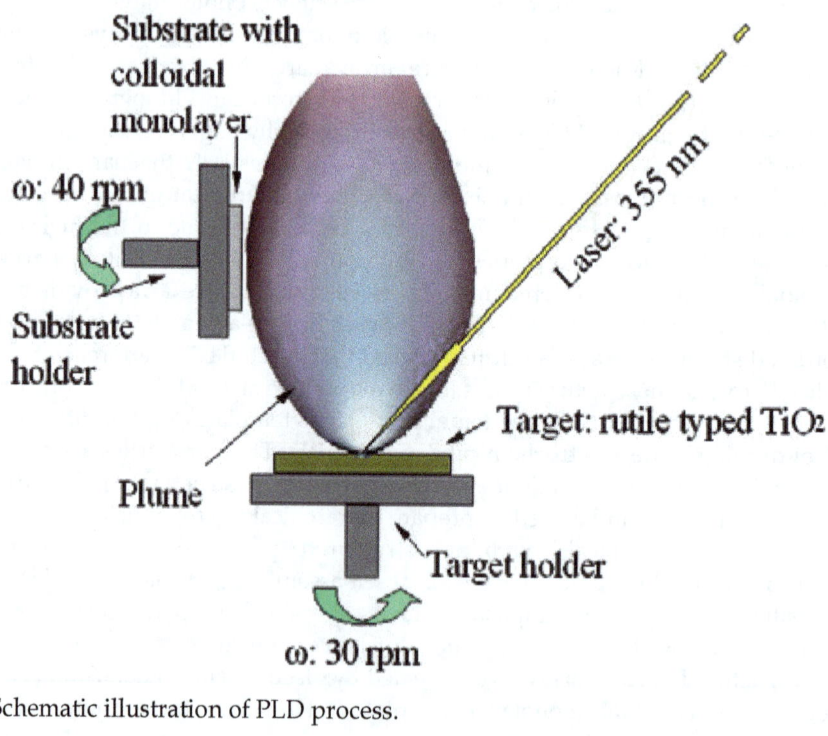

Fig. 1. Schematic illustration of PLD process.

Commercial monodispersed PS spheres dispersed in water with a certain size were purchased from companies. The PS colloidal monolayers were first fabricated on cleaned Si substrates by self-assembly using spin coating. The colloidal monolayer with its supporting substrate was placed in a deposition chamber of PLD, close to the target and at an off-axial position with respect to the target, as shown in Figure 1.A laser beam with a 355 nm wavelength from a Q-switched Nd:YAG laser (Continuum, Precision 8000), operated at 10 Hz with 100 mJ/pulse and a pulse width of 7 ns was applied and focused on the target surface with a diameter of about 2 mm. The desired target, for example, rutile typed titanium dioxide was used for deposition. The substrate and target were rotated at 40 and 30 rpm, respectively. PLD was carried out at a base pressure of 2.66×10^{-4} Pa and a background O_2 pressure of 6.7 Pa.

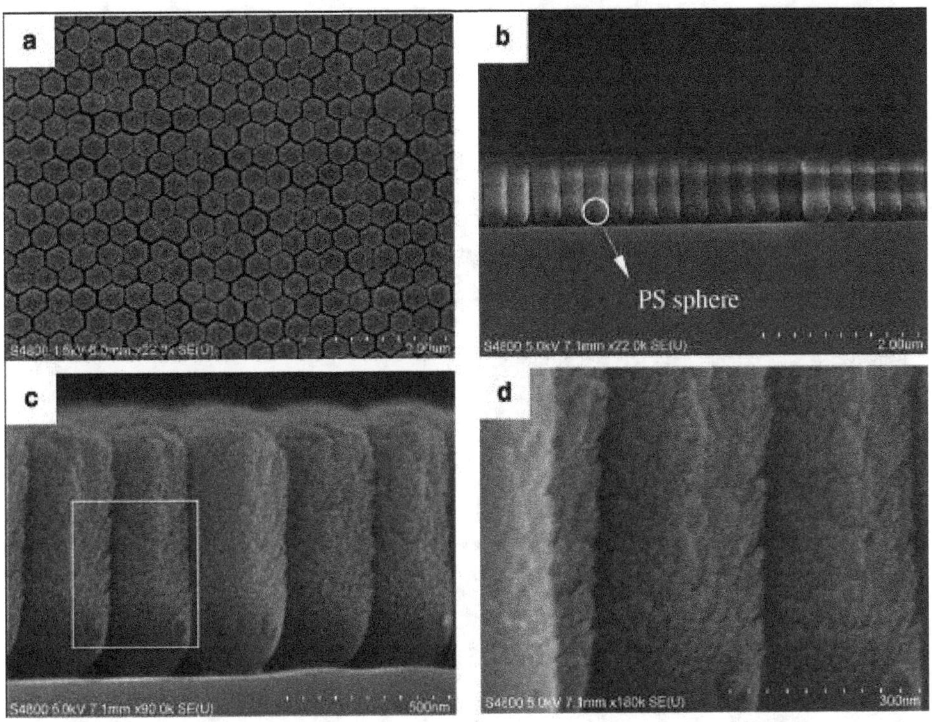

Fig. 2. Morphology of a sample obtained by PLD using a Si substrate with a PS colloidal monolayer coating (PS sphere size: 350 nm; deposition time: 70 min). (a) FESEM image from top view and (b) FE-SEM image of cross-section. (c) and (d) are high-resolution images observed from the side. (d) much higher magnification image of (c).

After deposition, the sample demonstrated a periodic hierarchical micro/nanorod array with an hcp arrangement, as reflected from Figure 2a. Each nanorod consists of two parts: a PS sphere at the bottom and a vertical nanorod on the top of the PS sphere (Figure 2b). The diameter of the nanocolumn was almost the same as that of the PS sphere, 350 nm, and its height was about 870 nm. The nanorod had a very rough structure on the surface and was composed of many nanobranches, according to the high-resolution images of the side view (Figure 2 c, d). TEM observation from the top of the nanorod arrays reflects that each nanorod consists of radiation-shaped nanobranches emanating from the center (Figure 3a).

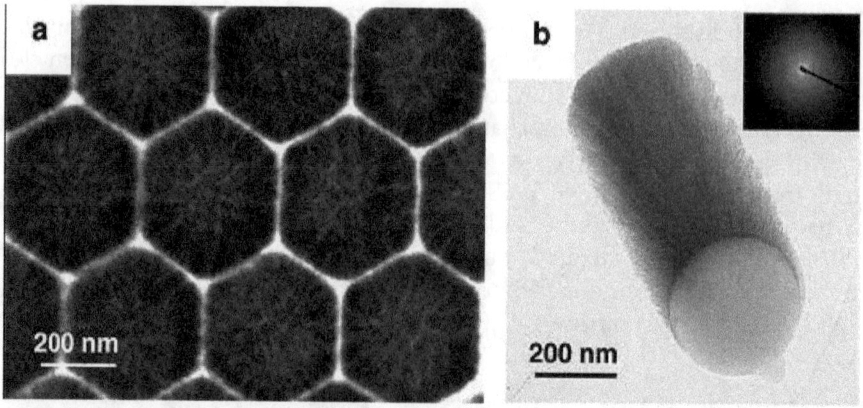

Fig. 3. Corresponding TEM images of the sample in Figure 2. (a) Periodic nanorod array observed from the top. (b) Single nanorod observed from the side. The inset in (b) is the corresponding electron diffraction pattern.

Fig. 4. a, b: FE-SEM images of a CuO hierarchical micro/nanostructured array obtained by combining the PS colloidal monolayer and PLD process. (PS sphere size 350 nm, deposition time 2 h, ambient oxygen pressure during deposition 6.7 Pa). (a) top-view image; (b) section view. Scale bars in parts a and b indicate 500 nm. c, d: TEM images of a CuO hierarchical micro/nanostructured array: (c) TEM image from the top; (d) TEM image of several separated units from the periodic array and the corresponding selected area electron diffraction (SAED) pattern.

The TEM image of a single nanorod also clearly displays that the nanorod consists of a PS sphere at bottom and a nanocolumn on the sphere surface. The nanorod possesses nanobranch-structures, which grow almost vertically on the PS sphere surface (Figure 3b). The nanobranch-structures indicate that the nanorod has a hierarchical, porous structure and hence has a high surface area. The selected area electron diffraction (SAED) pattern shows that the deposited materials on PS sphere surfaces by PLD are amorphous. Besides TiO_2 amorphous hcp nanocolumn arrays, the presented strategy can be extended to the fabrication of similar amorphous structures of SnO_2, WO_3, C, and so forth, just by changing the corresponding target in the PLD process.

Additionally, some materials, e.g. CuO, Fe_2O_3, ZnO are easily crystalline by PLD at room temperature. If the colloidal monolayer is applied as a template, the crystalline CuO, Fe_2O_3, ZnO etc. hierarchical micro/nanostructured arrays can be also obtained. Figure 4 shows the SEM and TEM image CuO crystalline hierarchical micro/nanostructured arrays using colloidal monolayers as templates by PLD. Each arrayed unit is composed of PS sphere at bottom and deposited materials at top. Deposited materials are well crystalline, they do not exhibit round shapes but radially aligned nanocolumns having tips with trigonal pyramidal shapes on the PS sphere.

The deposited CuO nanostructures can be tuned by varying ambient gas pressures during the PLD process. Figure 5 shows the FE-SEM and TEM images of samples achieved by PLD under higher ambient gas pressures during the PLD process using the colloidal monolayers as substrates. When oxygen pressure increased from 6.7 to 26.7 Pa, the morphology did not appreciably change and exhibited similar hierarchical structures as before (Figure 5a). However, when the oxygen pressure increased to 53.3 Pa, the morphology completely changed and was very different from those at lower pressures. The nanocolumn tips on the PS sphere demonstrated imperfect trigonal pyramid shapes, and the tip sizes became much smaller (Figure 5b). According to the corresponding TEM image and SAED pattern (Figure 5c), we find that hierarchical micro-/nanostructures were still observed at such high oxygen pressure, but the crystallization of deposited aligned nanocolumns on the PS sphere becomes worse than that obtained at lower oxygen pressure. When the gas pressure increased to as high as 79.8 Pa, similar hierarchical micro-/nanostructured array was not obtained, and many aggregates of small particles were produced on the colloidal monolayer template (Figure 5d). The XRD spectra of the samples obtained under different oxygen pressures are shown in Figure 6. Strong preferential orientation growth along (002) was observed at the gas pressure of 26.7 Pa. Increasing oxygen pressure led to weakening of this preferential orientation and broadening of X-ray diffraction peaks. This result reflects that deposited materials gradually changed to small nanoparticles from aligned nanocolumn arrays and the particles became much smaller with increasing oxygen pressure during PLD, agreeing with FE-SEM images. When the oxygen pressure increased to very high value, 79.8 Pa, the deposited material completely consisted of small nanoparticles or the aggregates of small nanoparticles, and there was no preferential orientation growth. Because when the gas pressure increases to a high value, the plume is compressed into a smaller space in PLD process, and the possibility of collision among ions or atoms in plasma is greatly enhanced, further resulting in a kinetic energy decrease of ions or atoms, which leads to less crystallization and smaller nanoparticle formation[75].

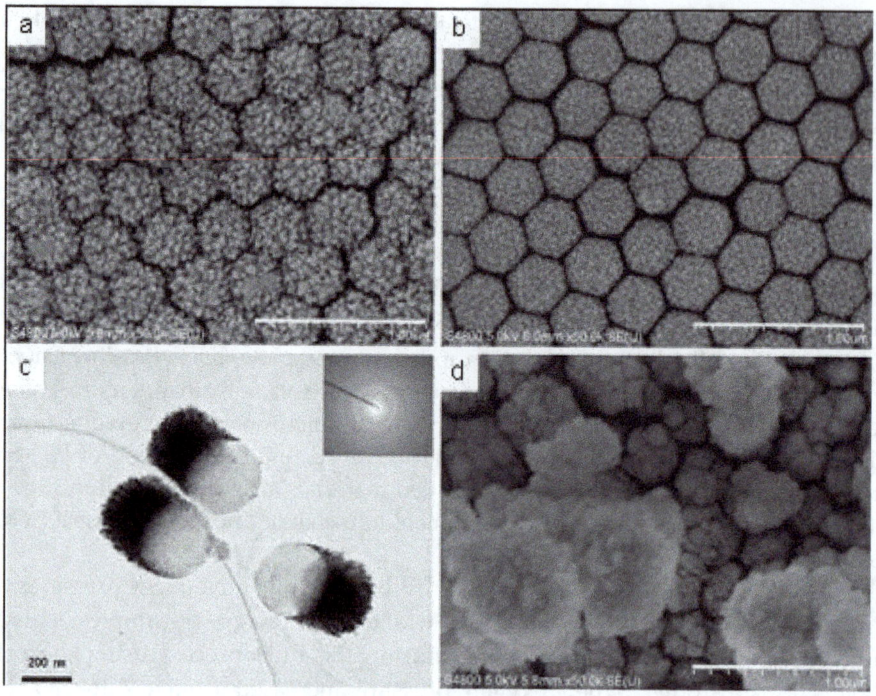

Fig. 5. Images obtained by the different ambient oxygen pressures: (a, b, d) FE-SEM images of the samples obtained under ambient oxygen pressure of 26.7, 53.3, and 79.8 Pa, respectively; (c) TEM image of the sample obtained at 53.3 Pa and the corresponding SAED pattern of several units. Scale bars in (a), (b), and (d): 1 μm.

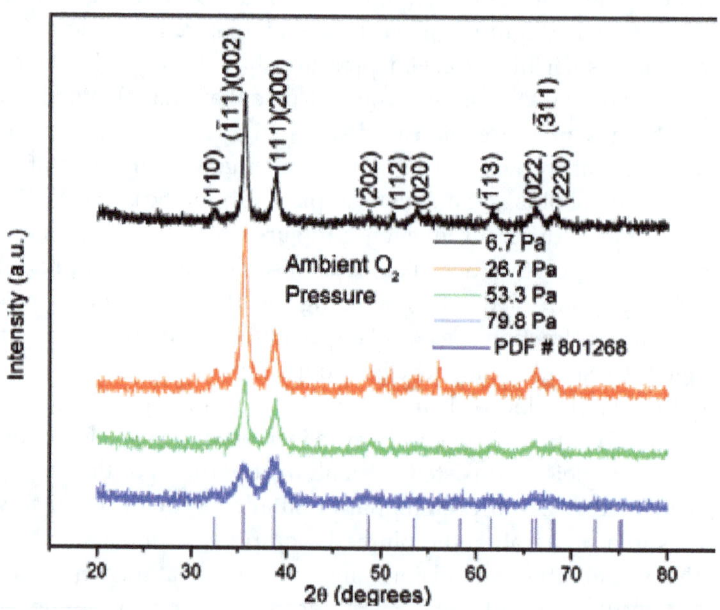

Fig. 6. XRD patterns of the samples obtained under different oxygen pressures.

Similar crystalline hierarchical micro-/nanostructured arrays of Fe_2O_3 and ZnO can be also created by the same route, as shown in Figure 7. Fe_2O_3 nanobelts or ZnO nanocolumns were well aligned on the PS sphere tops, like those of CuO. However, the Fe_2O_3 nanobelt or ZnO nanocolumn tops were not like those of CuO. The slight differences among CuO, Fe_2O_3, and ZnO fine nanostructures are determined mainly by their various chemical and physical properties: a crystal facet of the interface with different energies etc.

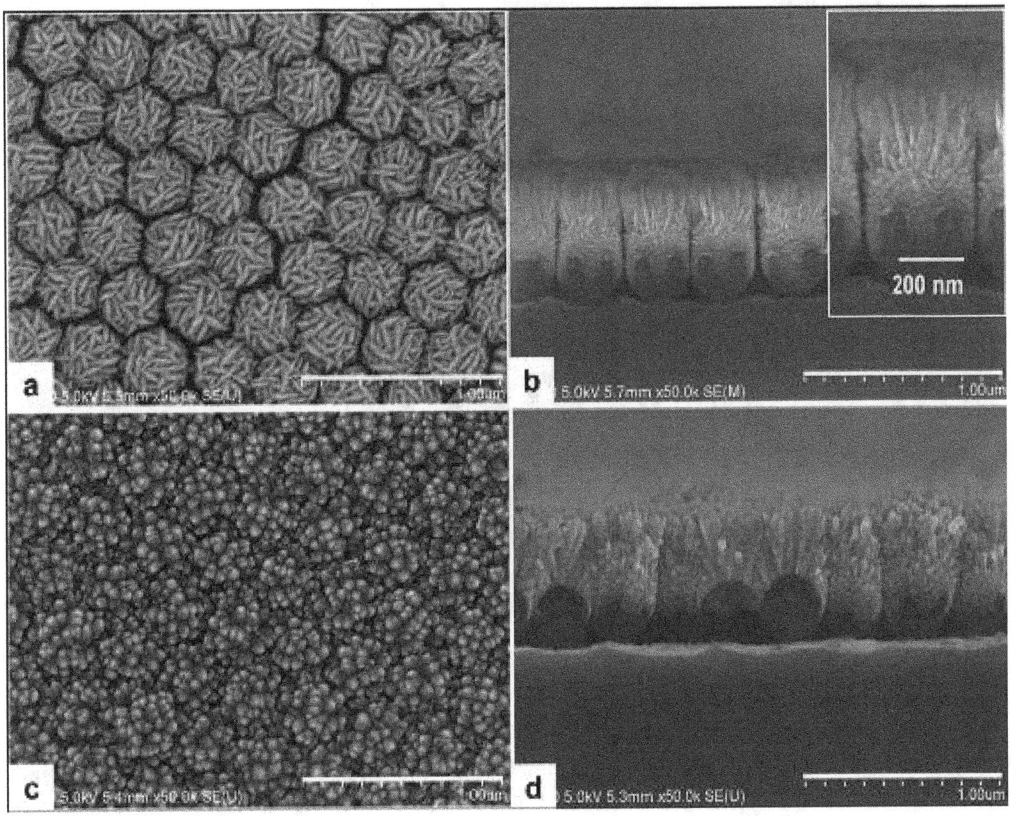

Fig. 7. FE-SEM images of hierarchical micro-/nanostructured arrays of Fe_2O_3 and ZnO. (a, b) Fe_2O_3, oxygen pressure 6.7 Pa, deposition time 1.5 h; (c, d) ZnO. a and c: top views; b and d: side views (oxygen pressure 6.7 Pa and deposition 40 min). The inset in (b): the high magnification image of a single Fe_2O_3 hierarchical micro-/ nanostructure.

In this strategy, the height of micro/nanostructured unit can be obviously controlled by varying deposition time during PLD process, the height will increase with increase of PLD time. From the Figure 8, it can be found that the unit height increases by increasing deposition time from 30 min to 60 min. However, if the deposition time is too long, to say, 180 min, the tops of micor/nanostructured units will aggregate with each other due to strong Van de Waals attraction among units in the deposition process, as shown in Figure 8 (c), (c′) and (c″). Additionally, the top of micro/nanostructured unit gradually flattens from convex shape with increasing deposition time, resulting in a weakening shadow effect. Therefore, a continuous film might be formed at top of hierarchical micro/nanostructured array if further increasing deposition time after 180 min.

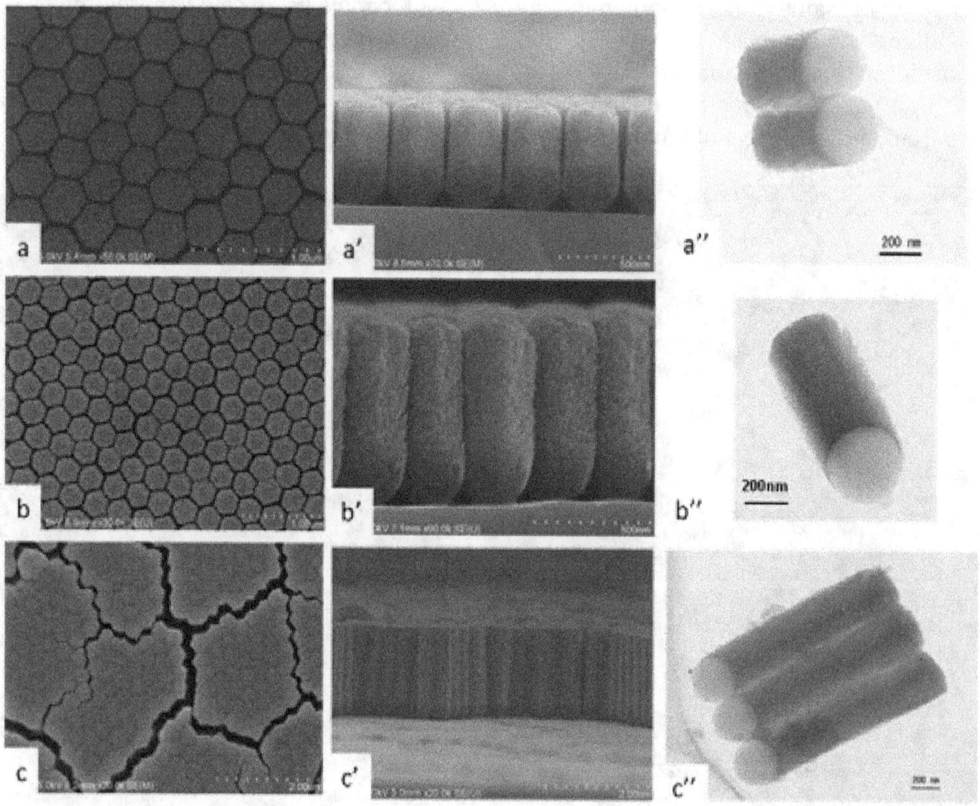

Fig. 8. The height changes of micro/nanostructured unit with increase deposition time. Deposition time: 30 min in (a), (a') and (a''); 60 min in (b), (b') and (b''); 180 min in (c), (c') and (c''). (a), (b) and (c) are SEM images of top view; (a'), (b') and (c') are SEM images of cross-section; (a''), (b'') and (c'') are TEM images of micro/nanostructure units.

The as-prepared hierarchical micro/nanostructured units in periodic arrays are composed of a PS sphere at the bottom and a micro/nano- particle or rod on the top of the PS sphere. If the PS colloidal template is dissolved by an organic solution (CH_2Cl_2), this periodic array could retain its integrity while being peeled from the substrate due to the van der Waals force between the neighboring micro/nanostructured units suspended in the solution. It could then be transferred to any desired substrate (e.g., TEM copper grid) by picking it up using another substrate, as illustrated in Figure 9 and Figure 10. The transferability avoids restrictions on substrates in the fabrication process of hierarchical micro/nanostructured arrays, which is helpful in the design and fabrication of new micro-/nano- devices on any desired substrates.

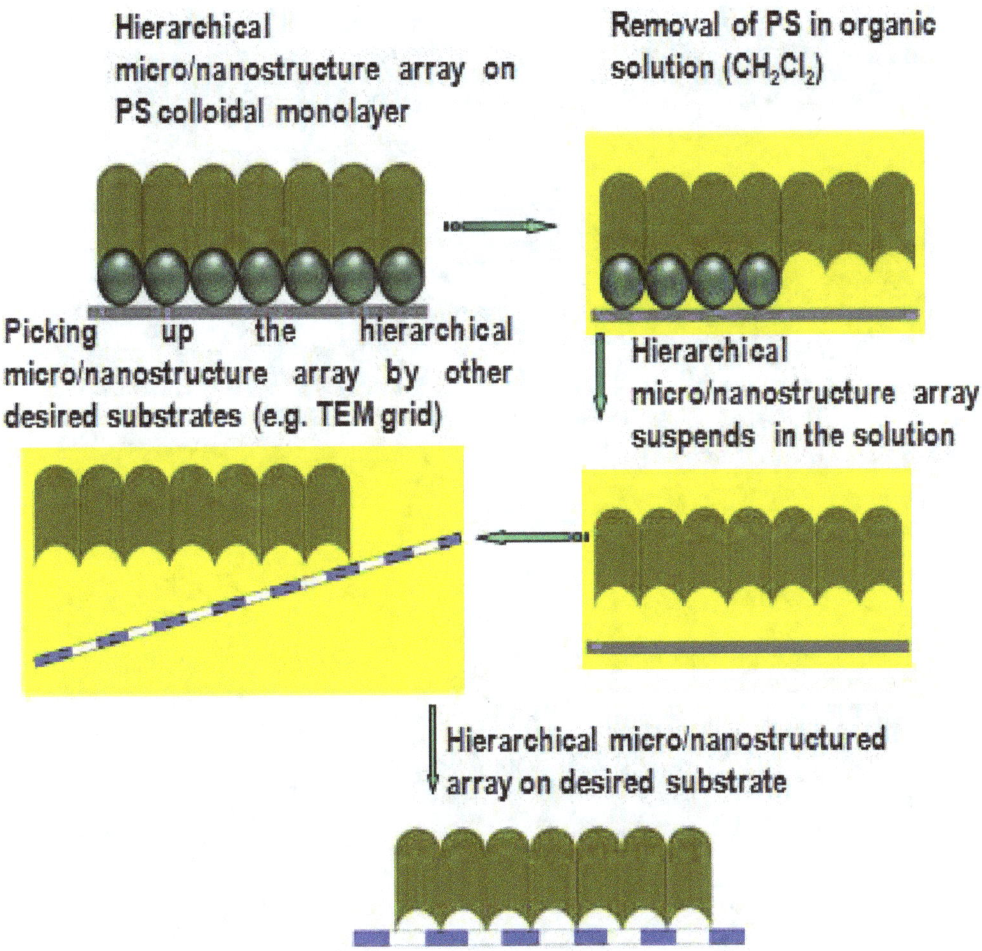

Fig. 9. Schematic illustration of transferability of hcp hierarchical micro/nanostructured arrays

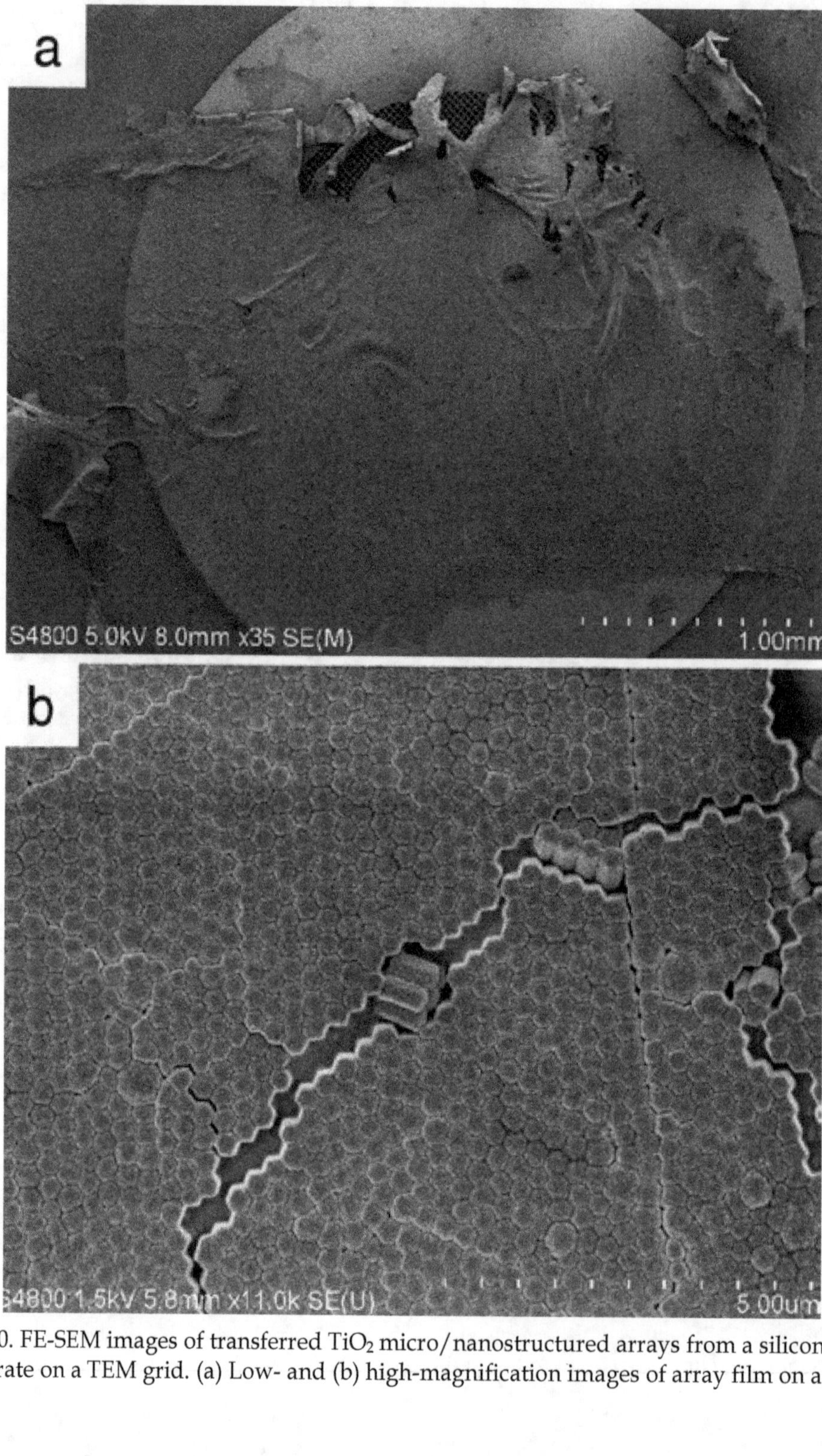

Fig. 10. FE-SEM images of transferred TiO_2 micro/nanostructured arrays from a silicon substrate on a TEM grid. (a) Low- and (b) high-magnification images of array film on a TEM grid.

2.2 Formation mechanism of hierarchical micro/nanostructured arrays by PLD

The formation process was traced by PLD using colloidal clusters with different PS spheres as templates. Herein, the TiO_2 was selected as desired material and colloidal monolayer with PS sphere size of 350 nm as template to demonstrate the formation process of hierarchical micro/nanostructured arrays. The colloidal clusters with different PS sphere were fabricated by spin-coating with a higher rotation speed (2000 rpm) and lower concentration (1.0 wt%) of PS colloidal microsphere suspension. For example, a single PS sphere or PS sphere clusters with different sphere numbers (2, 3, 4,...) can be easily created on the substrate by above route, as indicated in column A of Table 1. After PLD, morphologies observed from the top compared to those before PLD, as demonstrated in column B of Table 1. For a single PS microsphere, the shape kept spherical but that the size increased from 350 nm (PS sphere size) to 500 nm after PLD. For the PS sphere-clusters with sphere number from two to six, each unit size in the sphere-cluster still increased, but could not maintain the spherical shape after PLD. Growth of deposited TiO_2 was restricted at the contact point of two neighboring PS spheres, the contact between the neighboring units changed from a quasidot contact to a facet contact before (PS sphere-cluster) and after (PS sphere-cluster with deposited materials on the surface) PLD. If a PS sphere in sphere-cluster was completely surrounded by others, e.g., the central sphere in a hexagonal close packed (hcp) sphere-cluster of seven, its size after deposition was almost the same as before PLD and the morphology was slightly changed from spherical shape to hexagonal one. A section of a PS sphere-cluster of 10 spheres with hcp arrangement after PLD displays that hierarchical micro/nanorods have formed on the two spheres completely surrounded by the others and that hierarchical rod cannot be formed on the spheres at the edge of the sphere-cluster. This implies that a hierarchical micro/nanostructured array will be easily produced after PLD if a colloidal monolayer with a large-scale is applied in the PLD process. Additionally, if the desired materials are deposited on a bare silicon substrate without any PS spheres by PLD, nanocolumns grow vertically on the Si substrate, as seen in Figure 11.

Generally, nanocolumns prefer to grow in the normal direction on the substrate during the PLD process.[75] In the PLD process, the desired target (TiO_2) is irradiated by a laser beam using an energy level exceeding its threshold in vacuum environment, plasma including ions (Ti^{4+}, O^{2-}, etc.), molecules, electrons and clusters are released into the PLD chamber from the target. However, if a background gas with high pressure is introduced into the chamber, the movement direction of ions or electrons will be changed from an almost uniform direction to multidirection due to collisions between the ions, electrons, molecules and clusters of the ejected species and the background gas. According to the above facts, the formation mechanism of hierarchical hcp nanocolumn arrays can be easily understood, as displayed in Figure 12. If a substrate without PS spheres is used in the PLD process, a film consisting of vertical nanocolumns of small diameter will be formed. If a single PS sphere exists on the substrate, a composite of a PS sphere at the bottom and a shell composed of TiO_2 radiation-shaped nanobranches on sphere top will turn up, due to preferential vertical growth along the normal direction of the supporting surface and multidirectional deposition. For a PS sphere cluster (more than one sphere) on the substrate, a shadow effect will be produced in the deposition between any two neighboring spheres. If one sphere in the sphere-cluster is completely surrounded by six other spheres as in the case of hcp arrangement, one rod with hierarchical micro/nanostructure will grow on this sphere top. If a colloidal monolayer with a large scale is adopted, this route can easily fabricate an hcp hierarchical micro/nanostructured array. In this strategy, an off-axis configuration is adopted where the target and substrate are perpendicularly placed. It is similar to the glancing angle deposition (GLAD) or oblique angle deposition in which there is a large angle between the deposition direction and the normal direction of the substrate.[76-84] In the traditional GLAD method, atoms from the target obliquely arrive and condense on the substrate, and the tilted and separated nanowire or nanopillar array with a porous structure are gradually produced due to the shadow effect of the initial deposited nanoparticles under high-vacuum conditions. The critical difference between this route and GLAD is the background gas pressure during deposition, which converts the directional flow of ejected species in a vacuum into a multidirectional one at higher pressure. Therefore, this multidirection deposition and shadow effect are a principal reason why a vertical hierarchical micro/nanostructured array with hcp alignment is formed on the colloidal monolayer. This can be further verified by varying the angles between substrates and target in PLD process, as seen in Figure 13 and 14. If these experiments were carried out in a vacuum, tilted rod-like structured arrays with different angles would be obtained on the different substrates. However, from these results, the rod-like morphologies are independent of the angle between the substrate and target but the growth rates are different for different angles because of the plume shape in PLD , they always grow vertically on the substrate due the multiple direction deposition combined with shadow effect of neighboring colloidal sphere (Figure 14).

Number of PSs	(A) Before PLD	(B) After PLD
One	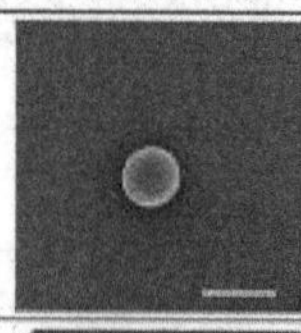	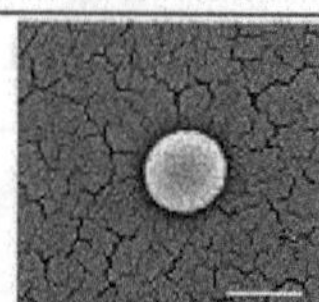
Two	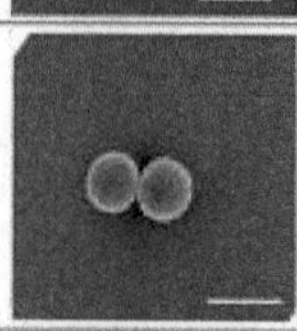	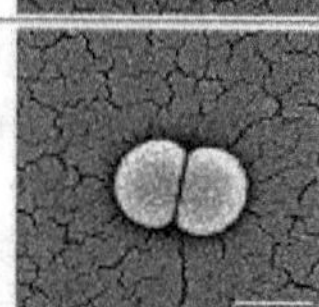
Three	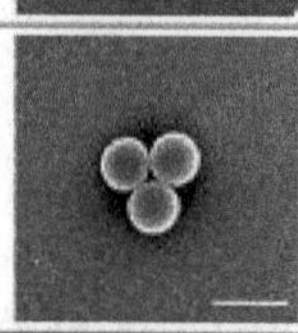	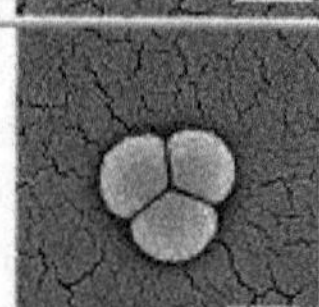
Four	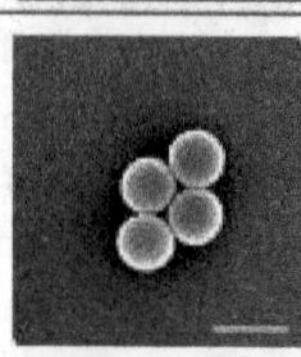	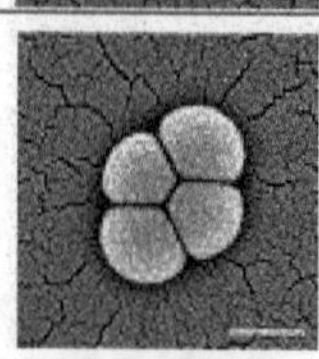
Five	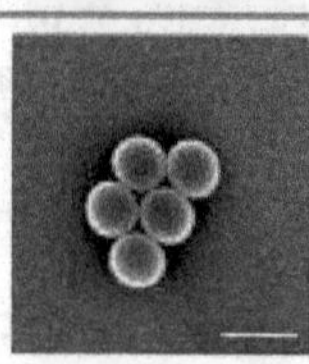	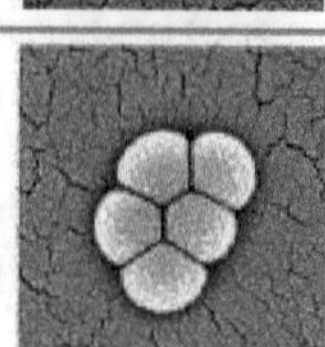
Six	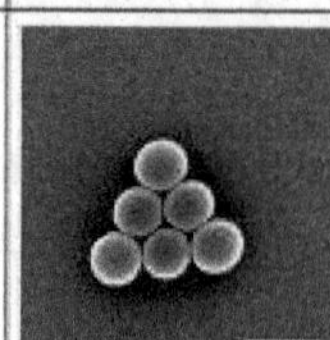	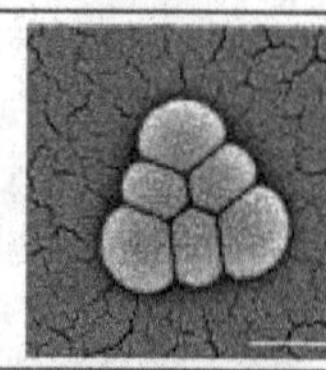
Seven	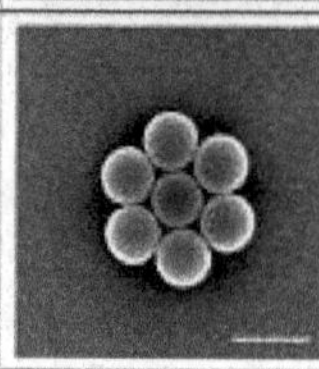	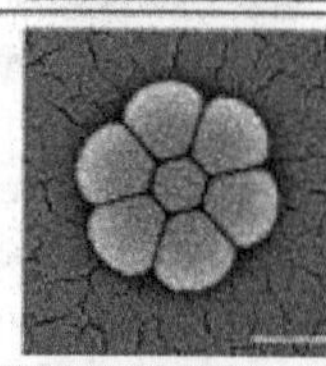
Ten	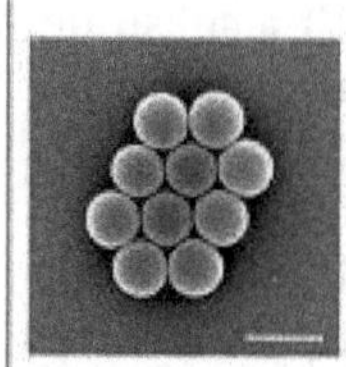	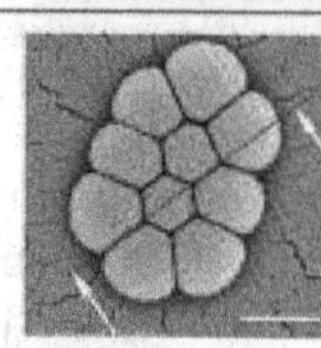 Observation from top 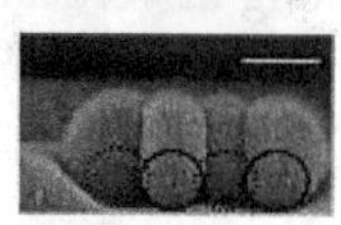 Section view along the red dot line

Table 1. Morphologies of before and after PLD on the PS sphere surface (Scale bars are 500 nm)

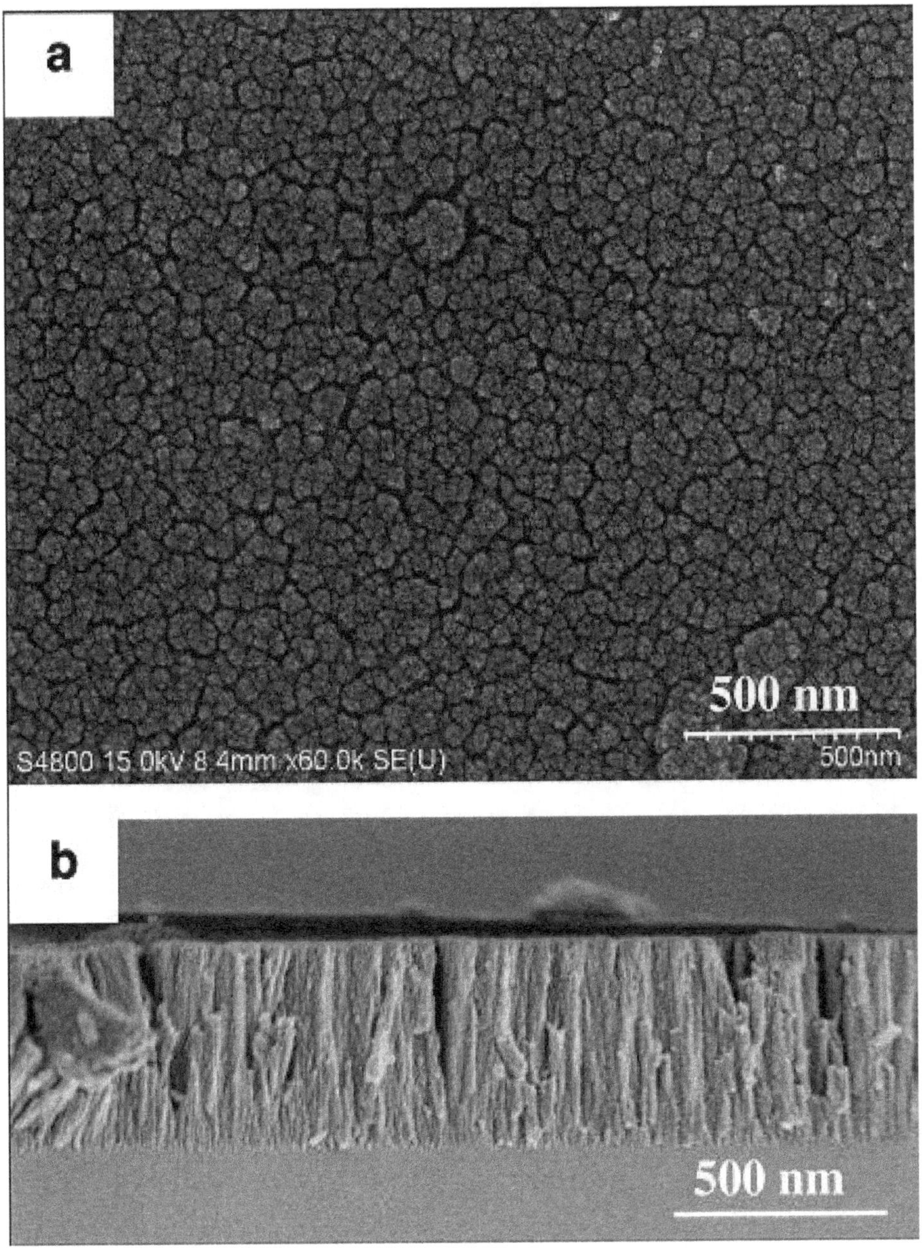

Fig. 11. FE-SEM images of the TiO_2 nanocolumns deposited by PLD directly on a bare substrate without colloidal spheres. (a) Observation from top. (b) Section view.

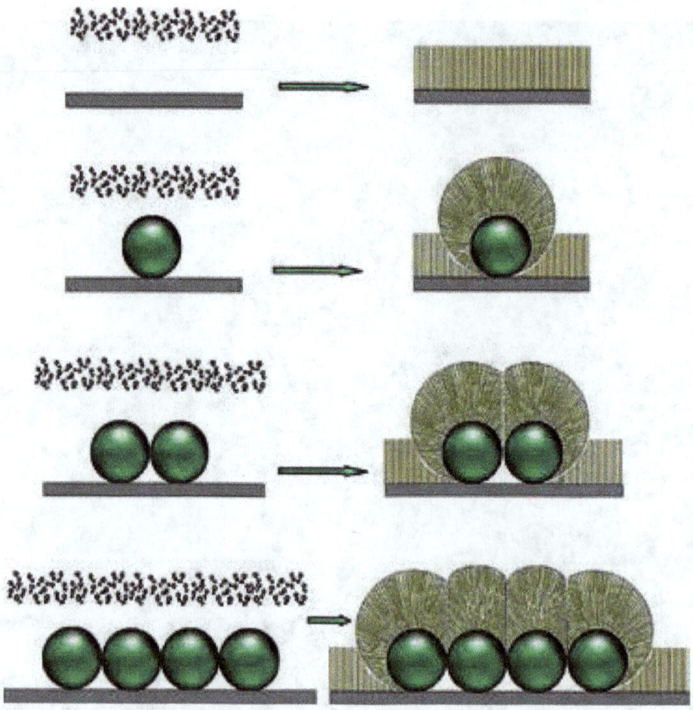

Fig. 12. Schematic illustration of formation mechanism of hcp hierarchical micro/nanostructured arrays.

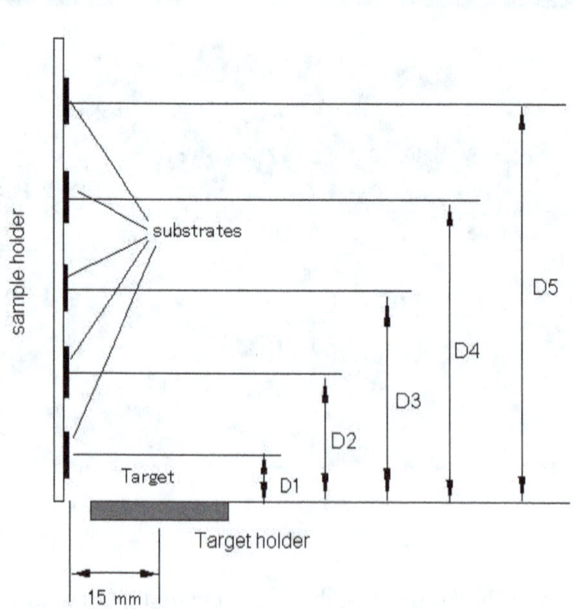

Fig. 13. Schematic illustration of multi-substrate experiment. D1: 8 mm. D2: 23 mm. D3: 35 mm. D4: 50 mm. D5: 64 mm. In this experiment, substrate rotation: 0 rpm. PS sphere size in colloidal monolayer: 350 nm.

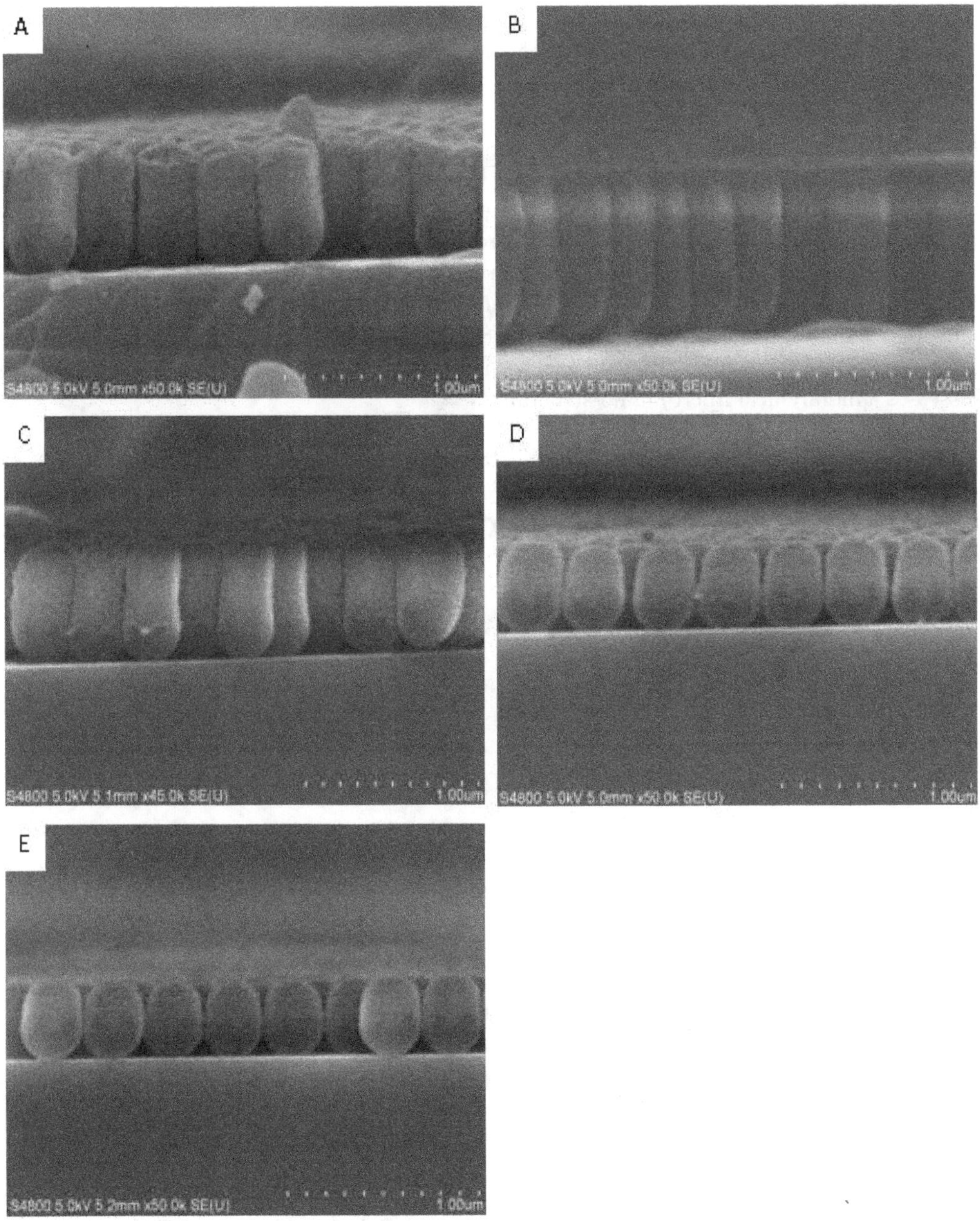

Fig. 14. FESEM cross-section images of the samples at different positions in Figure 10. Images A to E are from the samples at positions D1 to D5.

2.3 Extension of PLD assisted colloidal lithography

The hexagonal close-packed (hcp) micro/nanostructured arrays can be obtained by PLD using colloidal monolayer template. Such PLD assisted colloidal lithography can be extended to prepare hexagonal non-close-packed (hncp) micro/nano- particle or nanorod arrays by further annealing process.[85-88]

PS colloidal monolayer was first fabricated on Si substrates by a self-assembly process. It was then placed in a PLD chamber for deposition at room temperature. After deposition, the sample was moved to an oven from the PLD chamber and annealed in air. The hncp, hierarchically ordered micro/nano- particle arrays were thus prepared on the substrate, as illustrated in Figure 15.[85]

Figure 16 shows FE-SEM images of the TiO_2 hierarchical hncp micro/nano- particle array obtained by PLD assisted colloidal lithography and subsequently annealed at 650 °C for 2 h in air. One can clearly find that this particle-ordered array takes on hncp arrangement. Each particle in the periodic array exhibits a hemispherical shape with an average size of 240 nm (Figure 16c) and is composed of small nanoparticles (Figure 16d). Additionally, SEAD pattern (inset in Figure 16d.) indicates the deposited materials are changed from amorphous to anatase typed TiO_2 after annealing. These results reflect that this hncp particle array possesses a hierarchical micro/nano-structure.

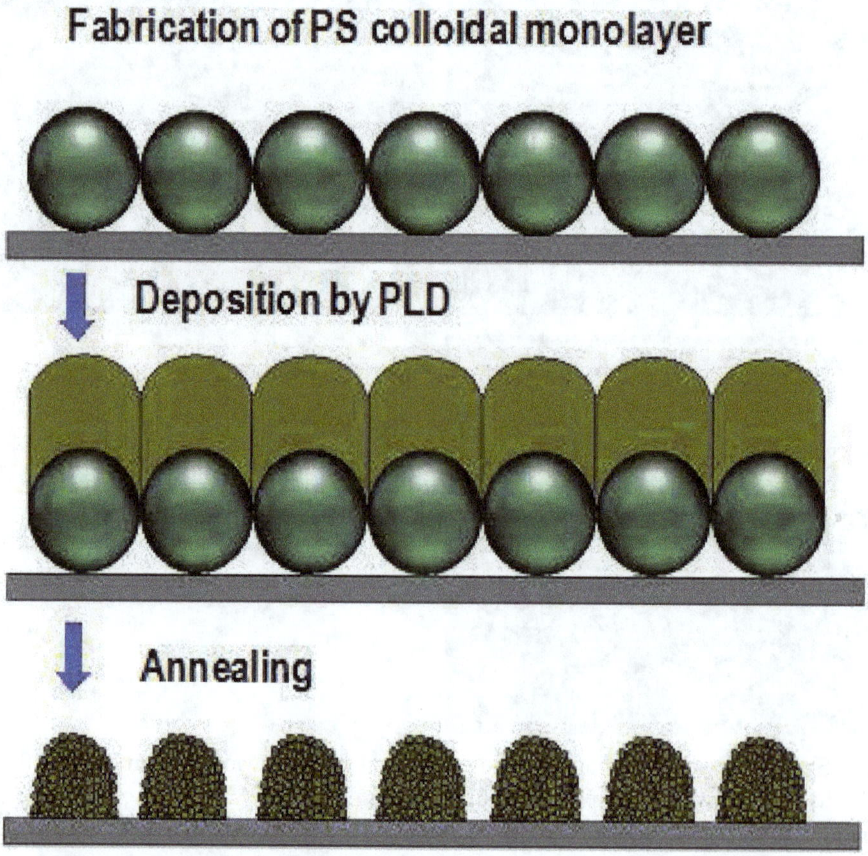

Fig. 15. Schematic illustration of fabrication process for the hierarchical hncp micro/nano-particle array.

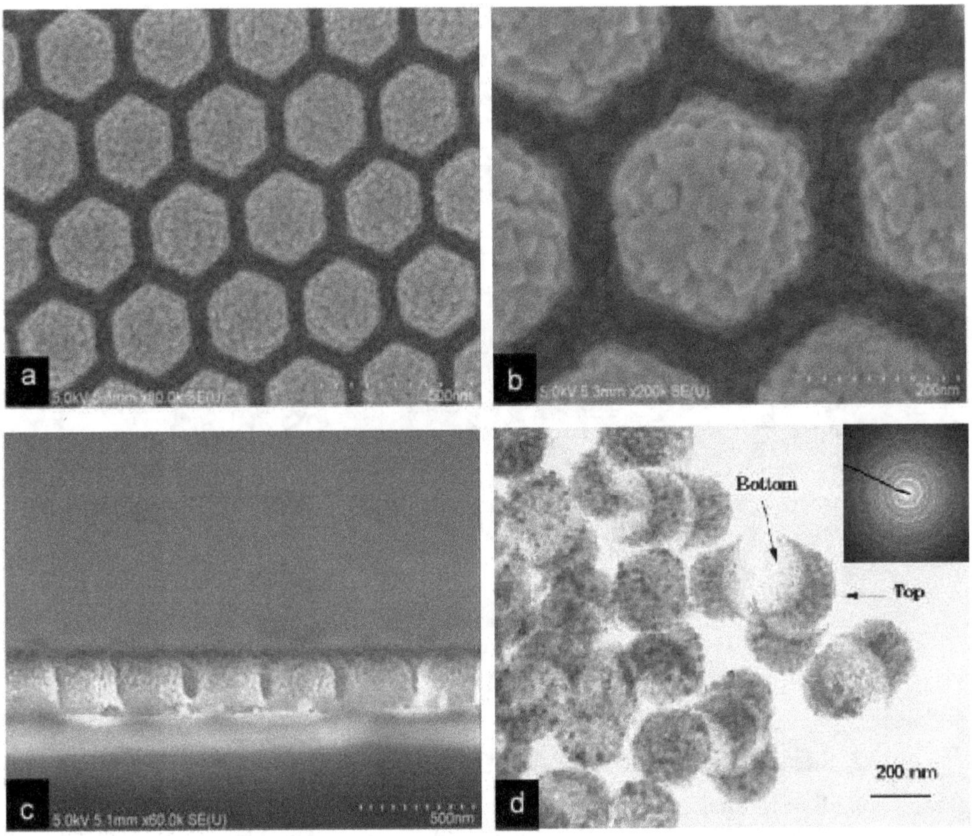

Fig. 16. FE-SEM and TEM images of the hierarchical hncp micro/nano particle arrays. a)-c): FE-SEM images. a) Large area hierarchical micro/nanostructured array b) Same at high magnification. c) Cross section d) TEM image of hierarchical particles (scratched from the supporting substrate by a knife and transferred to TEM copper grid for observation) and the corresponding SAED pattern (inset).

The formation of hncp, hierarchical micro/nano- particle array is discussed based on experimental results. Herein the explanation is given using TiO_2 as an example. The sample produced by PLD assisted colloidal lithography without further heating displayed an hcp alignment (Figure 17a). TiO_2 was deposited on the PS sphere surfaces and grew along the vertical direction (Figure 17b). Each particle is composed of two parts: the PS sphere at the bottom and an amorphous, porous TiO_2 layer consisting of smaller particles on the top (Figure 17c, d). The amorphous materials crystallize after being annealed at high temperature. In this case, when the amorphous TiO_2 with its supporting PS spheres was heated at 650 ^{o}C for 2 h, the PS spheres were burned out. Meanwhile, the TiO_2 particles on top of the PS sphere were changed to anatase polycrystals composed of smaller nanoparticles of ca.30 nm and were dropped vertically down to the original position of the PS sphere. Additionally, the volume of TiO_2 particles decreased during the change from the amorphous to the crystalline phase and hence an hncp hierarchical particle array was formed on the substrate, as illustrated in Figure 18. The hierarchical hncp mciro/nano-particle array film adhered tightly to the substrate after annealing and could not be detached from the supporting substrate even when it was washed ultrasonically in water for 30 min.

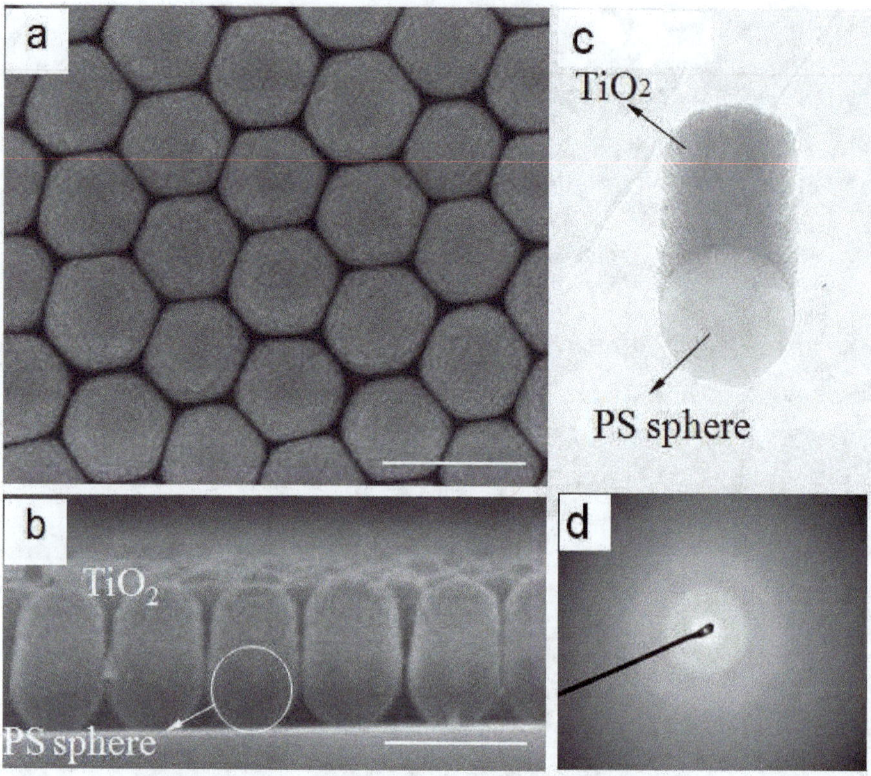

Fig. 17. (a, b): FE-SEM images of the as-deposited sample produced by the PLD in ambient atmosphere without heating. a) Top-view image. b) Cross-sectional image. c) TEM image of single unit. d) SAED pattern of deposited materials.

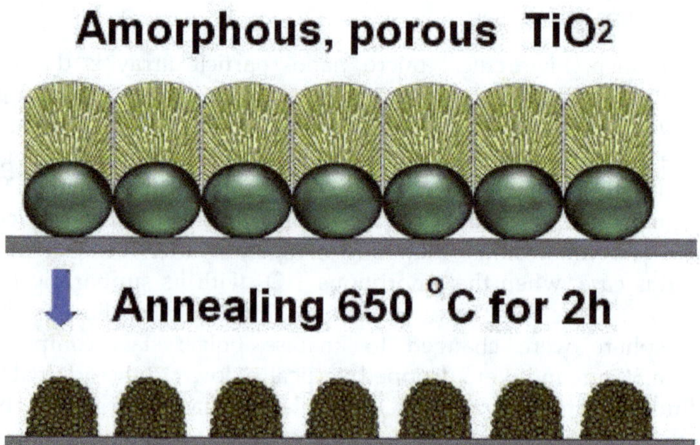

Fig. 18. Formation of hierarchical *hncp* micro/nanostructured particle arrays.

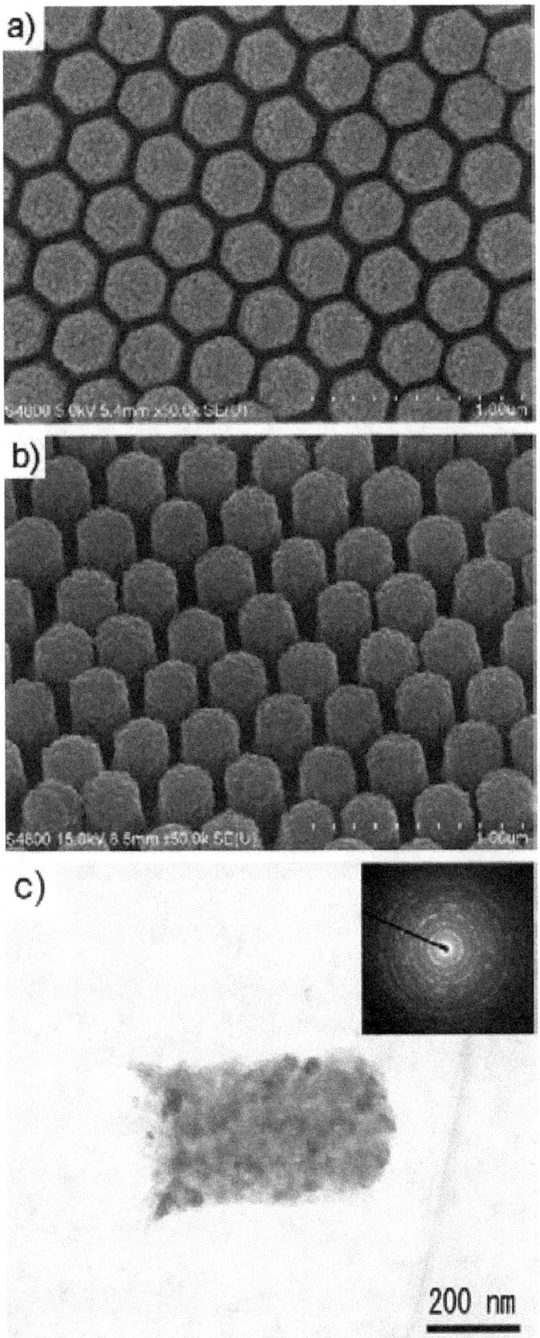

Fig. 19. (a, b) FE-SEM images of a nanorod array by PLD using a colloidal template monolayer with 350 nm PS spheres in O_2 at a pressure of 6.7 Pa for 60 min and subsequent annealing in air. a) A top-view FE-SEM image. (b) An FE-SEM image of a sample tilted to 45o. (c) TEM image of a nanorod and the corresponding selective area electron diffraction (SAED) pattern (inset).

In the PLD assisted colloidal lithography, if the deposition time increases to a appropriate time, an hncp micro/nano rod array can be obtained after PLD and subsequently annealing.[86] Figure 19 presents FE-SEM images of the hierarchical micro/nano rod array obtained by PLD assisted colloidal lithography (PS sphere size: 350 nm; background O_2 pressure: 6.7 Pa; longer deposition time: 60 min) after annealed at 650 °C for 2 h in air. Figure 19 a and b indicate that a periodic nanorod array takes an hncp arrangement and that each nanorod consists of many nanoparticles. The TEM image (Figure 19c) of a single nanorod shows that it has an aspect ratio of ca. 2:1 and is composed of small nanoparticles, and the PS sphere templates were entirely removed during annealing. The hncp micro/nano- rod arrays originated from amorphous, hcp micro/nanostructured array produced by PLD at room temperature without annealing, as displayed in Figure 20. Beside amorphous materials deposited by PLD, some crystalline materials, e.g. Fe_2O_3, their hncp micro/nanostructured arrays can be also achieved by PLD assisted colloidal lithography after annealing. Figure. 21a, b present SEM images of hierarchical, hcp, crystalline Fe_2O_3 micro/nanostructured arrays by PLD at oxygen pressures of 6 Pa at room temperature. Such sample was annealed in air at 450 °C for 3 h and, PS colloidal monolayer template was completely decomposed and hierarchical hncp micro/nano-structured array was formed, as shown in Figure 21 c-f.

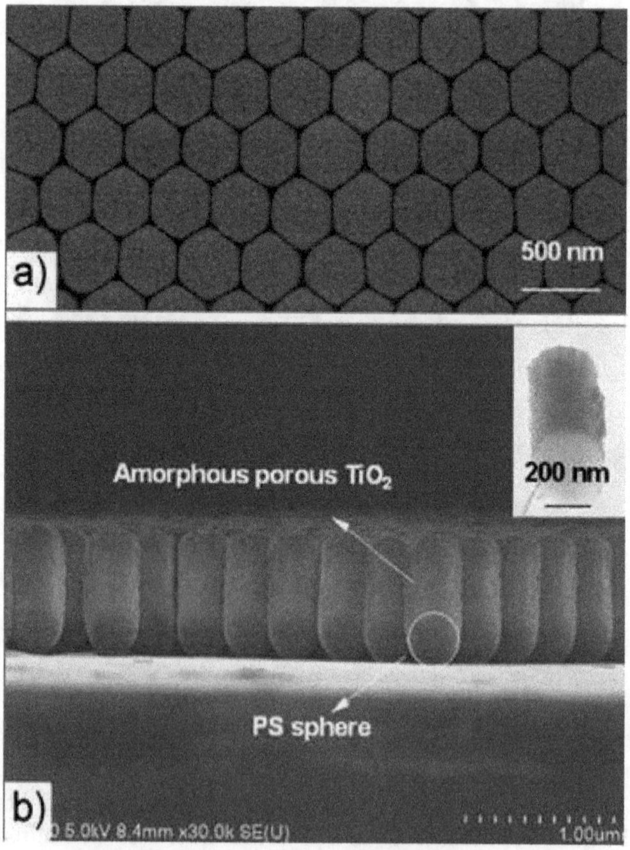

Fig. 20. FE-SEM images of an amorphous TiO_2 hcp nanorod array on the colloidal monolayer before annealing. a) Observation from top, b) Cross section. Inset in (b): a TEM image of a single amorphous TiO_2 nanorod with PS sphere.

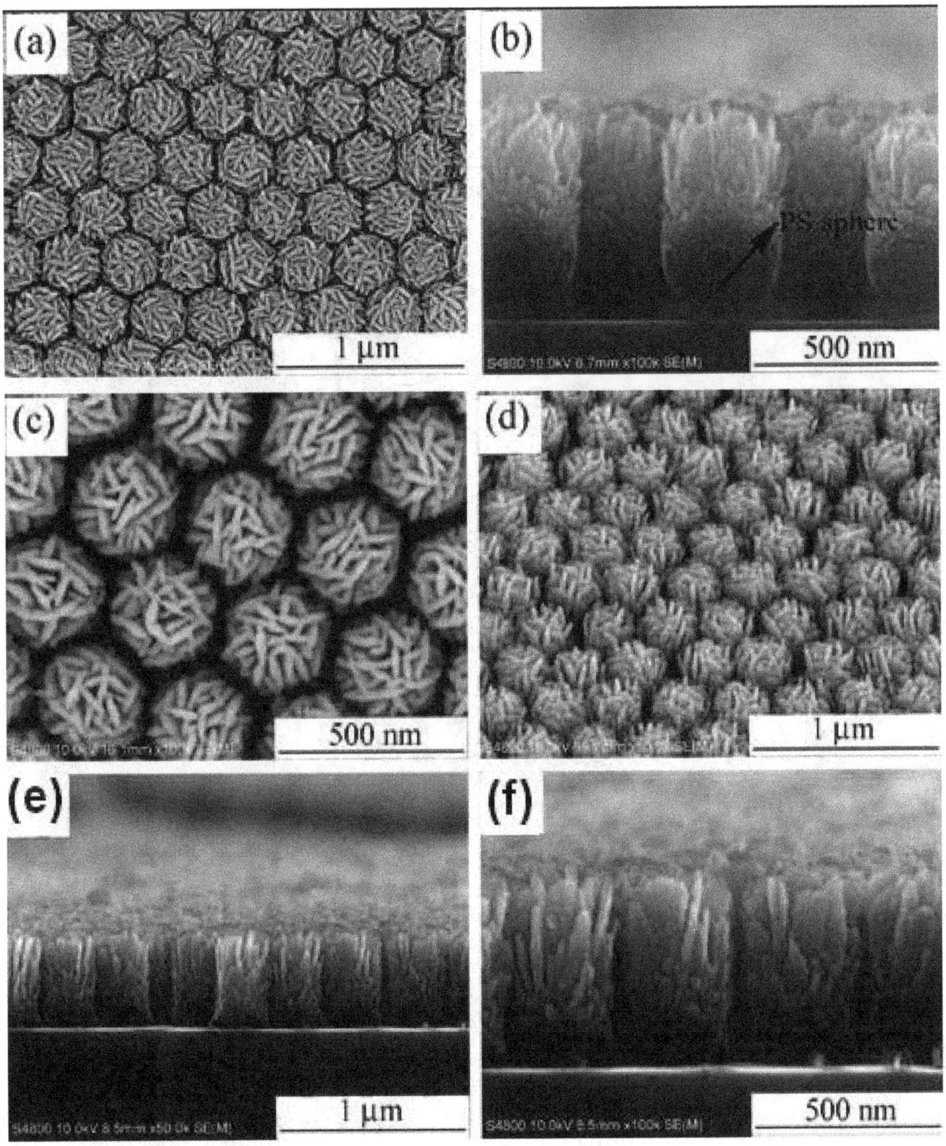

Fig. 21. SEM images of Fe_2O_3 micro/nanostructured array based on colloidal monolayer template and PLD at the oxygen pressure of 6 Pa. (a, b) Before (c-f) after annealing 450 °C for 3 h in air. (c) Top-surface image. (d) 45° tilted view. (e) Cross-section image. (f) Higher magnification.

In this route, the interspace between two neighboring rods can be controlled by changing the background gas pressures in PLD process. Usually, the porosity and specific surface area of the rods in arrays can be tuned by varying the background gas pressure in PLD deposition and they increase with increase of background gas pressure (Figure 22a–c). Therefore, the interspace between neighboring anatase rods can be tuned with different background gas pressures in an hncp array after annealing. The interspace will increase with increasing background gas pressure (Figure. 22a′–c′ and a″–c″).

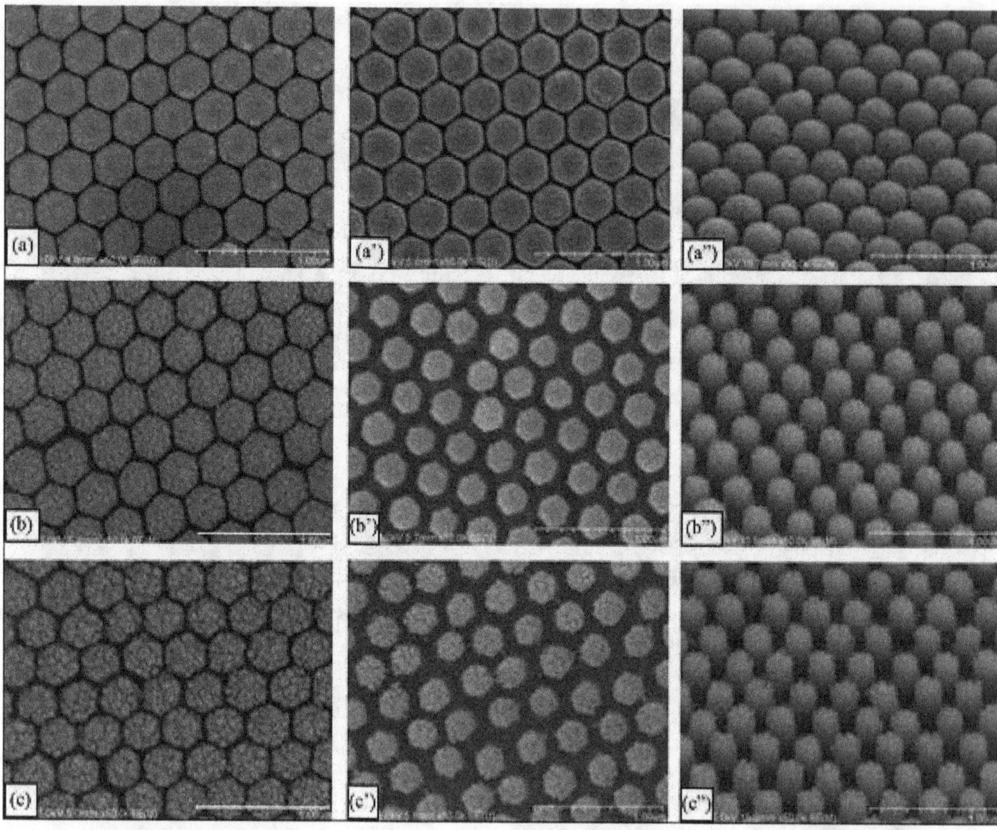

Fig. 22. FE-SEM images of an amorphous hcp TiO_2 nanorod array and anatase hncp TiO_2 nanorod arrays obtained by PLD under different background gas pressures and subsequent annealing. PLD was performed in oxygen (a) at 2.0 Pa for 200 min; (b) at 16.8 Pa for 43 min; (c) at 26.8 Pa for 30 min. (a–c) before annealing, (a′–c′) top views after annealing, (a″–c″) tilted view with 45 degrees after annealing.

3. Sputtering assisted colloidal lithography

In this method, a PS colloidal monolayer on a substrate is placed into the chamber of radio frequency (RF) magnetron sputtering for material deposition at room temperature and Ar is introduced as the background gas.[89] A unique hncp hierarchical micro/nanostructured array is formed due to PS-templated plasma etching/deposition in a relatively high vacuum (0.06 Pa). The features of this technique are (1) low-pressure sputtering; (2) PS-templated sputtering, which guarantees a periodic arrangement, and (3) plasma etching/deposition, which eventually produces the unique hncp hierarchical micro/nano-structure.

Figure 23 presents FE-SEM and TEM images of hierarchical alumina mciro/nanostructured arrays by sputtering target alumina using a PS colloidal template with a sphere size of 750 nm in a relatively high vacuum (0.06 Pa). They show the following three unique features. First, it is a periodic nanocolumn array cushioned by a semi-shell in an hncp arrangement. Each nanocolumn is composed of two parts: a semishell- shaped cushion at the bottom and a vertical nanocolumn on the top of the cushion. Such nanocolumn possesses a very rough structure on the surface and seems to be composed of many minicolumns, indicating that the sample possesses a hierarchical, porous structure and hence has a high surface area. Secondly, the periodicity was 750 nm, matching well with the initial size of the PS spheres. It is very evident that the sizes of the cushion and the central columns were reduced by about 15% and 45% compared to the original size of the PS template.

The formation of such hierarchical micro/nanostructured arrays is traced by the different sputtering deposition time, as demonstrate in Figure 24. With increase of deposition time (Figure 24 A–D), the PS sphere size gradually decreases and the alumina columns grow vertically in the center, finally the columnar structures and a salver-shaped semi-shell are formed. Generally, thin alumina continuous film is formed on bare substrate without PS spheres due to the strong ion energy and subsequent rapid surface migration under such a low sputtering pressure. In the case with PS sphere array on the substrate, alumina components sputtered from the target are impinged and implanted into the PS due to the strong ion energy and soft nature of PS. The part of PS sphere is also continuously etched away by argon ions and part of deposited alumina is also etched away, but remaining part will gradually form a structure. Additionally, the PS colloidal monolayer supplies the periodic array template. Merging these two aspects into one forms a unique hierarchical micro/nano-structured arrays. The PS spheres become smaller with plasma etching and a salver-shaped semi-shell gradually appear. Further sputtering causes the PS spheres to be etched more significantly, and the species generated from the target deposit perpendicularly onto the template (both the center and the semi-shell part), thus forming a column structure and salver-shaped semi-shell. Implanted components of aluminium and oxygen into PS sphere will be linked together by continuous etching of PS. But at the side edge of the spheres, the amount of PS is not much and easily etched away to form aluminium oxide film, resulted in the cushion shell. The amount of PS at the center part is much more and even by continuous etching a film cannot be formed and rod-like structures are generated. Further sputtering continues etching the PS sphere until the final unique hncp hierarchical structure forms. The formation process of this unique hierarchical mciro/nano-structure is schematically illustrated in Figure 25. In order to further confirm this process, the pressures of Ar were adjusted from low level (0.06 Pa) to 0.13 and final 6.7 Pa. The FE-SEM images of the samples are presented in Figure 26. With increase of the background gas pressure from 0.06 Pa to 0.13 and 6.7 Pa, the collision probability between the ejected species and Ar molecules increases, thus the PS spheres are more significantly etched and no semi-shell can form. Therefore, only columnar structures are obtained (Figure 26). The amounts of deposited materials in the inter-columnar structures are negligible probably due to the blocking effects of gaseous species emitted from decomposed PS spheres during sputtering. These results firmly prove that a relatively high vacuum condition subsequently induces mild plasma etching/deposition. Besides hncp alumina micro/nano-structured arrays with a periodicity of 750 nm, novel hierarchical arrays with periodicities of 350 nm, 1 μm, and 2 μm were also created by colloidal monolayers with different PS sphere sizes during sputtering at the same pressure of Ar as in Figure 27. Besides alumina, hierarchical arrays of other materials including Au/Al_2O_3 composite, CuO, and NiO can also be fabricated by the presented one-step plasma etching. Some of the results are presented in Figure 28. In this method, only the inorganic materials can be used as the deposited materials. Otherwise, the deposition cannot be guaranteed because of the subsequent etching.

Fig. 23. FE-SEM (A, B) and TEM (C) images of a sample obtained by sputtering using a PS colloidal monolayer as the substrate (PS sphere size: 750 nm; deposition time: 2 h). (A) Top view. (B) Titled view with 45° angle. The scale bar in A, B: 1 µm. (C) TEM image of one unit.

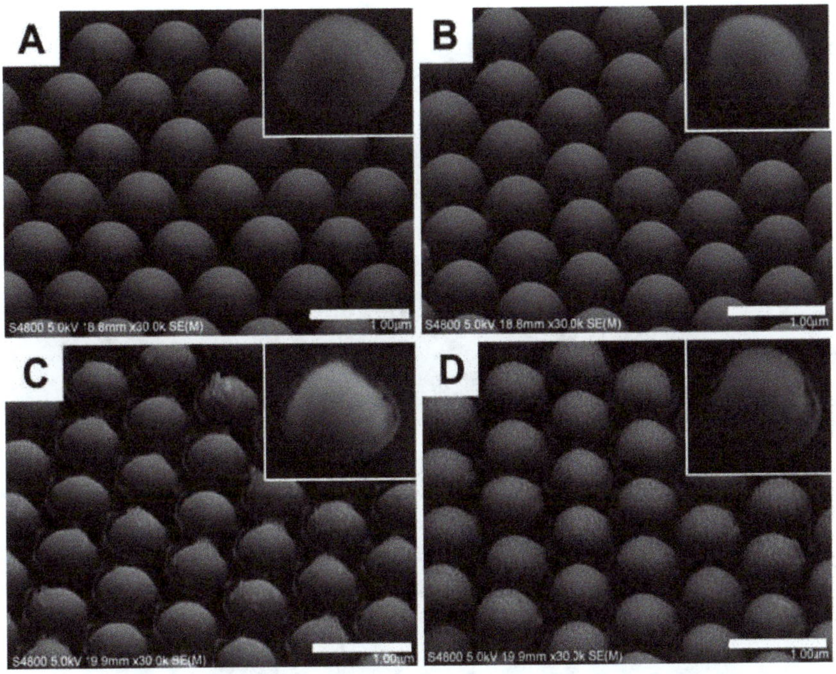

Fig. 24. FE-SEM images of samples obtained with different deposition times. (A) 10 min. (B) 25 min. (C) 30 min. (D) 60 min. The inset is high-magnification image of one unit in each figure.

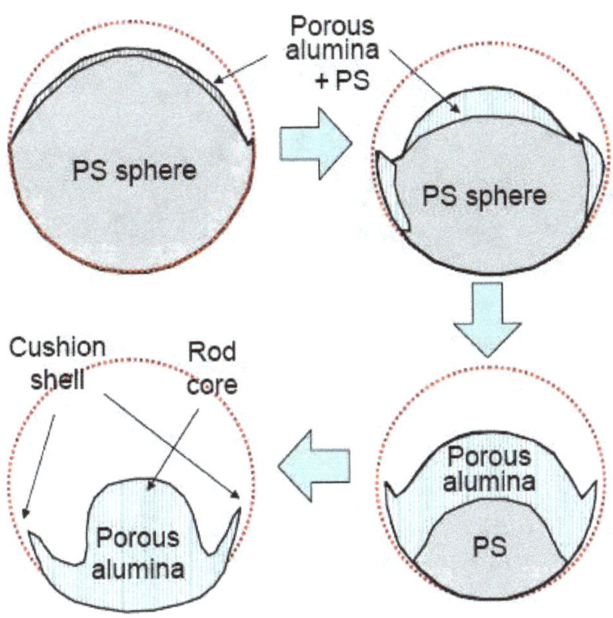

Fig. 25. Schematic illustration of the formation process of unique hncp hierarchical micro/nano-structure.

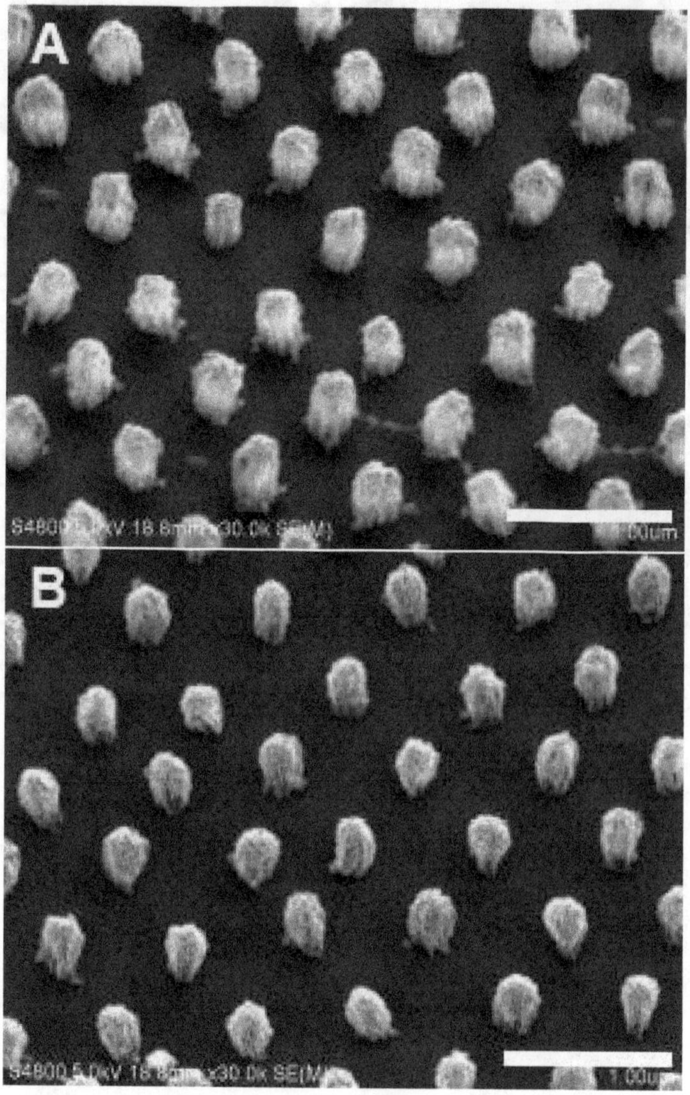

Fig. 26. FESEM images of a sample obtained by sputtering using a PS colloidal monolayer as the substrate at Ar pressures of (A) 0.13 Pa and (B) 6.7 Pa (PS sphere size 750 nm; deposition time 2 h). These images are observed with a tilt angle of 45°. The scale bar is 1 μm.

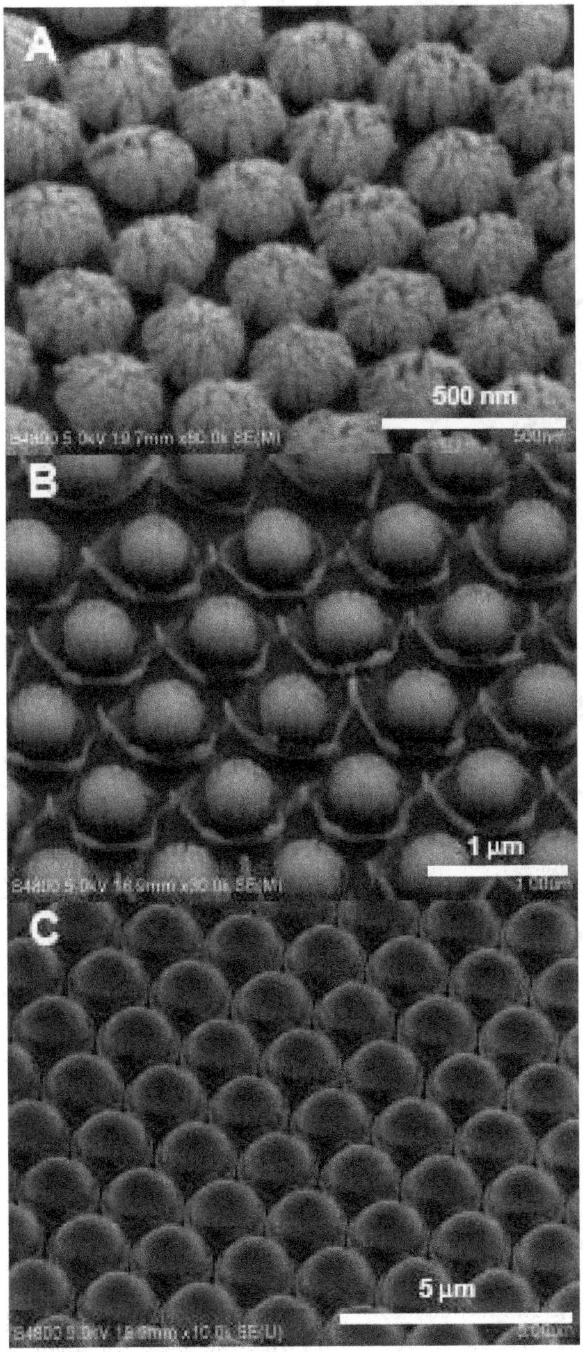

Fig. 27. FESEM images of a sample obtained by sputtering using a PS colloidal monolayer template with different PS sphere sizes at 0.06 Pa. (A) 350 nm. (B) 1 µm. (C) 2 µm.

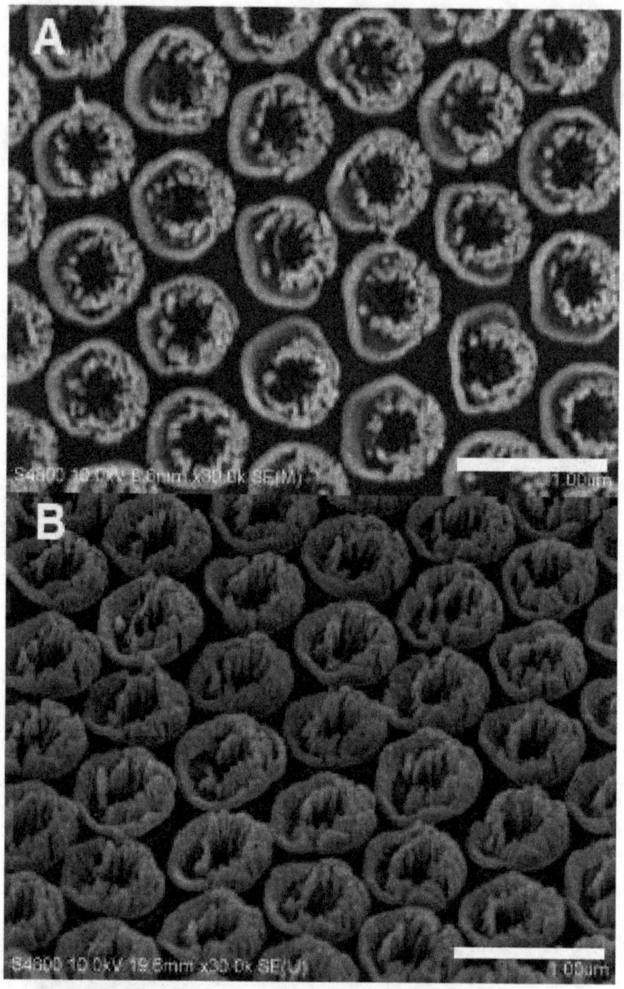

Fig. 28. FESEM images of periodic Au/Al_2O_3 nanocomposite arrays obtained by co-sputtering multiple targets consisting of an Al_2O_3 target and Au sheets and using a PS colloidal monolayer as the substrate (PS sphere size 750 nm; deposition time 2 h). (A) Image observed from the top. (B) Image with a tilt angle of 45°. The scale bar is 1 μm.

4. Applications of periodic micro/nanostructured arrays

Based on colloidal monolayer templates, the different hierarchical micro/nanostructured arrays can be fabricated by physical deposition and their morphologies and structural parameters (sizes and interspace of micro/nanostructured unit) can be facilely controlled by periodicities of colloidal monolayers, experimental conditions such as deposition time, background gas pressure etc. Some properties, such as the surface wettability, field emission, catalytic properties are closely morphology and structural parameter-dependent. These properties can be readily optimized by changing morphologies and parameters of the periodic arrays. Their investigations supply useful theoretic foundations and are highly valuable for designing micro/nanodevices based on these special arrays.

4.1 Wettability

Wettability is generally related to the surface morphologies, roughness and free energy of materials surface and it is evaluated by the water or oil contact angle. A special surface with self-cleaning effect is usually defined as a surface that has the ability to remove dirt or contaminants that are on it when water droplets slide along the surface. Self-cleaning is closely related to surface wettability[90-94]. The self-cleaning effect is normally attributed to superhydrophobicity (water contact angle (CA) exceeding 150° and sliding angle (SA) less than 10°) or superhydrophilicity (water CA less than 10°) of the surface. For superhydrophobicity with a self-cleaning effect, contaminants adhere to the water droplet surface and are removed after the water droplet slides off the solid surface with a small tilted angle, due to large water CA and low surface free energy. For superhydrophilic surfaces, contaminants can easily be swept away by adding water droplets on them, due to very low water CA. Wettability can be enhanced by increasing surface roughness, according to Wenzel's equation: [95]

$$\cos \theta_r = r \cos\theta \quad (1)$$

where r is the roughness factor, defined as the ratio of total surface area to projected area on the horizontal plane; θ_r is the CA of film with a rough surface; and θ is the CA of film with a smooth surface. Obviously, increased roughness can enhance the hydrophobicity and/or hydrophilicity of hydrophobic and/or hydrophilic surfaces. The hierarchical micro/nanostructured arrays based on colloidal monolayers are actually rough films at the micro/nano-scale level. It is expected that such hierarchical micro/nanostructured arrays could induce surface superhydrophilicity or superhydrophobicity with a self-cleaning effect, due to their high roughness. Amorphous, porous hierarchical TiO_2 micro/nanostructured arrays were prepared by PLD assisted colloidal lithography (Figure 3)[73]. These arrays exhibited strong superhydrophilicity. When a small water droplet was dropped on a hierarchical structured array, the droplet spread out rapidly on the surface and displayed a water CA of 0° in a 0.225 s (Figure 29). Additionally, this hierarchical array film exhibited superoleophilicity when a small oil droplet was placed on the nanorod surface and the oil CA became 0° in 0.5 s (Figure 30). These results suggest that this amorphous hierarchical micro/nano-structured array had superamphiphilicity with 0° of both water CA and oil CA. A TiO_2 film with superamphiphilicity can generally be obtained by UV irradiation, due to hydroxyl ions generated by oxygen defects or dangling bonds on its surface, induced by photochemical processes[96]. However, the TiO_2 hierarchical micro/nano- structured array film possessed superamphiphilicity without further UV irradiation. The ions (e.g., Ti^{4+}, and O^{2-}) and electrons are released into the PLD chamber and some oxygen species are lost in the vacuum environment in PLD after a TiO_2 target absorbs energy from laser irradiation by exceeding its threshold. Oxygen vacancies are produced in the deposited TiO_2 during PLD, converting relevant Ti^{4+} sites to Ti^{3+} sites that are favorable for dissociative water adsorption. Therefore, these defect sites microscopically form hydrophilic domains on the TiO_2 surface. However, the other parts surrounding the hydrophilic domain remain oleophilic on the surface. A composite TiO_2 surface having hydrophilic and oleophilic domains on a microscopically distinguishable scale demonstrates macroscopic amphiphilicity on the TiO_2 surface[96]. Additionally, a TiO_2 nanoparticle film prepared by PLD without a PS colloidal monolayer exhibited a water CA of 15° and an oil CA of 27° (Figure 31). The roughness of the hcp TiO_2 hierarchical micro/nano-structured array film was greatly increased compared with that of the nanoparticle TiO_2 film produced by PLD without using a colloidal monolayer. According to Wenzel's equation, wettability is enhanced from amphiphilicity to superamphiphilicity. Therefore, the superamphiphilicity of the amorphous micro/nano-structured array originates from the combination of the amphiphilicity produced by PLD and the special rough structures of hcp hierarchical arrays.

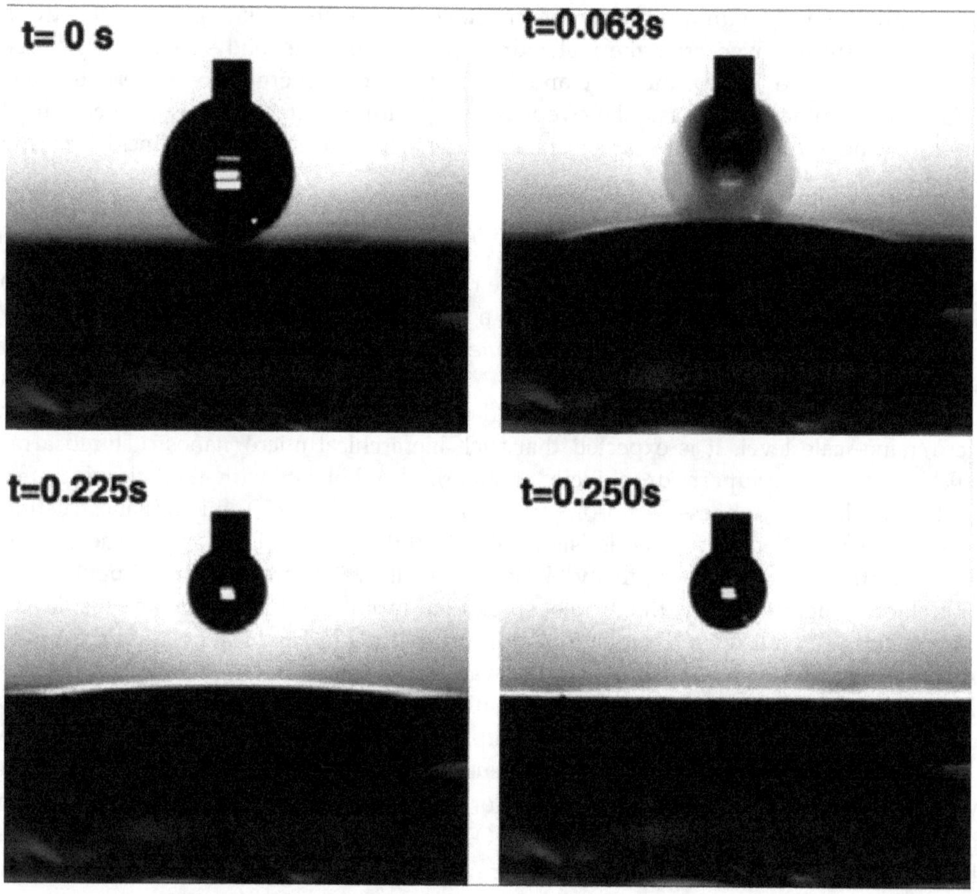

Fig. 29. Time course of water-contacting behavior on the amorphous TiO2 micro/nano-structured array film.

Fig. 30. Oil (rapeseed) droplet shape on amorphous TiO_2 micro/nano- structured array film. The oil contact angle becomes 0° in 0.5 s after it was dropped onto the surface.

More importantly, this amorphous TiO_2 hierarchical array demonstrated very good photocatalytic activity for organic molecular degradation (e.g., effective decomposition of stearic acid under UV illumination). A combination of superamphiphilicity and photocatalytic activity can yield a self-cleaning surface. For instance, an oily liquid contaminant spreads out on a surface due to superoleophilicity, which is helpful in improving the photocatalytic efficiency under light illumination. An organic contaminant including oil gradually degrades under sunlight irradiation (sunlight contains some UV light). The self-cleaning effect can be realized after washing away contamination from the superhydrophilic surface.

Additionally, superhydrophobic surfaces with large water CA and small SA have a self-cleaning effect. For superhydrophobic film, the surface should be sufficiently rough and have a chemical coating with low free-energy materials in order to trap the air on the rough surface. In this case, the area fraction of a water droplet in contact with the sample surface is very small, which helps obtain a small SA. Hierarchical periodic micro/nanostructured arrays based on colloidal templates provide surfaces with regularly ordered and well-defined roughness. They may lead to enhancement from hydrophobicity to superhydrophobicity on the surface after modification with low free-energy materials[97-99]. For instance, Co_3O_4 hierarchical, hncp micro/nano- rod arrays was created by PLD assisted colloidal lithography after annealing at 450 °C for 3 h (oxygen pressures: 93.1 Pa), as shown in Figure 32 a and b.[88] Such surface was chemical modification with fluorosilane, a kind of low free energy material, it presented superhydrophobicity with water CA of 152.6° and a very small SA, indicating self-cleaning effect. It can be explained by Cassie and Baxter equation,[100]

$$\cos\theta_r = f_1 \cos\theta - f_2 \tag{2}$$

Here, f_1 and f_2 are the surface area fractions of the projecting solid and air ($f_1 + f_2 = 1$). The large fraction of air trapped in the nanorod arrays forms a cushion at the film–water interface that prevents the penetration of water droplets into the grooves. In this case, θ_r is

152.6° and θ is 18.8°, so a value for f_1 of 0.06 is calculated from eqn (2) (i.e. f_2 is 0.94), implying that only 6% of the observed contact area beneath a water droplet is in contact with the water droplet. High f_2 of 0.94 means that the air was well trapped into the groove among nanorod arrays and hence the water droplet kept a spherical shape.

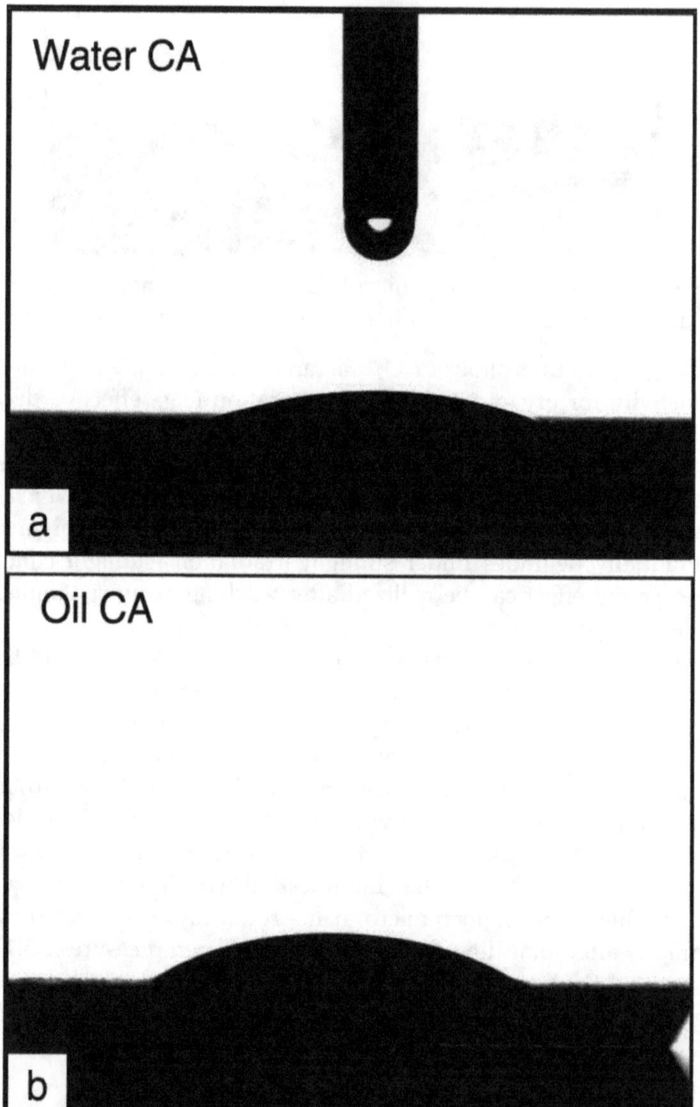

Fig. 31. Water and oil CAs on a TiO_2 film on a silicon wafer prepared by PLD without using a PS colloidal monolayer. (a) Water CA: 15 degrees. (b) Oil (rapeseed) CA: 27 degrees.

Fig. 32. FE-SEM images of Co_3O_4 hierarchical micro/nano- rod arrays obtained by PLD assisted colloidal lithography after annealing at 450 °C for 3 h (oxygen pressures: 93.1 Pa). (a) Top views; (b) Tilted at 45° (c) water CA after modification with low free energy materials.

4.2 Field emission

Field-emission (FE) properties have recently attracted so much attention due to great commercial interest in flat-panel displays and other microelectronic devices.[101] Besides carbon nanotubes, semiconductors have also attracted great interest in field emitters owing to their good mechanical stability, low work function, and high electrical and thermal conductivities.[102] FE properties are usually decided by the nature of the cathode materials as well as geometry and size of them. By well designing the geometry and size of cathode, for example, to introduce nanostructures on it, good FE performances have been achieved including faster turn-on time, compactness, and sustainability during the field emission compared to conventional bulky material forms. More importantly, researchers have found that cathodes composed of periodic regular arrays on the surface are very highly helpful for producing a low operating voltage and a stable current because of the elimination of the shield effect on densely packed 1D nanostructured arrays in field emission.

Periodic TiO_2 micro/nano-rod arrays with hncp arrangement can be synthesized by combining a colloidal monolayer template with pulsed laser deposition (PLD) followed by annealing in ambient air, as described before.[86] By this route, the periodicity of such special nanorods can be easily tuned by changing the colloidal sphere size in the colloidal monolayer template. While a distance between neighboring nanorods can be controlled by varying the background gas pressure during the PLD process if periodicity is fixed for a nanorod array. The well tunable periodicity and distance between neighboring nanorods are very useful for investigating and optimizing their FE performance.[86]

The periodic hncp TiO_2 nanorod array was fabricated by PLD using a colloidal monolayer template with 350nm PS spheres at 6.7 Pa O_2 for 60 min and subsequent annealing at 650 ^{o}C for 2 h in air (Figure 19). It demonstrated a low turn-on field of about 5.6 V μm^{-1} (here the turn-on field was defined as the value of electric field when an emission current density was 4.5 nA cm^{-2}) according to the FE current density-applied electric field curve (J–E) at a working distance of 60 μm from the anode to the nanorod array serving as the cathode (Figure 33). This FE current-voltage characteristics can be expressed by a simplified Fowler-Nordheim (FN) equation and a field-enhancement factor, β can be defined as $B\phi^{3/2}/\kappa$ according to the FN equation (here ϕ: the work function of cathode material; κ: slope in FN plot).[103] This hncp TiO_2 nanorod array showed a field-enhancement factor β of 8.38×10^2. However, a TiO_2 nanorod array with top aggregation for several neighboring nanorods caused by longer deposition displayed a much higher turn-on field of 15.8 V μm^{-1} and lower field-enhancement factor b 3.34×10^2 (Figure 34). This result indicates that the good FE properties of a periodic TiO_2 hncp nanorod array are mainly attributable to the aligned and periodic hncp nanorod morphology.

When a periodicity of hncp nanorod array increased from 350 nm to 750nm and 1 μm by choosing the colloidal monolayers with different PS sphere sizes during the PLD at the same background gas pressure as before and followed by the same annealing, as presented in Figure 35. The field-enhancement factor decreased with increasing periodicity of the hncp nanorod array (Figure 36a). This is mainly attributed to a decreasing number density of nanorods with an increase in the hncp array periodicity. When the periodicity of the hncp nanorod array was increased from 350 to 750 nm, the turn-on field also increased from 5.6 to 13.0 Vμm^{-1}. When a periodicity further increased to 1 μm, the turn-on field remained about 13.0 Vμm^{-1}. It is evident that the hncp nanorod array with the smallest periodicity of 350nm exhibited the best FE properties in this investigation.

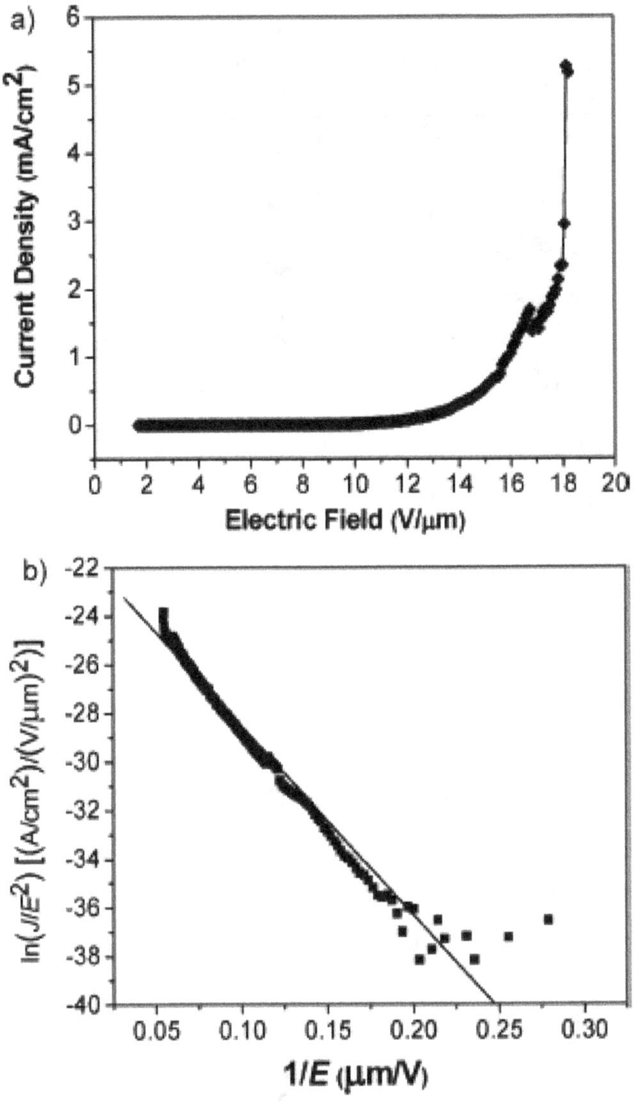

Fig. 33. FE properties of periodic hncp nanorod array by PLD using a colloidal monolayer template with 350nm PS spheres in O_2 at a pressure of 6.7 Pa for 60 min and subsequent annealing in air. a) FE current density-electric field (J-E) curves measured for an hncp TiO_2 nanorod array at an anode-cathode distance of 60 μm. b) Corresponding Fowler-Nordheim (FN) plot.

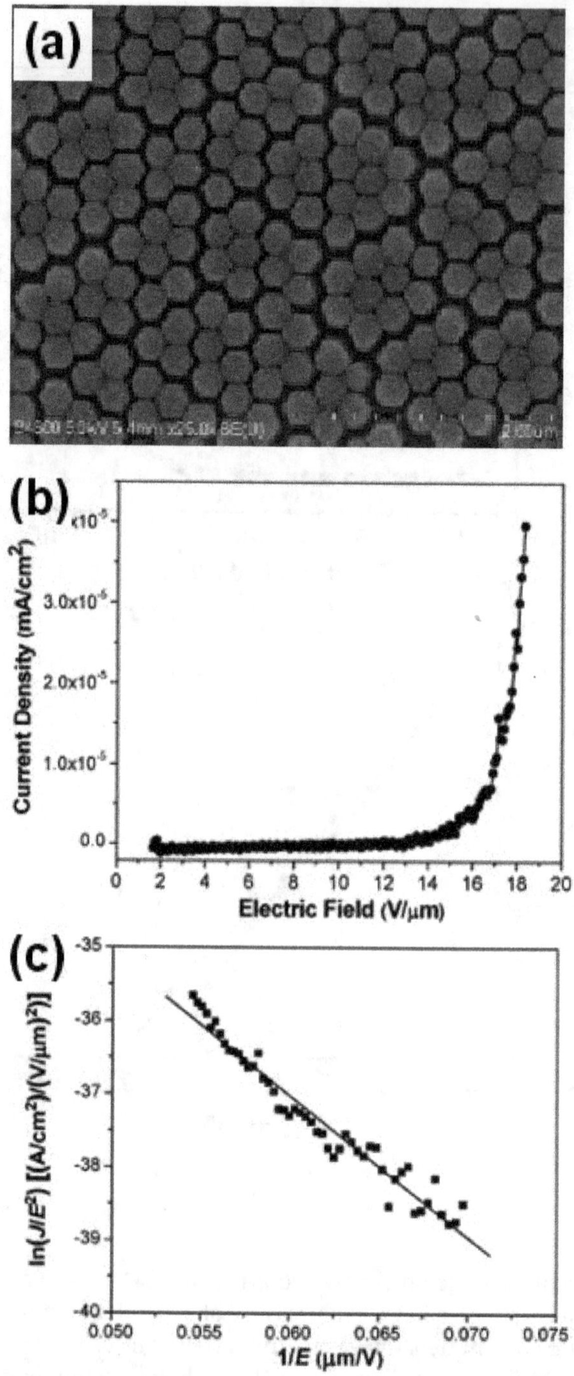

Fig. 34. Top aggregation of a TiO_2 nanorod array obtained PLD assisted colloidal lithography with a longer PLD time (80 min) and subsequent annealing. a) FE-SEM image; b) FE J–E curves measured at an anode–cathode distance of 60 mm for a top-aggregated TiO_2 nanorod array; c) Corresponding FN plot.

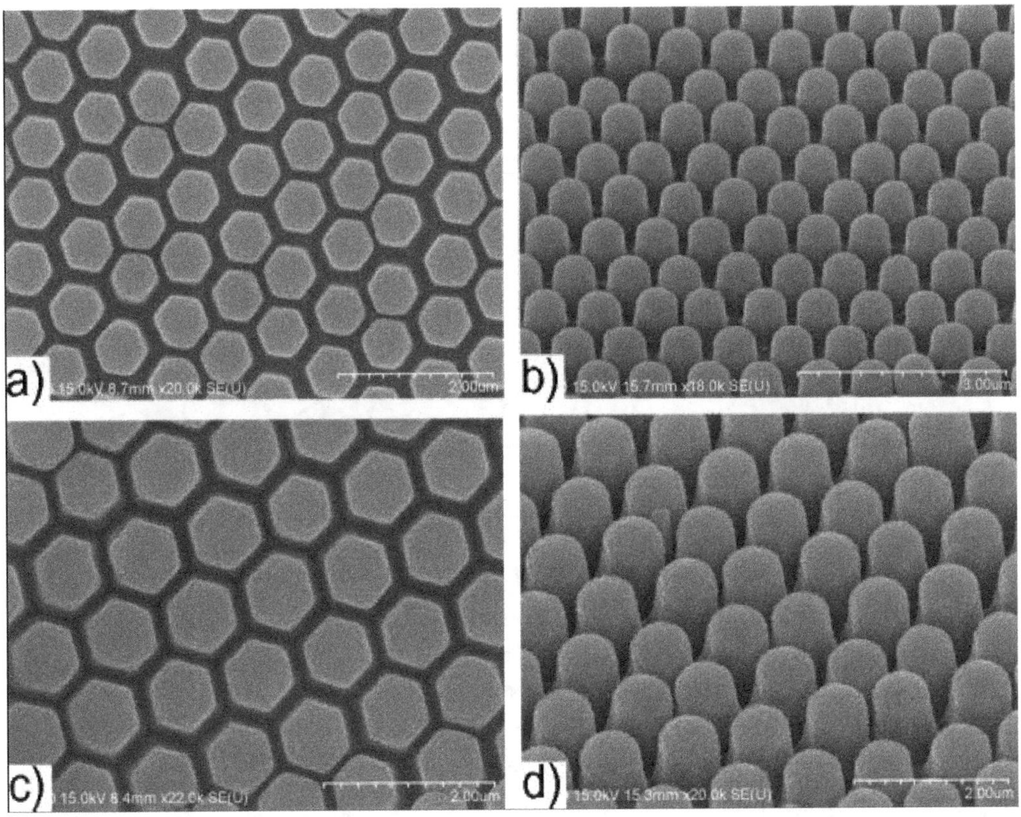

Fig. 35. FE-SEM images of hncp nanorod arrays with different periodicities: 750nm for (a) and (b); 1μm for (c) and (d). PLD was performed in 6.7 Pa of oxygen for 130 min. (a) and (c) are observed from the top, (b) and (d) are observed with a tilt angle of 458.

While a distance between neighboring nanorods can be tuned by changing the background gas pressure during the PLD process if periodicity is fixed to 350 nm for a nanorod array (Figure 22). With increasing this distance of neighboring ones, the field-enhancement factor increased and the turn-on field decreased (Figure 36a). The sample with a small nanorod distance of 20 nm exhibited a relatively low field-enhancement factor of 5.04×10^2 and a high turn-on field of 9.7 V μm^{-1}. When the nanorod distance increased to 50 nm, the FE properties showed enhanced performance with a high field-enhancement factor of 8.38×10^2 and a low turn-on field of 5.6 V μm^{-1}. When the nanorod distance further increased to 110 nm, the best FE properties with a field-enhancement factor of 9.39×10^2 and a turn-on field of 5.3 V μm^{-1} were obtained. (Figure 36b) The above results suggest that optimized FE properties can be achieved by increasing the nanorod distance by controlling the experimental parameters. The increased field-enhancement factor b with the increase of nanorod distance can easily be understood as follows if the periodicity of hncp nanorod array is fixed. The field enhancement factor β is generally related to geometry of an emitter and can be expressed as $\beta \propto h/r$, where h is the height and r is the curvature radius of an emitting center. With an increase in the nanorod distance, the effective diameter of an individual nanorod and a curvature radius r would decrease (Figure 37), resulting in an increase of β according to the above relationship.

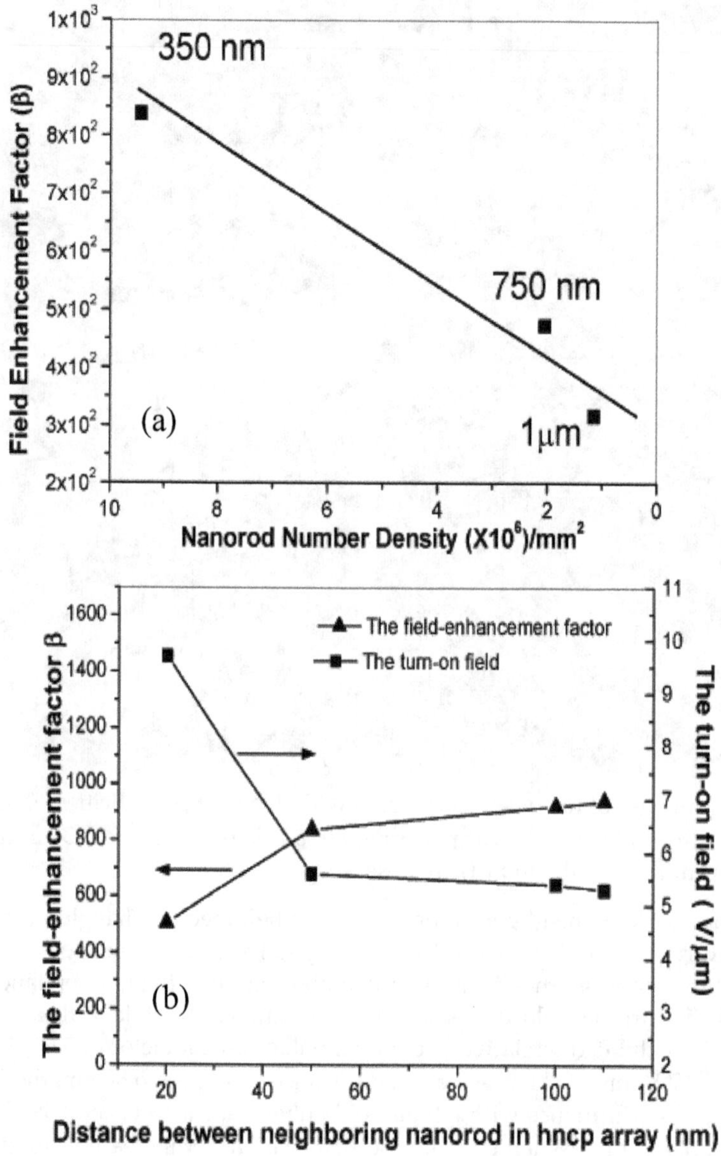

Fig. 36. (a) Field-enhancement factor β changing with increasing periodicity of an hncp nanorod array, (b) Change in field-enhancement factor and turn-on field with varying neighboring nanorod distance in an hncp array.

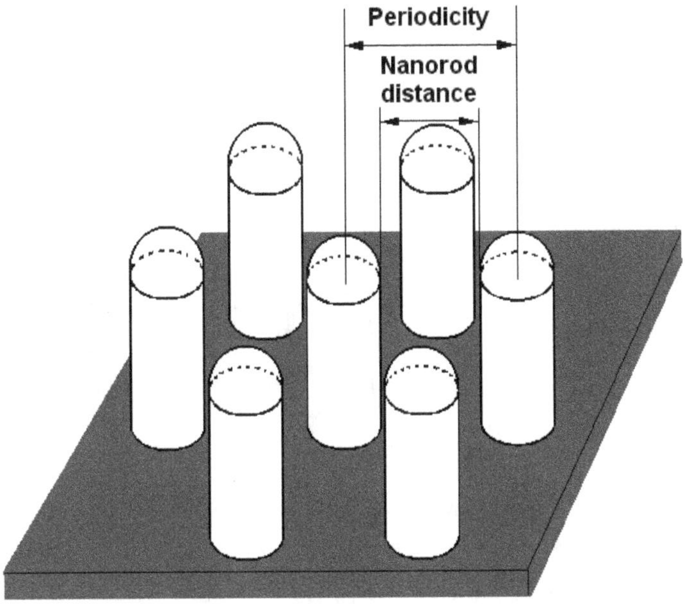

Fig. 37. Schematic illustration of defined nanorod distance in periodic hncp nanorod array.

4.3 Enhanced catalytic properties

The hierarchical micro/nanostructured arrays possess a large specific surface area and hence they might have important application in catalytic fields. For instance, the hcp amorphous TiO_2 micro/nanostructured array on a colloidal monolayer obtained by PLD assisted colloidal lithography demonstrated an enhanced photocatalytic activity (Figure 2). Its photocatalytic performance was estimated based on the decomposition of organic molecules, stearic acid under UV illumination by monitoring the FT-IR spectra.[104,105] The frequencies of 2919 and 2849 cm^{-1} reflect the methylene group asymmetric ($\nu_{asymmCH2}$) and symmetric ($\nu_{symmCH2}$) stretching modes of stearic acid. These values for the methylene group stretching mode are close to those of a crystalline alkane and are typically taken as evidence of the formation of a dense, well-ordered, self-assembled monolayer of stearic acid on the oxide surface.[106-108] Therefore, the photodegradation of stearic acid can be monitor by observing density of these two frequencies. With increasing the UV irradiation time, the vibrational bands of the methylene group gradually decreased and almost completely disappeared after 25 min, as shown in Figure 38a. The decrease in C-H vibrational bands reflects that the stearic acid is gradually photodegraded by such TiO_2 hierarchical micro/nanostructured array films under UV irradiation. Figure 38b shows that degradation curves of a stearic acid film on a silicon wafer, an amorphous TiO_2 film by PLD without using a colloidal monolayer, an hcp amorphous TiO_2 hierarchical micro/nano-rod array on the colloidal monolayer, and an anatase TiO_2 rod array (obtained by annealing hcp amorphous TiO_2 hierarchical micro/nano-rod array on the colloidal monolayer at 650 °C for 2 h. These results indicate that TiO_2 exhibited efficient degradation for stearic acid and that the hcp amorphous TiO_2 hierarchical micro/nano-rod array on a colloidal monolayer demonstrated the best performance compared to the amorphous film and the anatase rod array. Anatase is usually deemed to be more photocatalytically active than the rutile and

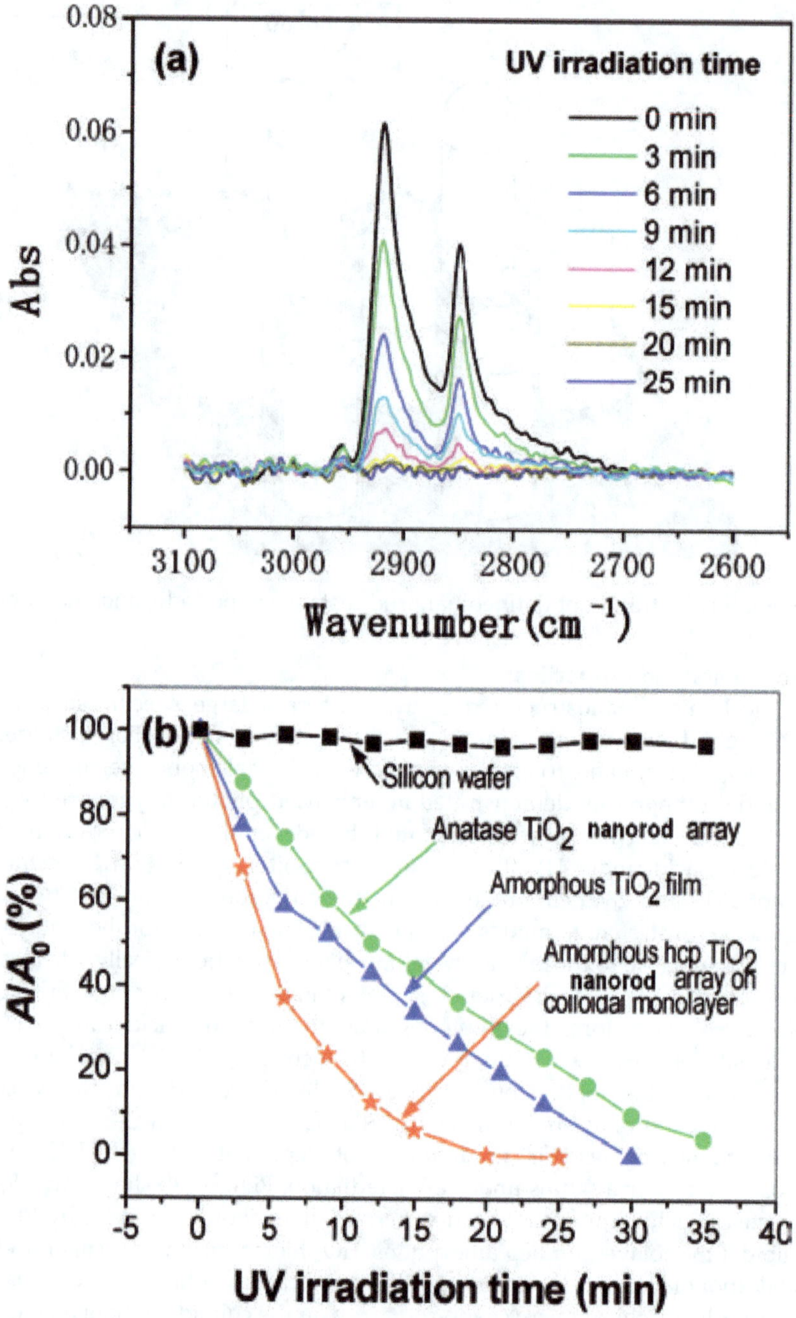

Fig. 38. (a) Photocatalytic activity of an hcp amorphous TiO_2 micro/nano-rod array with a PS colloidal monolayer. (b) Photocatalytic activity evaluation of different substrates based on the absorbance ratio A/A_0 as a function of UV irradiation time. A and A_0 are the absorbance after the UV irradiation and that from the initial surface, respectively.

amorphous TiO_2. However, besides the crystal phase, other factors, including the specific surface area, crystal composition, and material microstructures, also significantly affect the catalytic performance of TiO_2.[109-111] In this case, an amorphous hcp hierarchical micro/nano-rod array has porous structures and possesses a much higher specific surface area than that of an anatase rod array, which contributes to better photocatalytic properties. These results suggest that the surface area of TiO_2 is preferable to its crystal structure for enhancing photocatalytic activity. Additionally, a periodic structured array of amorphous TiO_2 can enhance photocatalytic activity better than an amorphous TiO_2 thin film produced by PLD without using a colloidal monolayer. This may be ascribed to special hierarchical structures composed of radiation-shaped nanobranches emanating from a center point on the PS sphere.[112] The combination of superamphiphilicity and photocatalytic activity is helpful in realizing a self-cleaning surface.

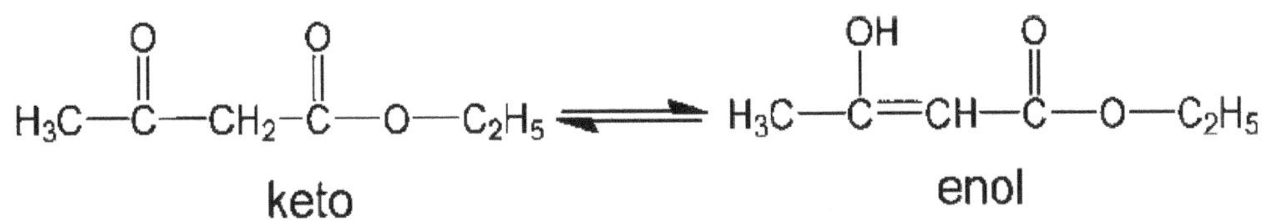

Fig. 39. Isomerization equilibrium of ethyl acetoacetate between the two kinds of isomers, keto and enol.

Additionally, hierarchical alumina mciro/nanostructured arrays demonstrate very excellent catalytic properties for some organic reactions. For example, as we know, ethyl acetoacetate has two kinds of isomers, keto and enol, because of the acidic hydrogen on the active methylene. The corresponding isomerization equilibrium can be depicted in Figure 39. Under common conditions, the keto isomer is more stable than the enol isomer. By the using hierarchical alumina mciro/nanostructured arrays as catalyst, keto isomer might be efficiently converted into the enol isomer. The isomerization can be commonly detected by on-line gas chromatography (GC) or on-line high-performance liquid chromatography (HPLC). Such technique is complicated and, most importantly, not cost-effective. An effective alternative is monitoring the UV-vis spectrum of reaction process. In the keto isomer configuration, there are only two isolated carbonyl groups, with an R absorption band having a quite small ε value in the UV-vis spectrum. For the enol isomer, there is a conjugation system between alkene and carbonyl group, with a K absorption band having a high ε value at around 244 nm in the UV-vis spectrum. This difference makes it convenient to analyze the content and the isomerization process using the UV-vis spectrum. Therefore, the catalytic activity of the hncp alumina was estimated based on the isomerization of ethyl acetoacetate by monitoring the UV-vis absorption spectrum (Figure 40A). The absorption at 244 nm indicates absorption of the conjugation system in the enol isomer as stated above. In the isomerization process, the absorption peak at 244 nm gradually increased and almost completely saturated after 40 min. The increase in absorption at 244 nm indicates that the proportion of enol isomer gradually increased, demonstrating the transition of the keto isomer to the enol isomer and verifying successful isomerization in the presence of hncp alumina. The corresponding control experiments were performed in order to confirm the

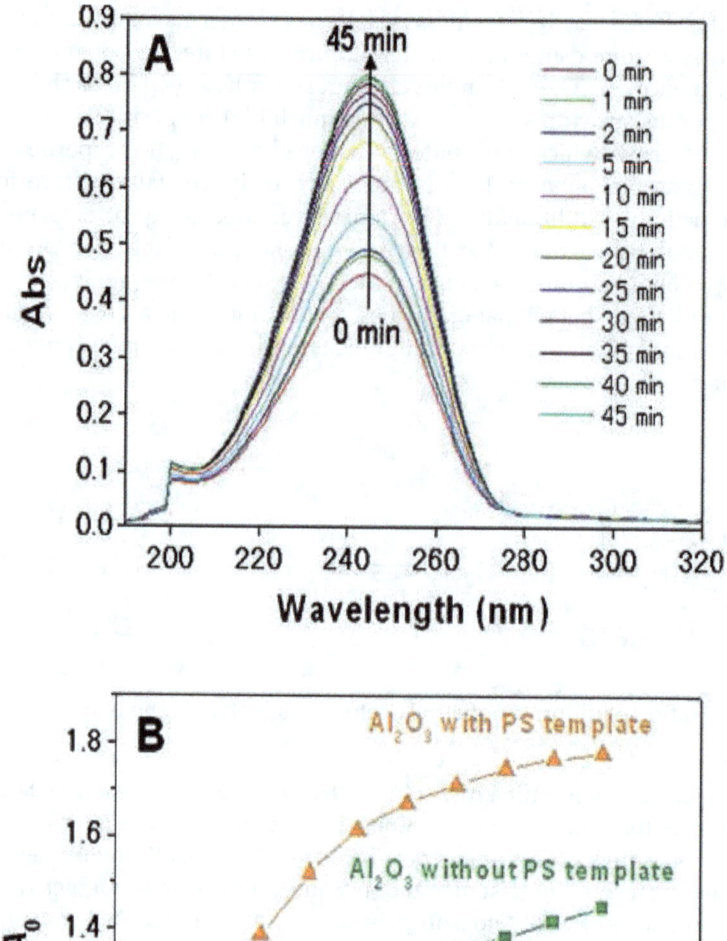

Fig. 40. (A) Catalytic activity of hncp Al_2O_3. (B) Catalytic activity evaluation of different substrates based on the absorbance ratio A/A_0 as a function of reaction time. A and A_0 are the absorbance after a given reaction time and that from the initial solution.

catalytic performance of the hncp alumina for such isomerization. Figure. 40B presents the isomerization process of ethyl acetoacetate in the presence of a silicon wafer and alumina film (by sputtering without using a colloidal monolayer) under the same deposition condition. These results indicate that the hncp alumina exhibited efficient catalytic activity for the isomerization of ethyl acetoacetate. γ-Alumina is generally deemed to be catalytically active. However, besides the crystal phase, other factors, including the specific surface area, crystal composition, material microstructures, and the absence of exotic species from the remnant starting materials, also significantly affect the catalytic performance.[113-116] In this case, the special hierarchical structures of the amorphous hncp alumina array have porous structures and possess a much higher specific surface area than the dense alumina film, which, together with the pure compositions, contributes to better catalytic properties.

5. Conclusions and remarks

The physical deposition assisted colloidal lithography has proven to be a facile, inexpensive, versatile route to construct hierarchical micro/nanostructured arrays with controlled morphologies, sizes, periodicities. The morphologies of these hierarchical micro/nanostructured arrays can be tuned by controlling the experimental conditions, including deposition time, background gas pressure in the vacuum chamber, periodicity of colloidal monolayer template etc. Compared with chemical routes, the physical deposition are more suitable for preparing high quality micro/nanostructured arrays with uniform morphologies. These special structures possess morphology- or size-dependent properties, such as superamphiphilicity, superhydrophobicity, photocatalytic activity, field emission etc., which have important applications in devices, microfluidic devices, field emitters, solar cells etc. Compared to development of fabrication strategies of micro/nanostructured arrays, investigation of morphology- or parameter- properties and micro/nanodevices dependent on them is not so much. The more micro/nanodevices based on these structures will be hoped, and it might be realized under researchers` efforts in the future.

6. Acknowledgements

This work was financially supported by the Natural Science Foundation of China (Grant Nos. 50831005, 10974203), provincial Natural Science Foundation of Anhui (Grant No. 11040606M62), and the National Basic Research Program of China (973 Program, Grant No. 2011CB302103).

7. References

M. Morariu, N. Voicu, E. Schäffer, Z. Lin, T. P. Russell, U. Steiner, *Nat. Mater,*. 2003, 2, 48.

C. H. Ye, L. D. Zhang, X. S. Fang, Y. H. Wang, P. Yan, J. W. Zhao, *Adv. Mater.*, 2004, 16, 1019.

S. V.Dorozhkin, *J. Mater. Sci.: Mater. Med.*, 2007, 18, 363.

G. Duan, W. Cai,Y. Luo, Y. Li, Y. Lei, *Appl. Phys. Lett.*, 2006, 89, 181918.

J. Y. Lao, J. G. Wen, D. Z. Wang, Z. F. Ren, *Nano. Lett.*, 2002, 2, 1287.

P. X. Gao, Y. Ding, Z. L. Wang, *Nano. Lett.*, 2003, 3, 1315.

R. Yan, D. Gargas, P. Yang, *Nat. Photonics,* 2009, 3, 569.

E. Roduner, *Chem. Soc. Rev.*, 2006, 35, 583.

C. M. Cobley, J. Chen, E. C. Cho, L. V. Wang, Y. Xia, *Chem. Soc. Rev.*, 2011, 40, 44.

Z. L., Wang, *Adv. Func Mater.*, 2008, 18, 3553.

R. A. Varin, L. Zbroniec, M. Polanski, J. Bystrzycki, *Energies,* 2011, 4, 1.

Smith, H. I.; Schat-tenburg, M. L. *IBM J. Res. DeV.* 1993, *37*, 319.

Stroscio, J. A.; Eigler, D. M. *Science* 1991, *254*, 1319.

Liu, G.-Y.; Xu, S.; Qian, Y. *Acc. Chem. Res.* 2000, *33*, 457.

Piner, R. D.; Zhu, J.; Xu, F.; Hong, S.; Mirkin, C. A. *Science* 1999, *283*, 661. Xia, Y. N.; Whitesides, G. M. *Langmuir* 1997, *13*, 2059.

Kumar, A.; Whitesides, G. M. *Appl. Phys. Lett.* 1993, *63*, 2002.

Quist, A. P.; Pavlovic, E.; Oscarsson, S. *Anal. Bioanal. Chem.* 2005, *381*,591. Kim, E.; Xia, Y.; Whitesides, G. M. *Nature* 1995, *376*, 581.

L. Gao, T. J. McCarthy, *Langmuir*, 2006, *22*, 2966.

Micheletto, R.; Fukuda, H.; Ohtsu, M. *Langmuir* 1995, *11*, 3333. Rakers, S.; Chi, L. F.; Fuchs, H. *Langmuir* 1997, *13*, 7121.

Denkov, N. D.; Velev, O. D.; Kralchevsky, P. A.; Ivanov, I. B.; Yoshimura, H.;Nagayama, K. *Langmuir* 1992, *8*,3183.

Denkov, N. D.; Velev, O. D.; Kralchevsky, P. A.; Ivanov, I. B.; Yoshimura, H.; Nagayama, K. *Nature* 1993, *361*, 1303.

Hulteen, J. C.; Van Duyne, R. P. *J. Vac. Sci. Technol.* 1995, *13*, 1553. Ozin, G. A.; Yang, S. M. *Adv. Funct. Mater.* 2001, *11*, 95.

Jiang, P.; McFarland, M. J. *J. Am. Chem. Soc.* 2004, *126*, 13778. Wang, D.; Möhwald, H. *Adv. Mater.* 2004, *16*, 244.

Mihi, A.; Ocana, M.; Mí guez, H. *Adv. Mater.* 2006, *18*, 2244.

Li, Y.; Cai, W. P.; Duan, G. T.; Sun, F. Q.; Cao, B. Q.; Lu, F. *Mater.Lett.* 2005, *59*, 276.

Antony, S.; Dimitrov, A. S.; Nagayama, K. *Langmuir* 1996, *12*, 1303.

Jiang, P.; Bertone, J. F.; Hwang, K. S.; Colvin, V. L. *Chem. Mater.*1999, *11*, 2132. Im, S. H.; Kim, M. H.; Park, O. O. *Chem. Mater.* 2003, *15*, 1797.

Kitaev, V.; Ozin, G. A. *Adv. Mater.* 2003, *15*, 75.

Choi, W. M.; Park, O. O. *Nanotechnology* 2006, *17*, 325.

G. Zhang, D. Wang, *Chem. Asian J.*, 2008, 4, 236.

Y. Li, N. Koshizaki, W. Cai, *Coord. Chem. Rev.*, 2010, 255, 357.

Y. Li, W. Cai, G. Duan, *Chem. Mater.*, 2008, 20, 615.

L. Li, T. Zhai, H. Zeng, X. Fang, Y. Bando, D. Golberg. *J. Mater. Chem.*, 2011, 21, 40.

J. Zhang, B. Yang, *Adv. Funct. Mater.*, 2010, 20, 3411.

S.-M. Yang, S.G. Jang, D.-G. Choi, S. Kim, H.K. Yu, *Small*, 2006, 2, 458.

Sun, F.; Cai, W.; Li, Y.; Duan, G.; Nichols, W. T.; Liang, C.; Koshizaki, N.; Fang, Q.; Boyd, I. W. *Appl. Phys. B:Lasers Opt.* 2005, *81*,765.

Burmeister, F.; Schäfle, C.; Matthes, T.; Böhmisch, M.; Boneberg, J.; Leiderer, P. *Langmuir* 1997, *13*, 2983.

Pacifica, J.; Gómez, D.; Mulvaney, P. *AdV. Mater.* 2005, *17*, 415.

Jensen, T. J.; Duval, M. L.; Kelly, K. L.; Lazarides, A. A.; Schatz, G. C.; Van Duyne, R. P. *J. Phys. Chem. B* 1999, *103*, 9846.

Malinsky, M. D.; Kelly, K. L.; Schatz, G. C.; Van Duyne, R. P. *J. Phys. Chem. B* 2001, *105*, 2343.

Hulteen, J. C.; Treichel, D. A.; Smith, M. T.; Duval, M. L.; Jensen, T. J.; Van Duyne, R. P. *J. Phys. Chem. B* 1999, *103*, 3854.

Tan, B. J. Y.; Sow, C. H.; Koh, T. S.; Chin, K. C.; Wee, A. T. S.; Ong, C. K. *J. Phys. Chem. B* 2005, *109*, 11100.

Sort, J.; Glaczynska, H.; Ebels, U.; Dieny, B.; Miersig, M.; Rybczynski, J. *J. Appl. Phys.* 2004, *95*, 7516.

Sun, F. Q.; Cai, W. P.; Li, Y.; Cao, B. Q.; Lei, Y.; Zhang, L. D. *Adv. Funct. Mater.* 2004, *14*, 283.

Sun, F. Q.; Yu, J. C.; Wang, X. C. *Chem. Mater.* 2006, *18*, 3774.

Li, Y.; Cai, W. P.; Duan, G. T.; Sun, F. Q.; Lu, F. *Appl. Phys. A: Mater. Sci. Process.* 2005, *81*, 269.

Li, Y.; Cai, W. P.; Cao, B. Q.; Duan, G. T.; Li, C. C.; Sun, F. Q.; Zeng, H. B. *J. Mater. Chem.* 2006, *16*, 609.

Sun, F. Q.; Cai, W. P.; Li, Y.; Cao, B. Q.; Lei, Y.; Zhang, L. D. *Mater. Sci. Technol.* 2005, *21*, 500.

Sun, F. Q.; Cai, W. P.; Li, Y.; Jia, L. C.; Lu, F. Adv. *Mater.* 2005, *17*, 2872.

Sun, F. Q.; Cai, W. P.; Li, Y.; Cao, B. Q.; Lu, F.; Duan, G. T.; Zhang, L. D. Adv. *Mater.* 2004, *16*, 1116.

Cao, B. Q.; Cai, W. P.; Sun, F. Q.; Li, Y.; Lei, Y.; Zhang, L. D. *Chem. Commun.* 2004, 1604.

Cao, B. Q.; Sun, F. Q.; Cai, W. P. *Electrochem. Solid-State Lett.* 2005, *8*, G237.

Duan, G. T.; Cai, W. P.; Li, Y.; Li, Z. G.; Cao, B. Q.; Luo, Y. Y. *J. Phys.Chem. B* 2006, *110*, 7184.

Duan, G. T.; Cai, W. P.; Luo, Y. Y.; Sun, F. Q. *Adv. Funct. Mater.* 2007, *17*, 644

Wang, X.; Lao, C.; Graugnard, E.; Summers, C. J.; Wang, Z. L. *Nano Lett.* 2005, *5*, 1784. Yan, F.; Goedel, W. A. *Nano Lett.* 2004, *4*, 1193.

Duan, G. T.; Cai, W. P.; Luo, Y. Y.; Li, Z. G.; Lei, Y. *J. Phys. Chem.B* 2006, *110*, 15729.

Li, Y.; Cai, W. P.; Duan, G. T.; Cao, B. Q.; Sun, F. Q. *J. Mater. Res.* 2005, *20*, 338.

Venkatesh, S.; Jiang, P. *Langmuir* 2007, *23*, 8231.

Li, Y.; Huang, X. J.; Heo, S. H.; Li, C. C.; Choi, Y. K.; Cai, W. P.; Cho, S. O. *Langmuir* 2007, *23*, 2169.

Huang, X. J.; Li, Y.; Im, H. S.; Yarimaga, O.; Kim, H. J.; Jang, D. Y.; Cho, S. O.; Cai, W. P.; Choi, K. Y. *Nanotechnology* 2006, *17*, 2988.

Correa-Duarte, M. A.; Kosiorek, A.; Kandulski, W.; Giersig, M. *Small* 2006, *2*, 220. Li, Y.; Lee, E. J.; Cho, S. O. *J. Phys. Chem. C* 2007, *111*, 14813.

Li, Y.; Li, C. C.; Cho, S. O.; Duan, G. T.; Cai, W. P. *Langmuir* 2007, *23*, 9802.

Duan, G. T.; Cai, W. P.; Luo, Y. Y.; Li, Y.; Lei, Y. *Appl. Phys. Lett.* 2006, *89*, 181918.

Qi, L. M. Li, Y.; Sasaki, T.; Shimizu, Y.; Koshizaki, N. *J. Am. Chem. Soc.* 2008, *130*, 14755.

Li, Y.; Koshizaki, N.; Shimizu, Y.; Li, L.; Gao, S. Sasaki, T. *ACS Appl. Mater. & Interfaces.* 2009, *1*, 2580.

Nam, H. J.; Sasaki, T.; Koshizaki, N. *J. Phys. Chem. B* 2006, *110*, 23081. Brett, M. J.; Hawkeye, M. M. *Science* 2008, *319*, 1192.

Robbie, K.; Beydaghyan, G.; Brown, T.; Dean, C.; Adams, J.; Buzea, C. *Rev. Sci. Instrum.* 2004, *75*, 1089.

Zhao, Y.-P.; Ye, D.-X.; Wang, G.-C.; Lu, T.-M. *Nano Lett.* 2002, *2*, 351.

Hrudey, P. C. P.; Szeto, B.; Brett, M. J. *Appl. Phys. Lett.* 2006, *88*, 251106. Kesapragada, S. V.; Victor, P.; Nalamasu, O.; Gall, D. *Nano Lett.* 2006, *6*, 854.

He, Y. P.; Fu, J. X.; Zhang, Y.; Zhao, Y. P.; Zhang, L. J.; Xia, A. L.; Cai, J. W. *Small* 2007, *3*, 153.

Zhou, C. M.; Gall, D. *Appl. Phys. Lett.* 2006, *88*, 203117.

Robbie, K.; Sit, J. C.; Brett, M. J. *J. Vac. Sci. Technol., B: Microelectron. Nanometer Struct.sProcess., Meas., Phenom.* 1998, *16*, 1115.

Vick, D.; Tsui, Y. Y.; Brett, M. J.; Fedosejevs, R. *Thin Solid Films* 1999, *350*, 49. Li, Y.; Sasaki, T.; Shimizu, Y.; Koshizaki, N. Small, 2008, 4, 2286.

Li, Y.; Fang, X.; Koshizaki, N.; Sasaki, T.; Li, L.; Gao, S.; Shimizu, Y.; Sasaki, T.; Bando, Y.; Golberg, M. Adv. Funct. Mater. 2008, 19, 2467.

Li, L.; Koshizaki, N. J. Mater. Chem., 2010, 20, 2972

Li, L.; Li, Y.; Gao, S.; Koshizaki, N. J. Mater. Chem., 2009, 19, 8366.

Gao, S.; Koshizaki, N.; Li,Y.; Li, L. J. Mater. Chem., 2011, 21, 2087.

A. Nakajima, K. Hashimoto, T. Watanabe, Monatsh. Chem. 132 (2001) 31.

T. Sun, L. Feng, X. Gao, L. Jiang, Acc. Chem. Res. 38 (2005) 644.

X.-M. Li, D. Reinhoudt, M. Crego-Calama, Chem. Soc. Rev. 36 (2007) 1350. X.-J.Huang,J.-H. Lee, J.-W.Lee, J.-B.Yoon, Y.-K. Choi, Small4 (2008) 211.

X.-J. Huang, D.-H. Kim, M. Im, J.-H. Lee, J.-B. Yoon, Y.-K. Choi, Small, 2008, 5, 90

R.N. Wenzel, J. Phys. Colloid Chem. 1949, 53, 1446.

R. Wang, K. Hashimoto, A. Fujishima, M. Chikuni, E. Kojima, A. Kitamura, M. Shimohigoshi, T. Watanabe, Adv. Mater. 1998, 10, 135.

W.-L. Min, B. Jiang, P. Jiang, Adv. Mater. 2008, 20, 3914.

Y. Li, J. Zhang, S. Zhu, H. Dong, Z. Wang, Z. Sun, J. Guo, B. Yang, J. Mater. Chem. 2009, 19, 1806.

Y.-R. Lin, H.-P. Wang, C.-A. Lin, J.-H. Hea, J. Appl. Phys. 2009, 106, 114310.

A. Cassie and S. Baxter, Trans. Faraday Soc., 1944, 40, 546.

S.S. Fan, M.G. Chapline, N.R. Franklin, T.W. Tombler, A.M. Cassell, H.J. Dai, Science 283 (1999) 512.

X.S. Fang, Y. Bando, U.K. Gautam, C.H. Ye, D. Golberg, J. Mater. Chem. 18 (2008) 509.

R.H. Fowler, L.W. Nordheim, Proc. R. Soc. London, Ser. A 119 (1928) 173. Wang,K.X.;Yao,B.D.;Morris,M.A.;Holmes,J.D.*Chem.Mater.*2005,*17*,4825.

Zhang, X.T.; Jin, M.; Liu, Z.Y.; Tryk, D.A.; Nishimoto, S.; Murakami, T.; Fujishima, A. *J. Phys. Chem. C* 2007, *111*, 14521.

Gao, W.; Dickinson, L.; Grozinger, C.; Morin, F. G.; Reven, L. *Langmuir* 1996, *12*, 6429. Nuzzo, R. G.; Dubois, L. H.; Allara, D. L. *J. Am. Chem. Soc.* 1990, *112*, 558.

Gawalt, E. S.; Avaltroni, M. J.; Koch, N.; Schwartz, J. *Langmuir* 2001, *17*, 5736. Fujishima, A.; Zhang, X. T. *C. R. Chimie* 2006, *9*, 750.

Macak, J. M.; Zlamal, M.; Krysa, J.; Schmuki, P. *Small* 2007, *3*, 300. Wu, C. Z.; Lei, L. Y.; Zhu, X.; Yang, J. L.; Xie, Y. *Small* 2007, *3*, 1518.

Li, H. X.; Bian, Z. F.; Zhu, J.; Zhang, D. Q.; Li, G. S.; Huo, Y. N.; Li, H.; Lu, Y. F. *J. Am. Chem. Soc.* 2007, *129*, 8406.

J. Li, R. Yan, B. Xiao, D. T. Liang and L. Du, Environ. Sci. Technol., 2008, 42, 6224; L. B. Sun, J. Yang, J. H. Kou, F. N. Gu, Y. Chun, Y. Wang, J. H. Zhu and Z. G. Zou, Angew. Chem., Int. Ed., 2008, 47, 3418;

C. Misra, American Chemical Society Monograph, in Industrial Alumina Chemicals, American Chemical Society, Washington, DC,1986, vol. 184;

B. R. Baker and R. M. Pearson, J. Catal., 1974, 33, 265.

8

Application of Nanoimprint Lithography to Distributed Feedback Laser Diodes

Masaki Yanagisawa
Sumitomo Electric Industries, Ltd.
Japan

1. Introduction

The recent growth of information-communication facilities such as the Internet, mobile telecommunication and video-on-demand services has led to explosion of worldwide communication traffic, which increase demand for faster and denser communication infrastructures including optical communication networks. Distributed feedback laser diodes (DFB LDs) have been widely used as optical sources in networks because of their high selectivity and stability of wavelength. Although they had limitedly been used in long-haul and high-speed network at one time, they are now increasingly required in metro and end-user fields because of the increasing traffic. Thus, necessity for inexpensive DFB LDs increases rapidly.

The characteristics of a DFB LD with uniform (constant period) gratings depend on the grating phase at the cleaved facet (Matsuoka et al., 1984). The variation of characteristics with the facet phase is a serious issue in view of productivity and usability of the LDs. One of the most effective ways of reducing the facet phase effect is to adopt phase-shifted gratings instead of uniform gratings (Kaden et al., 1992). The uniformities of the LD characteristics such as mode-stability and output power are improved by adopting phase-shifted gratings, thus the yield of LDs increase and their production cost is effectively reduced.

In general, there are various fabrication methods for diffraction gratings of DFB LDs, for example, interference exposure, electron beam lithography (EBL), and optical projection exposure. Interference exposure cannot feasibly be used for fabricating phase-shifted gratings, because it exclusively generates exposure patterns with a uniform bright-and-dark period. Although EBL has sufficient resolution to be used for phase-shifted gratings, exceedingly expensive apparatus is necessary for volume production with sufficient throughput. For the optical projection method, a forefront stepper having sufficient resolution for gratings is also expensive, and the cost is too high for fabricating DFB LDs, of which production volume is relatively small compared to that of such semiconductor devices as LSIs.

Nanoimprint lithography (NIL) has been studied by many organizations since the middle of the 1990s. Chou et al. indicated that sub-10-nm features could be formed by imprint, which started the era of NIL technology (Chou et al., 1995). A novel method of NIL using a UV-curable resin was introduced by Haisma et al (Haisma et al., 1996), and Bailey et al.

demonstrated a step-and-repeat imprint method named SFIL (Bailey et al., 2000). NIL is a simple method applicable to forming fine patterns smaller than 100 nanometers, so it is studied as a candidate for a next-generation fabrication technique in many application fields such as storage devices, displays, optical devices, MEMS, semiconductor and so on.
Since 2004, we have investigated the use of NIL for fabricating diffraction gratings, because the new technology have been considered as an attractive solution to the above issues concerning the fabrication of phase-shifted gratings because of its high resolution, throughput, and low cost. In 2009, we have reported that our NIL method have been successfully applied to fabricating quarter-wavelength shifted DFB LDs (Yanagisawa et al., 2009). To the best of our knowledge, it is the unprecedented demonstration of NIL application to DFB LDs in wafer-scale fabrication having the potential of near-future volume production.

2. Fabrication process

Our fabrication method is based on our conventional production process of DFB LDs except for the formation of diffraction gratings by NIL, which is a strong point because we make the best use of the mature process and minimize the risk of unpredictable difficulties induced by adopting new methods.
From an early stage of this study, we have used UV-NIL rather than other imprint methods such as thermal NIL and soft lithography, because UV-NIL has advantages of high over-lay alignment accuracy, low imprint pressure and high throughput. Over-lay alignment is especially important in view of mixing and matching the NIL method with the conventional LD fabrication. Imprint pressure is desirable as low as possible, because it directly influences crystal damage of a substrate as described later.

2.1 Mold design

One of the major issues of applying NIL process to DFB LDs is preparation of molds (templates). In general, NIL molds are fabricated by utilizing the forefront method such as electron-beam lithography (EBL), whose throughput is much lower than other optical lithography methods. Thus, long delivery time and high cost for mold fabrication are serious problems. DFB LDs generally used in optical networks vary in emission wavelengths from 1.3 μm to 1.6 μm, thus corresponding periods of diffraction gratings accordingly vary from 200 nm to 250 nm. If a mold were designed with containing the sole type of grating period, we would have to prepare a large number of molds for covering all wavelength range. It would lead to insurmountably high initial costs for DFB LD fabrication.
We have developed a novel NIL method with a new concept of mold design named "VARI-mold." In the new process, one mold can be used for fabricating various types of DFB LDs with different wavelengths. The prepared mold contains more than 9000 grating patterns in an imprint field, consisting of about 900 subfields, and each subfield contains more than 10 different types of grating patterns with the period from 194 nm to 248 nm exemplified in Fig. 1. As described in the next section, various types of LDs are fabricated by utilizing one VARI-mold, leading to reducing the cost for mold preparation.
The molds used in this study are fabricated by a reticle fabrication method and an etching procedure.

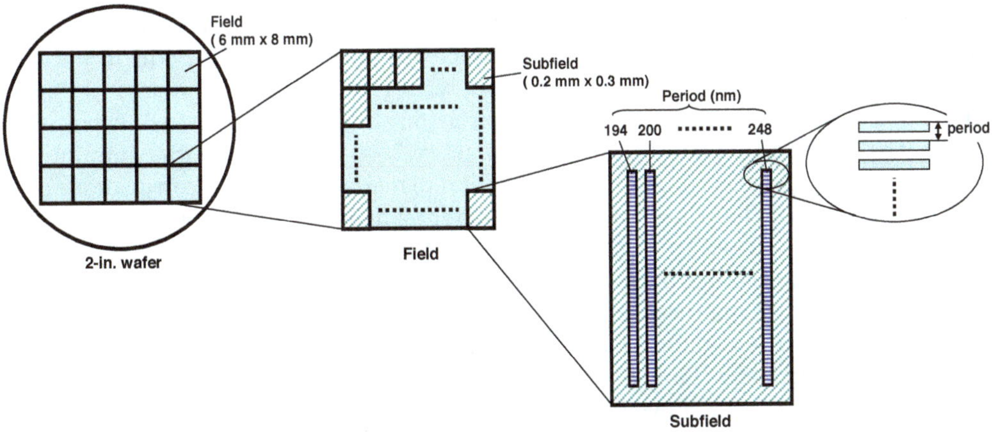

Fig. 1. Schematic structure of a "VARI-mold"

2.2 Wafer fabrication

In general, commercially available compound semiconductor substrates used for optical devices, such as GaAs, InP, and GaN, have large undulations compared with Si substrates widely used for electronic devices such as LSIs and memories. For example, 2-in. diameter InP substrates, which are generally used for fabricating DFB LDs, typically have over 3 μm in total thickness variation. If such undulating substrates are applied to NIL, the mold probably come in contact with a limited portion of the substrate, in which case severe nonuniformity of the residual layer thickness will lead to large variations of the transferred figures in the imprinted area. Thus, we have formed diffraction gratings by using a reverse-tone nanoimprint in order to suppress the variation of residual layer thickness resulting from the undulation of substrates (Miller et al., 2005).

A schematic structure and a layer structure of the LD are shown in Fig. 2. We have prepared a 2-in. InP wafer with epitaxial layers including a grating layer, an active layer, a buffer layer and a lower cladding layer grown by metalorganic vapor phase epitaxy (MOVPE).

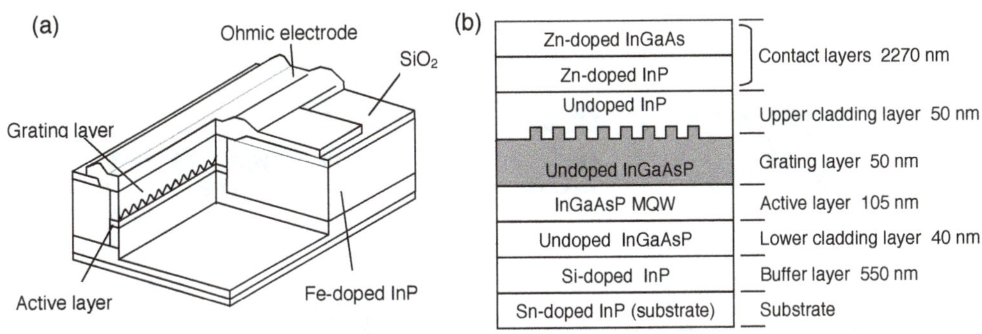

Fig. 2. Schematic structure of a DFB LD (a) and crystal layer structure of the prepared wafer (b).

The fabrication procedure is shown in Fig. 3. First, a 50-nm-thick SiN film is deposited on the wafer by plasma-enhanced chemical vapor deposition (PE-CVD). Next, a primer material is spin-coated in order to increase adhesion between UV-curable resin and the SiN film. Then, UV-NIL using a VARI-mold is conducted to form grating patterns in the UV-curable resin layer (Figs. 3(a) and 3(b)). In this study, imprinting was performed in 16 fields on the wafer with a step-and-repeat equipment. The imprint pressure before UV exposure is approximately 0.1 MPa, and the exposure time is 20 seconds. After the imprint, Si-containing resin is spin-coated in order to cover and planarize the grating corrugations (Fig. 3(c)). The thickness of the planarization layer is approximately 200 nm. Subsequently, the planarization layer is etched by reactive ion etching (RIE) until the tops of the corrugations are revealed (Fig. 3(d)). After that, the revealed layer is selectively etched and penetrated until the SiN masks are revealed (Fig. 3(e)). This penetration etching is one of the essential techniques of this process, so detailed in next section. The formed resin patterns are used as masks for the subsequent etching, transferring the grating patterns to the SiN film (Fig. 3(f)). Next, the resin layers are removed by O_2 plasma etching. After that, we use inductively coupled plasma RIE (ICP-RIE) with CH_4 / H_2 gas for etching of the crystal layer (Fig. 3(g)). Finally, the SiN masks are stripped by a wet chemical process using HF solution, and the diffraction grating structure is achieved. After the formation of the gratings, an upper cladding layer and contact layers are formed on the grating layer by MOVPE. The contact layers consist of InP and InGaAs layers. Then, stripe patterns of SiO_2 are formed on the contact layer by using chemical vapor deposition (CVD) and conventional photolithography method in order to define the cavities of the DFB LDs. In this step, cavity stripes are overlaid to the grating patterns having a specific period corresponding to the required wavelength of LDs. For example, when we would fabricate DFB LDs with the wavelength of 1310 nm, stripe patterns had to be aligned onto 200 nm-period gratings (Fig. 4). As a matter of course, different types of LDs with various wavelengths can be achieved simultaneously on a wafer provided that we adjust the alignment of the cavity stripe layer in each imprint field. The stripe patterns of SiO_2 are used as masks for subsequent crystal etching by ICP-RIE with CH_4 / H_2 gas. In this etching step, all unused grating patterns (excepting the selected one under the stripe) are removed. After that, Fe-doped InP is selectively grown onto the etched area as an insulating layer by MOVPE. Subsequently, a SiO_2 film is deposited as a passivation layer, in which contact holes are formed by selective etching by RIE. Finally, metal electrodes are formed by high-vacuum evaporation and a lift-off method.

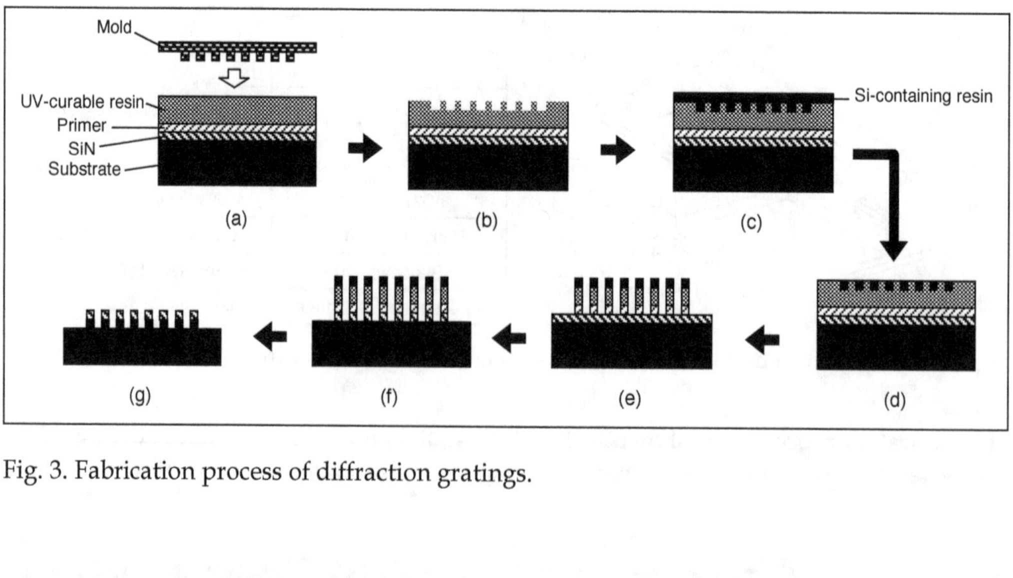

Fig. 3. Fabrication process of diffraction gratings.

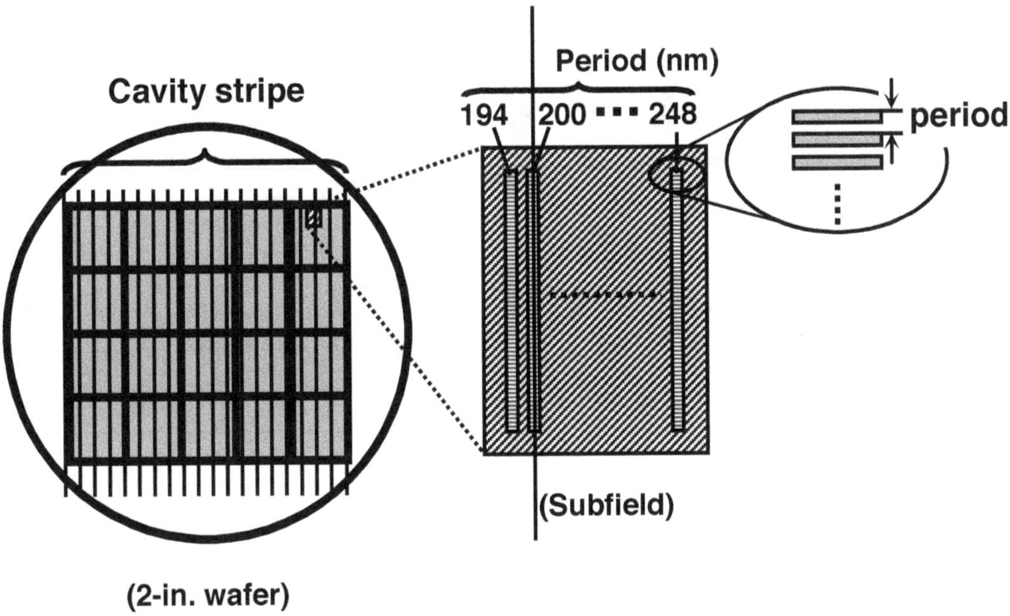

Fig. 4. Concept of cavity stripe delineation.

2.3 Etching process

As mentioned above, the resin etching is one of the key techniques of our fabrication process. Nonuniformity of the linewidth of corrugations after the penetration etching leads to inhomogeneity of the grating figures (Fig. 3(e)), resulting in yield reduction of DFB LDs.

We have developed the etching method using a low-temperature ICP-RIE system in order to achieve highly-uniform and highly-repeatable grating fabrication (Tsuji et al., 2011). The etching method is characterized by the etching gas and the low-temperature substrate stage. Oxygen and nitrogen are used as the etching gas, and the substrate stage is controlled with the temperature from 260 K to 270 K. The both features contributes to suppress the undercut of the UV-curable resin during the penetration etching, resulting from the sidewall effect produced by the optimized plasma condition and substrate temperature (Kure et al., 1991; Kinoshita et al., 1999). The optimized etching condition is shown in Table 1, and a cross-sectional view of the grating after the penetration etching observed by a scanning electron microscope (SEM) is shown in Fig. 5.

Gas flow rate	N_2	9 sccm
	O_2	1 sccm
Bias power		70 W
ICP power		250 W
Substrate temperature		263 K

Table 1. Etching condition of the penetration etching.

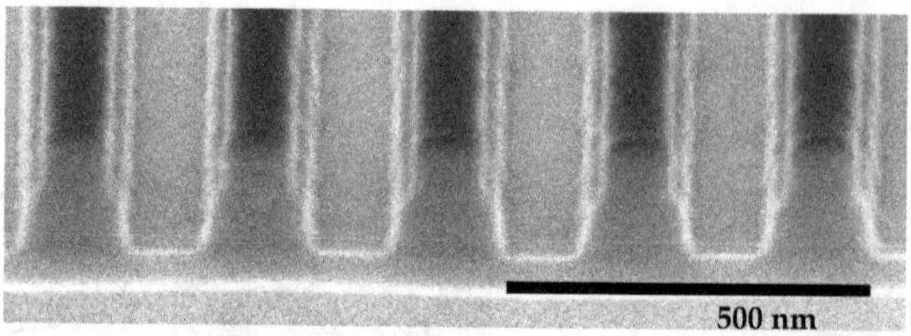

Fig. 5. SEM image of the diffraction grating after the penetration etching.

We have evaluated the linewidth uniformity of the corrugations within 6 wafers. Figure 6 is the histogram of the linewidth, indicating the standard deviation is less than 4 nm.

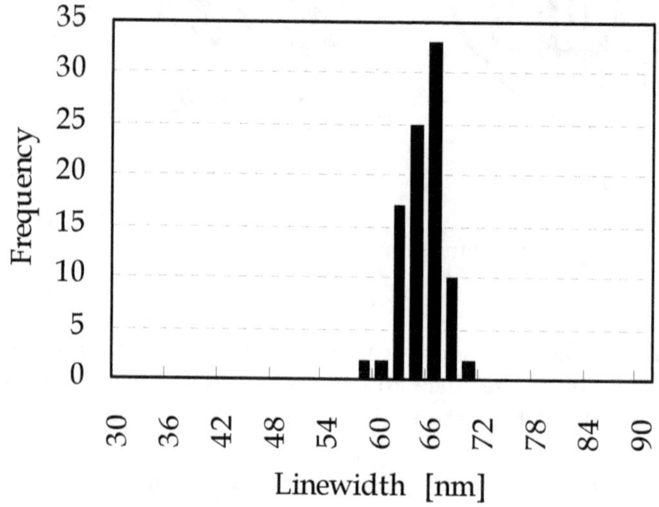

Fig. 6. Histogram of the corrugation linewidth within 6 wafers after the penetration etching.

3. Results

When we use NIL for fabricating DFB LDs, we have paid attention to mechanical damage in epitaxial layers by imprint pressure, because compound semiconductor crystals used for LDs are easily damaged by mechanical stress, leading to severe deterioration in long-term reliability of device characteristics (Fukuda et al., 1985). In this study, photoluminescence (PL) method have been used for evaluation of the crystal damage, and simple 2-D simulation is conducted as qualitative analysis of distribution of the PL intensity.

Furthermore, we have also focused on the accuracy of the grating period, because it dominantly influences the emission wavelength which is the essential parameter of LD characteristics. The emission wavelength directly impacts on the yield of DFB LDs, so its repeatability is required to be sufficiently high for volume production. We have verified the accuracy of the grating period by measuring diffraction angles described below.

Finally, we have fabricated quarter-wavelength shifted DFB LDs by utilizing the new fabrication process combining NIL and the conventional LD process, and evaluated their characteristics including long-term reliability.

3.1 Influence of imprint pressure

3.1.1 Photoluminescence intensity

In order to evaluate the mechanical damage in semiconductor crystal induced by imprinting pressure, we have investigated deterioration of PL intensities from the epitaxial layers. We prepared two indium-phosphide substrates with epitaxial layers and a blank (with no patterns) mold. We compared photoluminescence intensities between the two samples: imprinted with UV-curable resin between the mold and the substrate, and imprinted without resin. The field size of the blank mold used here is 10 mm x 10 mm. The evaluation results of PL intensities are shown in Fig. 7. Field size of the blank mold used here is 10 mm x 10 mm. Imprinting pressure is 0.8 MPa for both samples. The sample without resin shows evident deterioration of PL intensities indicated in dark (green-like) colors in Fig. 7. The deteriorations of intensity are found mainly in the edge of the imprinted area. This means that the imprinting pressure concentrates near the edge of the mold. On the other hand, no evident deterioration is found in the sample with resin. These results indicate that the resin functions such as a cushion to prevent severe damage in epitaxial layers by imprinting pressure.

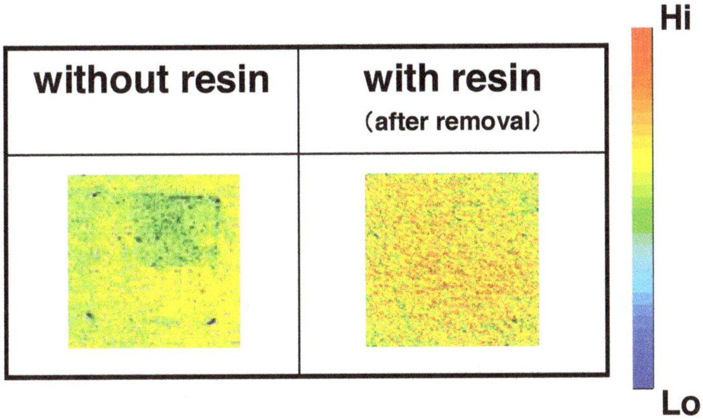

Fig. 7. PL intensities of the epitaxial layers after imprints. The field size of the imprint is 10 mm x 10 mm.

3.1.2 Simulation

We have calculated how resin between a mold and a substrate influences on mechanical stress induced in the substrate. We used a simple model for the finite element method (FEM) as described in Table 2 and Fig. 8. In this study, UV-curable resin is to be considered as liquid having nonlinear viscoelasticity; however, elastomer is substituted for resin because we do not have an adequate tool for dynamic simulation of viscoelastic material. Thickness and the Young's modulus of the elastomer are assumed of 50 nm and 1000 MPa, respectively. Imprinting pressure is 1×10^7 MPa. The elastomer and the other materials are connected with common FEM nodes at the boundaries.

Figure 8 shows two-dimensional distribution of von Mises stress in molds and substrates. For the sample without elastomer, mechanical stress is concentrated in periphery of the edge of the mold [Fig. 8(a)]. On the other hand, when the elastomer is supposed between the mold and the substrate, concentration of stress on the surface of the substrate is clearly suppressed [Fig. 8(b)]. These results are qualitatively consistent with the evaluation using PL described in above section.

		Young's modulus [MPa]	Poisson ratio
Mold	SiO_2	70000	0.17
Substrate	InP	60700	0.36
Liquid	Elastomer	1000	0.49

Table 2. Mechanical properties used for the stress simulation.

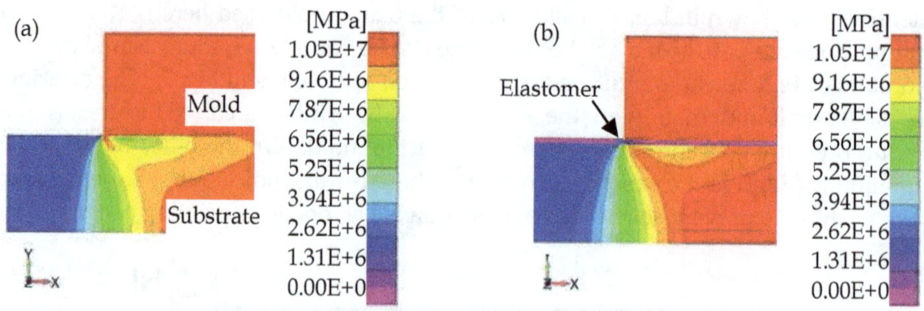

Fig. 8. Distribution of von Mises stress without (a) and with (b) elastomer between a mold and a substrate.

3.2 Grating figure

A SEM image of the fabricated diffraction gratings is shown in Fig. 9. It demonstrates that the line edge roughness (LER) of the gratings is markedly low, indicating that the LER of the grating corrugations in the master mold is sufficiently suppressed and that the master patterns are precisely transferred to the substrate by the imprinting and the subsequent etching process. The depth of the diffraction gratings was measured to be approximately 15 nm by an atomic force microscope.

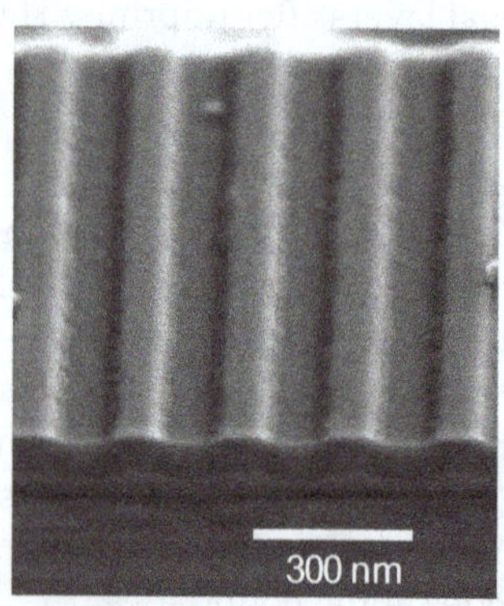

Fig. 9. SEM image of diffraction gratings.

3.3 Reproducibility of the grating period

The diffraction gratings with the period from 194 to 204 nm have been formed on a 2-in. InP wafer. The grating periods have been verified by measuring the diffraction angles of the transferred grating patterns. The wavelength of the incident light is 364.8 nm, and the measurement error of the grating period is estimated less than 0.03 nm. Figure 10 shows a correlation between the grating periods of the mold and those of the transferred patterns. The difference between both of the values is less than 1%, and the wafer-to-wafer variation of the periods is less than 0.2 nm through the 6 wafers. This shows that the NIL process has high reproducibility in transferring grating patterns.

3.4 Device characteristics

We have evaluated DFB LDs fabricated by our novel process in order to verify the basic characteristics and their uniformity in 2-in. wafer. The nominal wavelength of the measured LDs is 1490 nm, and the corresponding grating period is approximately 232 nm (excluding Fig. 12).

Figure 11 shows the dependence of the optical output and the slope efficiency on the supplied current for a typical phase-shifted DFB LD fabricated in this study. The threshold current and the slope efficiency at room temperature were measured to be 8 mA and 0.28 W/A, respectively, which are comparable to those of typical phase-shifted DFB LDs fabricated by utilizing the conventional EBL process.

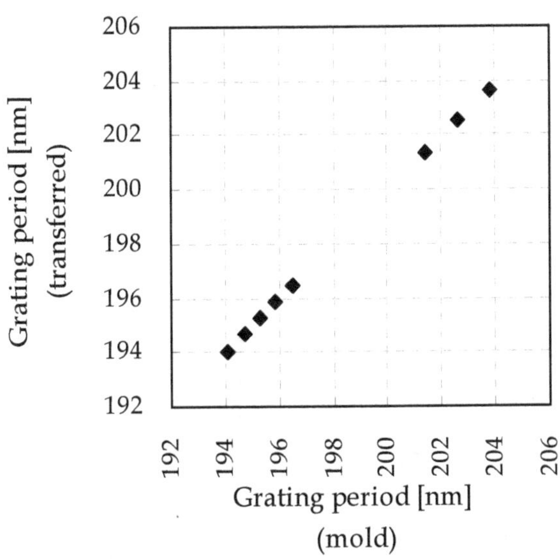

Fig. 10. Correlation between the designed (horizontal) and the measured (vertical) grating periods.

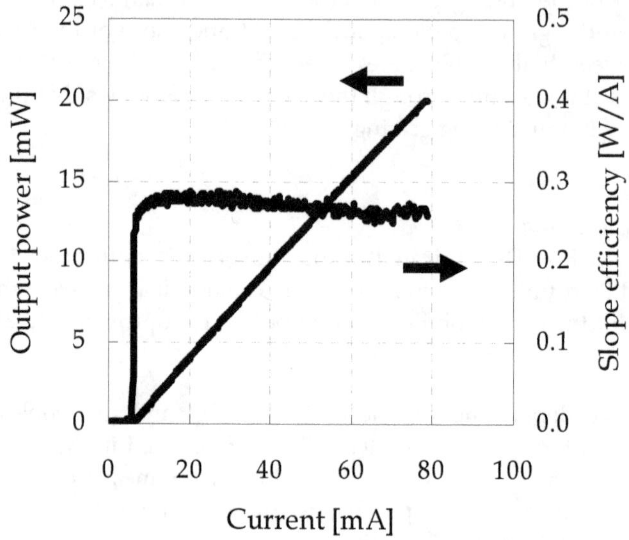

Fig. 11. Supplied current vs optical output and slope efficiency.

Figure 12 shows the oscillation spectrum of a phase-shifted DFB LD. The resolution of the wavelength is 0.02 nm. This demonstrates that the peak wavelength corresponds to the Bragg wavelength at the center of the stopband, indicating that the phase-shifted gratings function properly.

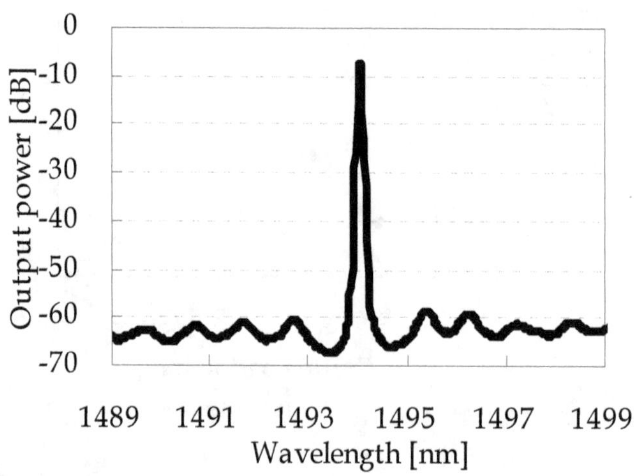

Fig. 12. Oscillation spectrum of the DFB LD.

Figure 13 shows the oscillation spectra of DFB LDs with the grating period of 195 nm and 202 nm, which are simultaneously fabricated on a wafer. This demonstrates that the peak wavelengths correspond to the respective Bragg wavelengths corresponding to each grating period.

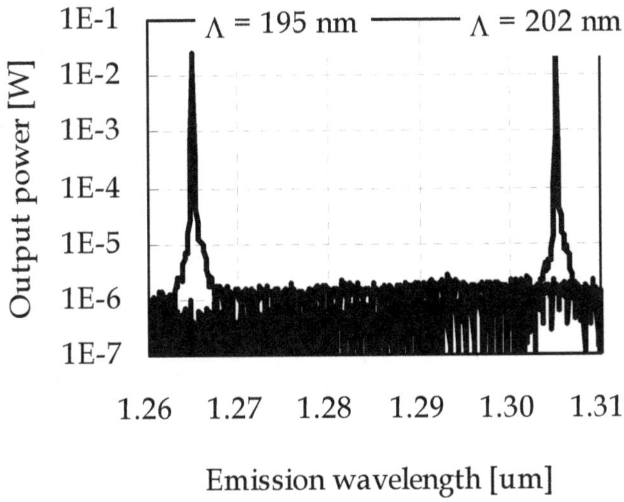

Fig. 13. Oscillation spectra of the DFB LDs with the grating period (Λ) of 195 nm and 202 nm.

We have compared the side-mode suppression ratio (SMSR) of phase-shifted LDs fabricated by NIL with those fabricated by EBL. SMSR is one of the parameters indicating the stability of the single-mode emission of DFB LDs. More than 300 LDs randomly sampled from respective 2-in. wafers have been evaluated. Histograms of SMSR for both types of LD are shown in Fig. 14, which demonstrate that they show comparable variations in SMSR.
These results indicate that DFB LDs fabricated here by utilizing NIL have comparable characteristics and their uniformities to ones by conventional EBL process.

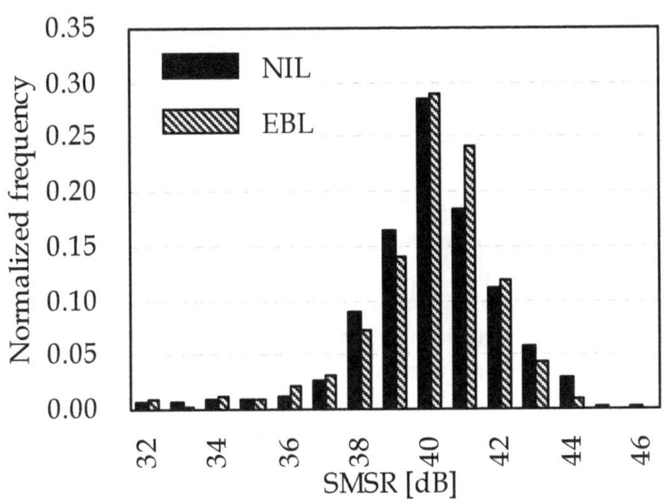

Fig. 14. Histograms of SMSR for phase-shifted LDs fabricated by NIL and by EBL. The standard deviations of SMSR are almost the same, 2.0 and 1.8, respectively.

3.5 Reliability

We have investigated long-term reliability of fabricated DFB LDs. Figure 15 shows the time-dependent change in the threshold current of phase-shifted LDs with the output power of 10 mW at the ambient temperature of 358 K. The number of samples is 18. The change in the threshold current after 5000 hours is less than ±1%, indicating that the phase-shifted DFB LDs fabricated in this study have high stability in lasing characteristics.

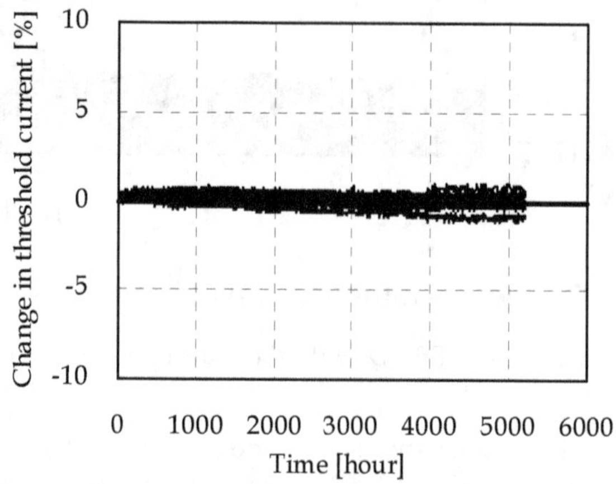

Fig. 15. Time-dependent change in the operation current of the DFB LDs with the output power of 10 mW at the ambient temperature of 358 K.

4. Conclusion

We have successfully demonstrated fabrication of phase-shifted DFB LDs by utilizing NIL process, which have comparable characteristics and their uniformities to ones fabricated by conventional EBL process. Fabricated DFB LDs have shown high stability of characteristics in long-term reliability test. We have also demonstrated the feasibility of the VARI-mold, which can be used for the fabrication of DFB LDs with various wavelengths, indicating that we can drastically reduce the cost of molds in our mass production phase in the near future. Considering the results above, we conclude that NIL is a promising candidate of the production technique for phase-shifted DFB LDs featuring low cost and high throughput.

NIL is expected to be used as a fabrication process for many applications. However, there are still some difficulties with its use as a mass-production process, for example, defects and low throughput in patterned media, defects and poor alignment accuracy in semiconductor lithography, and the necessary of increasing field size and throughput in displays. Although those difficulties may be common to the fabrication of DFB LDs, they would not be insurmountable problems. Even if defects in the imprinted pattern influence the yield of LDs, failed chips could be easily rejected because each dye is as small as approximately 300 μm. There is no need for a larger field size than a 3-in.-diameter circle, because the diameter of compound semiconductor substrates used for LDs is 3 inches or smaller. Regarding alignment accuracy, an error of up to approximately 5 μm is acceptable. Even if throughput is limited to less than one wafer per hour, NIL would still have a higher throughput than EBL.

As described above, NIL is an effective and promising method for fabricating phase-shifted DFB LDs and is expected to have the advantage of mass-production capability in the near future. We conclude that NIL has high potential for fabricating DFB LDs, and we expect that NIL will be used for fabricating various optical devices consisting of nanostructures.

5. Acknowledgment

We would like to appreciate Dr. Taniguchi and his laboratory members at Tokyo University of Science for cooperative works and fruitful discussion.

6. References

Bailey, T.; Choi, B. J.; Colburn, M.; Meissl, M.; Shaya, S.; Ekerdt, J. G.; Sreenivasan, S. V. & Willson, C. G. (2000). Step and flash imprint lithography: Template surface treatment and defect analysis, *J. Vac. Sci. Technol.,* B 18, pp. 3572-3577, ISSN 1071-1023

Chou, S. Y.; Krauss, P. R. & Renstrom, P. J. (1995). Imprint of sub-25 nm vias and trenches in polymers, *Appl. Phys. Lett.,* Vol. 67, pp. 3114-3116, ISSN 1077-3118

Fukuda, M. & Iwane, G. (1985). Degradation of active region in InGaAsP/InP buried heterostructure lasers, *J. Appl. Phys.*, Vol. 58, pp. 2932-2936, ISSN 0021-8979

Haisma, J.; Verheijen, M.; Heuvel, K. and Berg, J. (1996). Mold-assisted nanolithography: A process for reliable pattern replication, *J. Vac. Sci. Technol.,* B 14, pp. 4124-4128, ISSN 1071-1023

Kaden, C.; Griesinger, U.; Schweitzer, H.; Pilkuhn, M. H. & Stath N. (1992). Fabrication of nonconventional distributed feedback lasers with variable grating periods and phase shifts by electron beam lithography, *J. Vac. Sci. Technol.,* B 10, pp. 2970-2973, ISSN 1071-1023

Kinoshita, H.; Yamauchi, A. & Sawai, M. (1999). High-resolution resist etching for quartermicron lithography using O_2/N_2 supermagnetron plasma, *J. Vac. Sci. Technol.,* B 17, pp. 109-112, ISSN 1071-1023

Kure, T.; Kawakami, H.; Tachi, S. & Enami, H. (1991). Low-Temperature Etching for Deep-Submicron Trilayer Resist, *Jpn. J. Appl. Phys.,* 30, pp. 1562-1566, ISSN 1347-4065

Matsuoka, T.; Nagai, H.; Noguchi, Y.; Suzuki, Y. & Kawaguchi, Y. (1984). Effect of the grating phase at the cleaved facet on DFB laser properties, *Jpn. J. Appl. Phys.* 23, pp. L138-L140, ISSN 1347-4065

Miller, M.; Doyle, G.; Stacey, N.; Xu, F.; Sreenivasan, S. V.; Watts, M. & LaBrake, D. L. (2005). Fabrication of nanometer sized features on non-flat substrates using a nano-imprint lithography process, *Proc. SPIE 5751,* pp. 994-1002

Tsuji, Y.; Yanagisawa, M.; Yoshinaga, H.; Inoue, N. & Nomaguchi, T. (2011). Highly uniform fabrication of diffraction gratings for distributed feedback laser diodes by nanoimprint lithography, *Jpn. J. Appl. Phys.* 50, 06GK06, ISSN 1347-4065

Yanagisawa, M.; Tsuji, Y.; Yoshinaga, H.; Kono, N. & Hiratsuka, K. (2009). Application of Nanoimprint Lithography to Fabrication of Distributed Feedback Laser Diodes, *Jpn. J. Appl. Phys.*, Vol. 48, 06FH11, ISSN 1347-4065

Yanagisawa, M.; Tsuji, Y.; Yoshinaga, H.; Kono, N. & Hiratsuka, K. (2009). Evaluation of nanoimprint lithography as a fabrication process of phase-shifted diffraction gratings of distributed feedback laser diodes, *J. Vac. Sci. Technol.*, B 27, pp. 2776-2780, ISSN 1071-1023

Yanagisawa, M.; Tsuji, Y.; Yoshinaga, H.; Kouno, N. & Hiratsuka, K. (2011). Wafer-level fabrication of distributed feedback laser diodes by utilizing UV nanoimprint lithography, *Proc. SPIE*, 7970, pp. 797014, ISSN 0277-786X

9

High-Index Immersion Lithography

Keita Sakai
Canon Inc.
Japan

1. Introduction

The resolution capability of photolithography is given by Rayleigh's equation.

$$R=k_1 \cdot \lambda / NA\ , \quad (1)$$

where R is the half-pitch resolution of the image, k_1 is a constant that depends on the resist process and exposure method, λ is the exposure wavelength, and NA is the numerical aperture of the projection optic. According to Rayleigh's equation, there are three ways to enhance the resolution. The first is to shorten the exposure wavelength such as extreme ultraviolet lithography (EUVL). The second is to improve the k_1 value, for example, using the double-patterning technique. The third is to increase the numerical aperture (NA) as ArF immersion lithography. It has already realized the NA up to 1.35 and moreover can increase the NA using high-index materials. In this chapter, high-index immersion lithography with the NA over 1.45 is focused.

The NA is actually determined by the acceptance angle of the lens and the refractive index of the medium surrounding the lens and is given by eq. (2).

$$NA=n \cdot \sin\theta\ , \quad (2)$$

where n is the refractive index of the medium surrounding the lens and θ is the acceptance angle of the lens. Therefore, the NA can be enlarged by replacing the air (n=1) with a fluid (n>1) as a medium. In this immersion lithography, the film stack consists of a lens, a fluid layer, and a resist layer. The value of n·sinθ is invariant through the film stack because it obeys Snell's Law. Since sinθ is smaller than 1, the maximum NA (=n·sinθ) is limited by the layer with the minimum refractive index. For example, the refractive index of water is 1.44 at 193.4 nm, thus the NA over 1.44 cannot be established as shown in Fig. 1 (a) because of the total reflection. To realize the NA over 1.44, the water must be replaced with a fluid which has a higher refractive index than water (Fig. 1 (b)). In Fig. 1 (b), fused silica has the smallest refractive index in the film stack and it limits the maximum NA. For further increasing the NA, a high-index lens material must be used as a lens material.

As described above, high-index lens materials and high-index immersion fluids are indispensable to realize high-index immersion lithography. One of the candidates of a high-index lens material is lutetium aluminium garnet (LuAG), which has a refractive index of 2.14. Second-generation (G2) and third-generation (G3) fluids are saturated hydrocarbon fluids whose refractive indices are approximately 1.64 and 1.80, respectively.

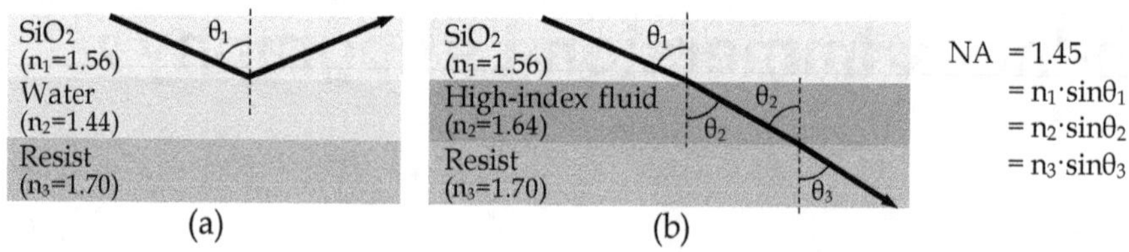

Fig. 1. Light propagation through the film stack at 1.45 NA. (a) The light does not reach the resist layer because of the total reflection at the lens-water boundary. (b) A high-index fluid enables the light to propagate into the resist.

High-index immersion lithography can be classified into three types by combining a lens material and an immersion fluid: fused silica and a G2 fluid (type 1), LuAG and a G2 fluid (type 2), and LuAG and a G3 fluid (type 3). With these types, the maximum NAs are estimated 1.45, 1.55, and 1.70, respectively.

Table 1 shows k_1 values for typical half-pitch and NA. The k_1 is calculated using Rayleigh's equation and needs to be at least 0.25 to resolve the patterns of the half-pitch for the theoretical limit. The resolution of 36 nm is achieved with 1.45 NA optic, which can be realized using only a G2 fluid. Those of 1.55 NA and 1.65 NA can achieve the resolutions of 34 and 32 nm, respectively.

Half-pitch	1.35 NA	1.45 NA	1.55 NA	1.65 NA	1.70 NA
39 nm	0.272	0.292	0.313	0.333	0.343
36 nm	0.251	0.270	0.289	0.307	0.316
34 nm	0.237	0.255	0.272	0.290	0.299
32 nm	0.223	0.240	0.257	0.273	0.281

Table 1. k_1 values for typical half-pitch and NA.

Although an exposure tool of 1.45 NA does not need new materials except for a G2 fluid, customers have little interest in the tool because of its modest gain in resolution. On the other hand, a tool of over 1.65 NA seems attractive for resolution enhancement. However, it would be difficult to realize G3 fluids immediately because no materials meet the requirements. In such a situation, it is a realistic way to develop a tool of 1.55 NA using LuAG and a G2 fluid.

In the next subchapter, projection optics with LuAG are explained. Development of LuAG is a hard work because the specifications of lithography-grade lens materials are extremely stringent. According to the history of LuAG development by Schott Lithotec (Parthier et al., 2008), great progress was achieved in the absorbance but it does not reach the target. The key issue for the optical design with LuAG is a correction of intrinsic birefringence (IBR). An effective method has been developed for IBR correction and it reduces the wave-front aberration to a practical level.

In the third subchapter, an immersion system using a G2 fluid is described. It was demonstrated that fluid absorbance can be kept low enough through an in-line purification unit and an oxygen removal unit. Although lens contamination is an important issue in a G2 fluid system, it was found that contamination can be suppressed by addition of a small amount of water into a G2 fluid. With this water-addition and in-line purification, the necessity for lens cleaning decreases from three times per day to once a week. Fluid

confinement is also a challenge for a G2 fluid system because residual fluid is easy to remain on a wafer. Some issues arising with residual fluid, such as fluid darkening due to reentry of oxygen-rich residual fluid, were solved. By accepting residual fluid on a wafer, the scanning speed and the throughput can be raised.
Finally, the remaining challenges to realize high-index immersion lithography are discussed.

2. Projection optics with LuAG

Key parameters for high-index lens materials are refractive index, absorbance, and intrinsic birefringence (IBR). High-index lens materials must have a refractive index to permit NA scaling sufficient to justify the development cost. The absorbance must be sufficiently low to avoid the image degradation by thermal aberrations. The intrinsic birefringence must be minimal to allow a correction to avoid introducing unacceptable aberrations in the final aerial image.
The National Institute of Standards and Technology (NIST) has searched for high-index lens materials that meet the above requirements such as barium lithium fluoride ($BaLiF_3$) and LuAG (Burnett et al., 2006). $BaLiF_3$ developed by Tokuyama is available in various sizes with low absorbance (Nawata et al., 2007). However, the refractive index of $BaLiF_3$ is not high enough to achieve a sufficient NA for the enhancement of the resolution. Only LuAG remains as a candidate for a high-index lens material. LuAG has the intrinsic birefringence over 30 nm/cm and still has a high absorbance caused by impurities. Thus, the status of LuAG should be paid attention to and an IBR correction method should be developed.

2.1 Status of LuAG

Lithography-grade LuAG has been aggressively developed by Schott Lithotec since 2005. The absorbance, which is the biggest issue for LuAG, was largely improved down to 0.035 /cm by purifying the raw material and optimising the crystal growing process as shown in Fig. 2. Since it was found that the intrinsic absorbance of LuAG is 0.00118 /cm (Letz et al., 2010), the absorbance will be less than 0.005 /cm by further reduction of impurities.

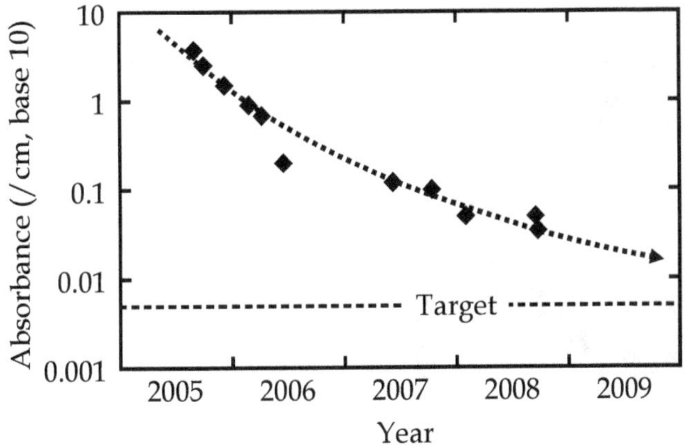

Fig. 2. History of LuAG development by Schott Lithotec. The absorbance of LuAG has been improved down to 0.035 /cm but it has not reached the target (0.005 /cm).

Table 2 shows the target and the status of each requirement. The stress birefringence (SBR) and the homogeneity seem to have gotten closer to the targets. However, these were achieved with a small crystal of yttrium aluminium garnet (YAG). Therefore, it is necessary to confirm the SBR and the homogeneity of large-sized LuAG.
The remaining challenges, such as the absorbance, will be solved if the development is accelerated by strong supports from the industry.

	Target	Status
Absorbance (/cm, base 10)	< 0.005	0.035
Stress birefringence (RMS, nm/cm)	< 0.5	0.73 (40 mmϕ, YAG)
Index homogeneity (-Z36, ppm)	< 0.05	0.03 (45 mmϕ, YAG)

Table 2. Target and status of LuAG.

2.2 Intrinsic birefringence correction

Since the absorbance of LuAG is higher than that of fused silica, it is difficult to use many LuAG lenses in a projection optic. From the viewpoint of optical design, applying LuAG to the final lens is most effective. In the final lens through which light converging at an image plane passes, the range of incident angles to the optical axis is wide. Assuming that the bottom surface of the final lens, which touches an immersion fluid, is flat, the maximum ray angles in LuAG are 46.4 and 52.6° for the NAs of 1.55 and 1.70, respectively.
The IBR distribution depends on the direction of a crystal. The distribution of a (111)-oriented crystal has three-fold rotational symmetry. When the (111)-oriented crystal is used, the angle at the maximum IBR is 35.3° from the (111) axis. On the other hand, a (100)-oriented crystal has four-fold rotational symmetry, and the angle at the maximum IBR is 45° from the (100) axis. In short, the maximum ray angle exceeds both angles of the maximum IBR even in the case of NA 1.55.

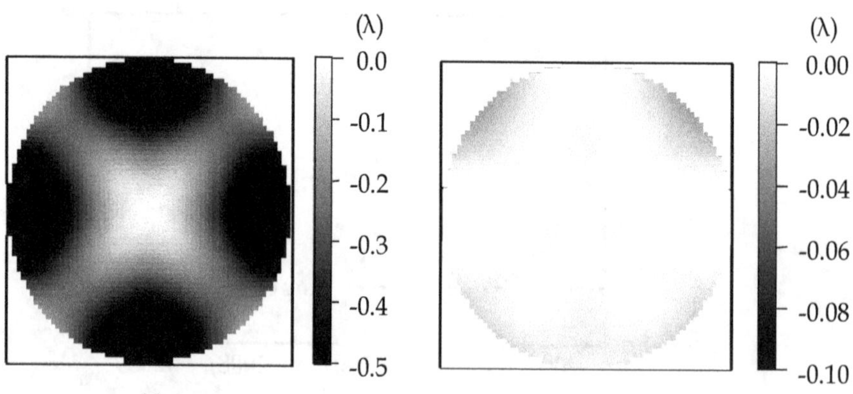

Fig. 3. Difference between radial and tangential polarized wave front for a projection optic of 1.55 NA; (a) without IBR correction, (b) with an original correction method.

Figure 3 shows wave front aberration maps for a projection optic of 1.55 NA. These show the difference between the radial and the tangential polarized wave fronts. The wave front difference without the correction has a large rms value of 217 mλ and a clear 4θ component. On the other hand, the original correction method reduces the rms difference to 4.2 mλ. This value is within the practical level.

In the case of 1.70 NA, the wave front without the correction has a large rms difference of 680 mλ and the rms difference after the correction is 21.2 mλ. Although this value is still large compared with the case of 1.55 NA, it is close to the practical level. Therefore, it can be concluded that the IBR correction is feasible even at 1.70 NA.

3. Immersion system using a G2 fluid

High-index immersion fluids are categorized as second-generation (G2) or third-generation (G3) fluids. G2 fluids have refractive indices of approximately 1.64 and enable an NA to increase up to 1.55 with LuAG as a final lens material. G3 fluids are targeted to have refractive indices over 1.80 for achieving an NA over 1.65 with LuAG.

DuPont, JSR, and Mitsui Chemicals have reported both G2 and G3 fluids (French et al, 2007, Furukawa et al., 2007, Kagayama et al., 2007). Their G2 fluids are saturated hydrocarbon and have sufficient performance as an immersion fluid. They also developed G3 fluids as some extension of organic G2 fluid materials. This type of G3 fluids has not yet met the requirements of refractive index or absorbance. The other types of G3 fluids, which include nanoparticles, are being developed, but they are still within the research phase (Zimmerman et al., 2008). In the above situation, an immersion system using a G2 fluid has been preferentially developed (Sakai et al., 2008).

To achieve the target cost of $1/layer for a G2 fluid, the recycling of a G2 fluid is required. G2 fluids are easily degraded by laser irradiation and the dissolution of oxygen from the atmosphere. Therefore, a fluid circulation system is desired to keep the fluid absorbance sufficiently low using purification and oxygen removal functions. In addition, lens contamination, bubble, and residual fluid on a wafer are also important issues. An immersion system composed of a fluid circulation system, an immersion nozzle and a cleaning unit should be established to accommodate a G2 fluid.

3.1 Fluid degradation

The absorbance of water does not change largely with dissolved oxygen, and immersion water is not used repeatedly as a G2 fluid. These are the reasons the degradation of water is not a concern as an immersion fluid.

On the other hand, the absorbance of a G2 fluid is easily increased by dissolved oxygen or laser irradiation. Moreover, the dn/dT of a G2 fluid is 5~6 times larger than that of water (French et al., 2006, Santillan et al., 2006, Furukawa et al., 2007). This means that a temperature change causes a larger change in the refractive index and results in larger thermal aberration (Sekine et al., 2007). To suppress a temperature rise with the absorption of exposure light, the fluid absorbance has to remain sufficiently low.

3.1.1 Dissolved oxygen

Although the absorbance of a G2 fluid is lower than that of water (French et al., 2005, Wang et al., 2006), the degradation of a G2 fluid with dissolved oxygen is larger than that of water. The first reason for this large degradation is the high oxygen solubility. It has been reported that the solubility of oxygen in G2 fluids are 55~70 ppm in air atmosphere (Furukawa et al., 2006). These values are 7~9 times higher than that in water. According to the research at Columbia University (Gejo et al., 2007), dissolved oxygen forms a charge-transfer complex with cycloalkane of a G2 fluid and the complex has strong absorption in the ultraviolet wavelength range. This is the second reason for the large degradation induced by dissolved oxygen.

The induced absorbance against dissolved oxygen in a G2 fluid has been experimentally obtained. Figure 4 shows the induced absorbance of HIL-203 (JSR) compared with water. It was confirmed that the oxygen-rich HIL-203 exhibits strong absorption. Therefore, a removal function of oxygen is necessary to keep the fluid absorbance low in a circulation system.

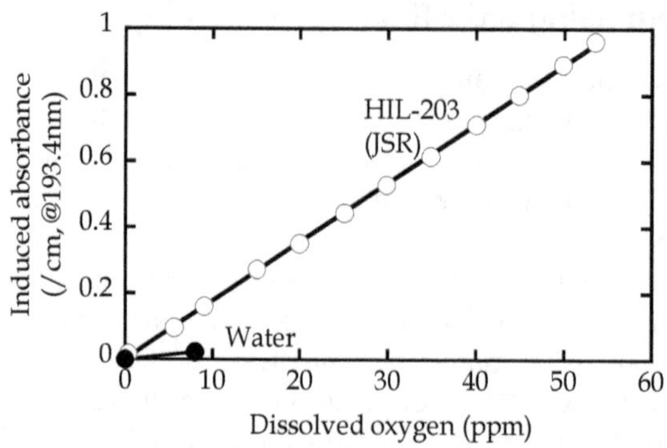

Fig. 4. Experimental results of induced absorbance with dissolved oxygen. Open circles show the induced absorbance of HIL-203 and filled circles show that of water.

An immersion nozzle supplies a G2 fluid under a final lens and sucks the fluid with the neighbouring gas. If the atmosphere around a wafer is air, the oxygen concentration in the recovered fluid increases with the sucked air. Under such a condition, two types of oxygen removal functions were evaluated using an experimental system as shown in Fig. 5. The fluid used in this experiment was HIL-203 (JSR) and the fluid flow rate was 400 mL/min.

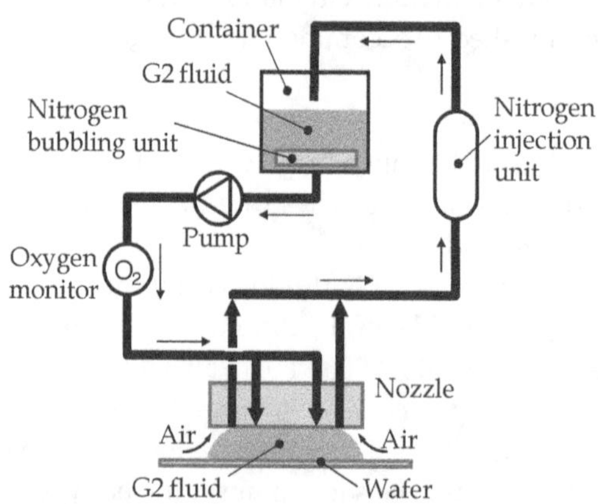

Fig. 5. Schematic view of an experimental system to investigate two types of oxygen removal functions. One is a nitrogen bubbling unit in the fluid container, and the other is a nitrogen injection unit attached to the fluid line. Oxygen concentration in the fluid is measured at the position just before the immersion nozzle.

Figure 6 shows dissolved oxygen using two types of oxygen removal functions. In the case with the nitrogen bubbling unit alone, the oxygen concentration was 4.5 ppm. It corresponds to the absorption of 1.83 %/mm, which can not be permitted. By using the nitrogen bubbling and the nitrogen injection units simultaneously, the oxygen concentration was reduced to 0.2 ppm. Further reduction was achieved increasing the nitrogen flow rate into the injection unit from 5 to 10 L/min. The oxygen concentration of 0.1 ppm is the permissible level because it corresponds to only 0.04 %/mm degradation. The nitrogen injection unit removes the dissolved oxygen from the oxygen-rich fluid just after recovery and prevents the oxygen-rich fluid from going back to the container. That is why the injection unit is efficient for the reduction of dissolved oxygen.

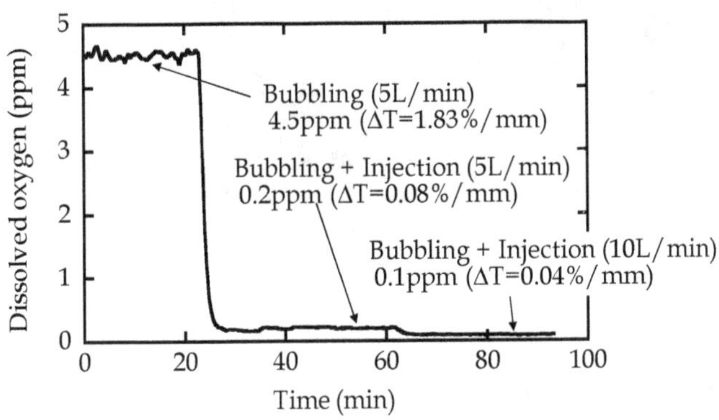

Fig. 6. Experimental results of dissolved oxygen using oxygen removal functions. At first, the dissolved oxygen was not sufficiently low with the nitrogen bubbling unit alone. Then, the large reduction of dissolved oxygen was achieved with the simultaneous use of the nitrogen bubbling and the nitrogen injection units.

3.1.2 Laser irradiation

The absorbance of a G2 fluid increases with the photodecomposed materials of a G2 fluid induced by ArF laser irradiation. To achieve the target cost of $1/layer, it is necessary to remove the photodecomposed materials using an in-line purification unit.

Figure 7 shows the instruments used for a laser irradiation test. ArF excimer laser G41A3 (Gigaphoton), which can irradiate with the rep. rate of up to 3 kHz, is used for irradiation. The irradiation chamber has two beam lines and is purged with nitrogen gas to keep the oxygen concentration less than 1 ppm. The fluid circulation system with the oxygen removal functions can maintain dissolved oxygen at 0.1 ppm and below. The fluid absorbance monitor contains four flow-through cells which can evaluate a number of fluids simultaneously.

The experimental conditions are shown in Table 3. Fluid suppliers have developed in-line purification units for their G2 fluid. Therefore, fluid degradation was evaluated with the combination of a fluid and a purification unit made by the same supplier, such as HIF-A and Unit-A.

The experimental results are shown as the induced absorbance against fluid dose. The fluid dose is defined as an incident dose modified by the volume dilution factor as shown in eq. (3). It has been suggested as the appropriate dose metrics (Liberman et al., 2007).

$$D_{fluid}=I \cdot N \cdot V_{irr}/V_{total}, \quad (3)$$

where D_{fluid} is the fluid dose, I is the fluence per pulse, N is the total pulse count, and V_{irr} and V_{total} are the irradiated volume and the total fluid reservoir volume, respectively.

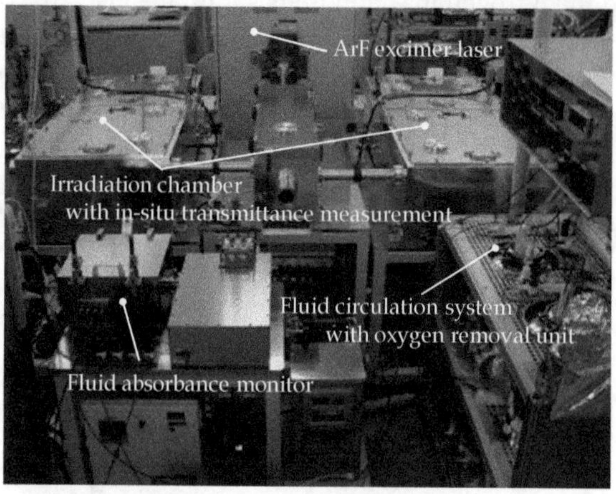

Fig. 7. Instruments for a laser irradiation test. The fluid circulation system consists of a fluid container, a pump, and oxygen removal units.

Immersion fluid / Purification unit	HIF-A / Unit-A HIF-B / Unit-B HIF-C / Unit-C
Fluence	0.5 mJ/(cm^2 · pulse)
Rep. rate	2000 Hz
Irradiation area	8.0 mmϕ
Fluid gap	1.0 mm
Fluid flow rate	100 ~ 350 mL/min

Table 3. Experimental conditions for evaluation of in-line purification units.

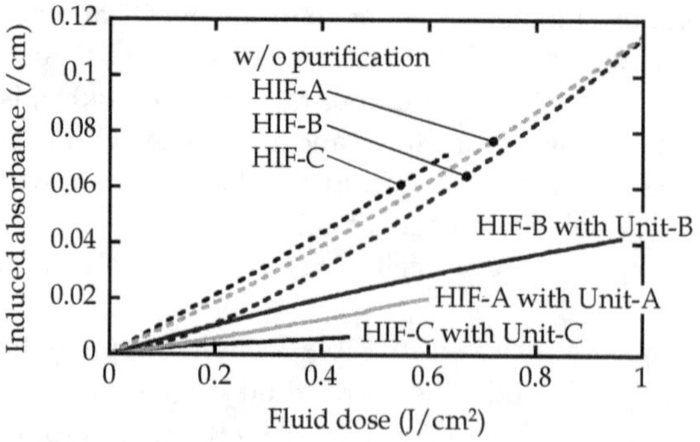

Fig. 8. Induced absorbance of a G2 fluid against the fluid dose. The dashed line shows the induced absorbance without a purification unit. The solid line shows the result with a purification unit. The fluid dose is calculated using eq. (3) and the conditions as shown in Table 3.

Figure 8 shows the induced absorbance of each G2 fluid with and without an in-line purification unit. For example, the degradation rate of HIF-C with Unit-C is one-seventh-fold smaller than that without Unit-C. Using a purification unit, fluid lifetime will be up to 1 week before the fluid absorbance will reach an unacceptable level. Therefore, it can be concluded that recycling of a G2 fluid is feasible using an in-line purification unit.

3.2 Lens contamination

It has been reported that the exposed surface of a final lens is polluted by photodecomposed materials of a G2 fluid (French et al., 2007). This lens contamination diminishes the uniformity of exposure dose on a wafer and would be a source of particles. Thus, the suppression of lens contamination is a serious issue for an immersion system using a G2 fluid.

The suppression of lens contamination was examined by three procedures. The first is to select an appropriate fluid, which has a lower deposition rate. The second is to use an in-line purification, which can remove photodecomposed materials. The third is water-addition into a G2 fluid, which is an original suppression method.

Figure 9 shows a schematic diagram of the experimental setup. The cell absorbance, which consists of the absorbance of windows and a fluid of 1 mm path length in the cell, was measured using two power meters. In addition, the fluid absorbance was measured by a fluid absorbance monitor. Then, the absorbance of windows was calculated by subtracting the fluid absorbance from the cell absorbance as shown in eq. (4).

$$A_{window} = A_{cell} - A_{fluid}, \qquad (4)$$

where A_{window} is the absorbance of windows, A_{cell} is the cell absorbance, and A_{fluid} is the fluid absorbance through 1 mm path length.

The laser fluence was 1.0 mJ/(cm^2·pulse), which is almost the same as in an actual exposure tool. The dose estimation on an actual tool is approximately 60 kJ/(cm^2·day). Although the effective cleaning method was proposed (Liberman et al., 2007), the necessity for lens cleaning should be less than once a week for minimizing tool downtime.

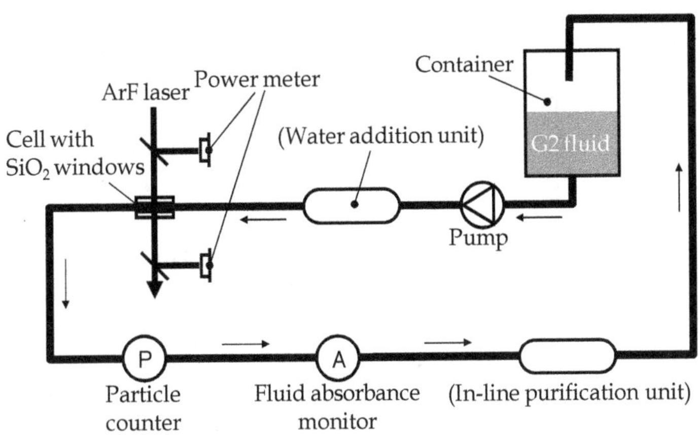

Fig. 9. Schematic diagram of the experimental setup. G2 fluid flows through the 1 mm gap space between the SiO_2 windows of the irradiation cell. The particle counter can detect particles larger than 60 nm in diameter.

3.2.1 Appropriate fluid

Induced absorbance of windows, which correspond to a final lens of a projection optic, was evaluated by using various G2 fluids. Figure 10 shows induced absorbance of windows against exposure dose. HIF-C exhibits a higher deposition rate than HIF-A and D. Therefore, HIF-A or D should be selected from the point of view of lens contamination. However, lens contamination occurs in a few hours on an actual exposure tool, even if using HIF-A or D. It is necessary to introduce some methods for suppressing contamination.

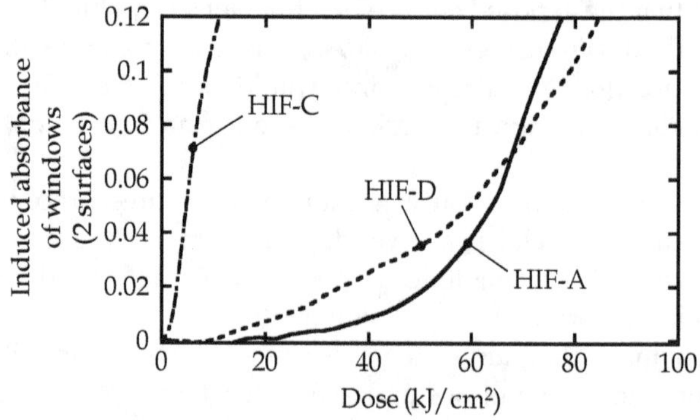

Fig. 10. Induced absorbance of windows using various G2 fluids. The deposition rates of HIF-A (solid line) and HIF-D (dashed line) are lower than that of HIF-C (dashed-dotted line).

3.2.2 In-line purification

It is expected that an in-line purification unit can suppress lens contamination because it can reduce photodecomposed materials. The experiments with several G2 fluids and in-line purification units were carried out using the experimental setup as shown in Fig. 9.

Figure 11 shows experimental results of window absorbance with and without a purification unit. It was confirmed that each purification unit can improve lens contamination. But the performance of purification units are not enough and a final lens needs to be cleaned more than once a day under the practical use conditions. Therefore, the additional method is desired to extend the cleaning interval.

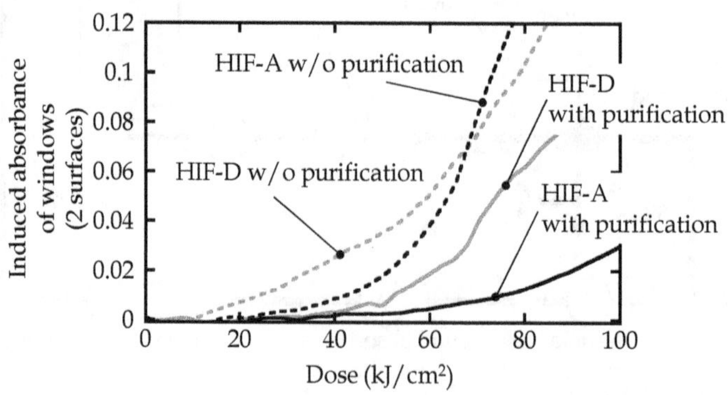

Fig. 11. Induced absorbance of windows with and without a purification unit. The deposition rate with a purification unit (solid line) is lower than that without a purification unit (dashed line).

3.2.3 Water-addition

It was found that lens contamination can be suppressed by addition of a small amount of water into a G2 fluid (Sakai et al., 2008). Since water in a G2 fluid is removed by an oxygen removal unit or an in-line purification unit, a water-addition unit is necessary for a fluid circulation system to keep water in a G2 fluid. Figure 12 shows a schematic diagram of a water-addition unit. Using the water-addition unit, a sufficient amount of water can be solved into a G2 fluid.

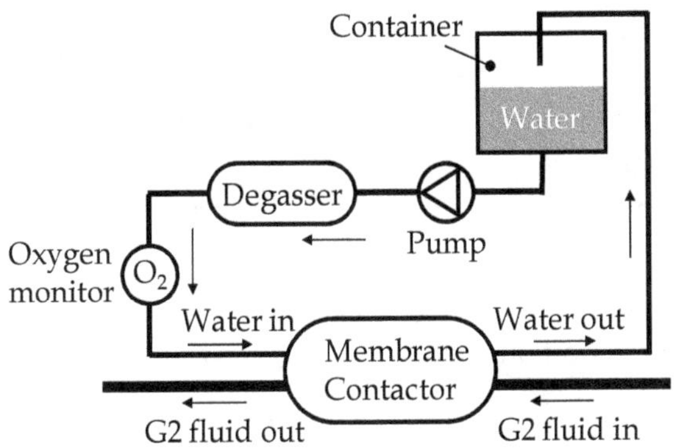

Fig. 12. Schematic diagram of a water-addition unit. The unit consists of a membrane contactor and a circulation unit of degassed water. Degassed water contacts a G2 fluid through the porous membrane with a large area.

Figure 13, 14, and 15 shows the experimental results using HIL-203 (JSR) as a G2 fluid. In the case without water-addition, the absorbance of windows increased gradually (Fig. 13) and many streak contaminations were observed on the window (Fig. 14 (a)). On the other hand, the window was not contaminated in the case with water-addition (Fig. 13, Fig. 14 (b)). A strong impact was also obtained in terms of particle. As shown in Fig. 15, particles increased with lens contamination in the case without water-addition. Some portions of lens contamination would be removed by laser irradiation and flow into the fluid as particles. Using the water-addition unit, there were no particles in the fluid made from lens contamination. It was also confirmed that the water-addition unit does not generate any particles.

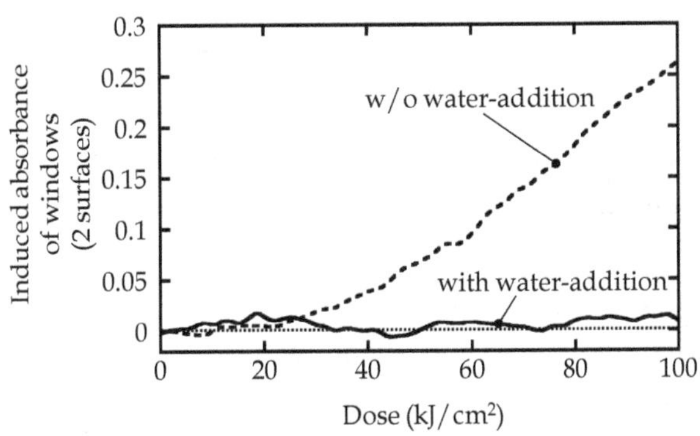

Fig. 13. Induced absorbance of windows against exposure dose. The absorbance of windows increased with the exposure dose in the case without water-addition (dashed line). In contrast, the water-addition unit suppressed contamination (solid line).

As described above, water-addition is the superior method for suppressing lens contamination. In addition, it seems to be a method without any disadvantages. For example, the refractive index of a G2 fluid does not change substantially with water-addition because the solubility of water into a G2 fluid is very low (several tens of ppm).

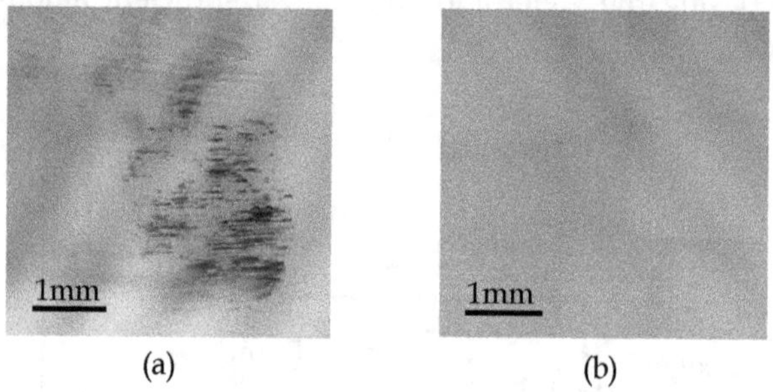

Fig. 14. Micrographs of window surfaces; (a) without water-addition after 100 kJ/cm^2 exposure, (b) with water-addition after 100 kJ/cm^2 exposure.

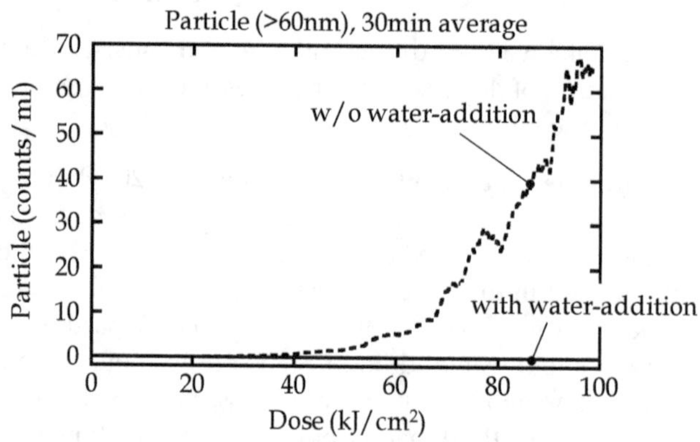

Fig. 15. Particles in a fluid against exposure dose. The number of particles increased with the exposure dose without water-addition (dashed line). In contrast, the water-addition unit suppressed particles induced by lens contamination (solid line).

3.2.4 Long-term test

A long-term test for lens contamination was done with an in-line purification unit and a water-addition unit simultaneously. IF132 (DuPont) and ARP (DuPont) were used as a G2 fluid and an in-line purification unit, respectively.

As shown in Fig. 16, the absorbance of windows started to increase at 20 kJ/cm^2 without any suppression units. The dose estimation on an actual exposure tool is approximately 60 kJ/(cm^2·day). Therefore, lens cleaning must be done three times per day in this case. Even if with ARP, the absorbance started to rise at 60 kJ/cm^2 and a lens needs to be cleaned once a day. The cleaning interval is not allowed in the case with ARP alone. On the other hand, lens contamination was suppressed until 420 kJ/cm^2 using ARP and a water-addition unit simultaneously. The necessity for lens cleaning is once a week and it is practical. Figure 17 shows micrographs of window surfaces after the experiments. The micrographs also exhibit that the simultaneous use of ARP and a water-addition unit has a sufficient performance. It is concluded that lens contamination is not a critical issue anymore.

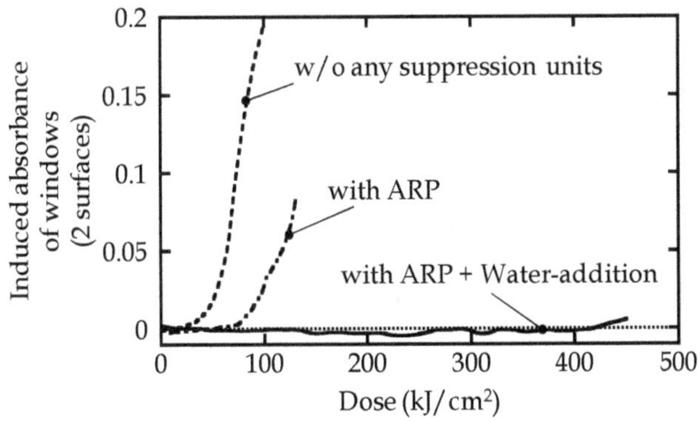

Fig. 16. Induced absorbance of windows against exposure dose. Simultaneous use of ARP and a water-addition unit suppressed lens contamination until 420 kJ/cm^2.

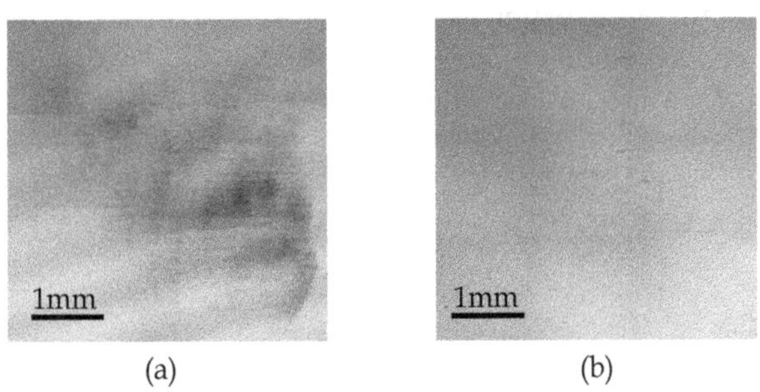

Fig. 17. Micrographs of window surfaces; (a) with ARP after 130 kJ/cm^2 exposure, (b) with ARP and a water-addition unit after 450 kJ/cm^2 exposure.

3.3 Bubble

A bubble is one of the origins of immersion defect. A bubble is entrapped on an air-liquid boundary and is translated with a wafer motion. If the bubble lifetime is sufficiently short, the bubble will be eliminated before reaching an exposure area. Therefore, the reduction of bubble lifetime is important.

The gas in a bubble diffuses from the surface of the bubble into the fluid. This diffusion causes the bubble to shrink with time and to eventually vanish. The bubble lifetime is approximated using eq. (5) (Honda et al., 2004).

$$\tau = \frac{\rho d^2}{8D(C_S - C_\infty)}, \quad (5)$$

where τ is the bubble lifetime, ρ is the density of the gas inside the bubble, d is the diameter of the bubble, D is the diffusion coefficient, C_S is the saturated concentration, and C_∞ is the concentration of the dissolved gas at a position far from the bubble. Figure 18 shows a schematic diagram of the experimental setup for the bubble lifetime measurement. By increasing the pressure of the fluid from 1 to 2 atm, the bubble vanishes at a certain lifetime. It corresponds to the lifetime in the half-degassed fluid. Figure 19 shows the bubble lifetimes in water and two types of G2 fluids. The experimental results in water correlated with the theory fairly well. The bubble lifetime in the G2 fluid was shorter than that in water. Hence, the concept of the bubble elimination method for a water immersion system, i.e., using degassed fluid, is also feasible in an immersion system using a G2 fluid.

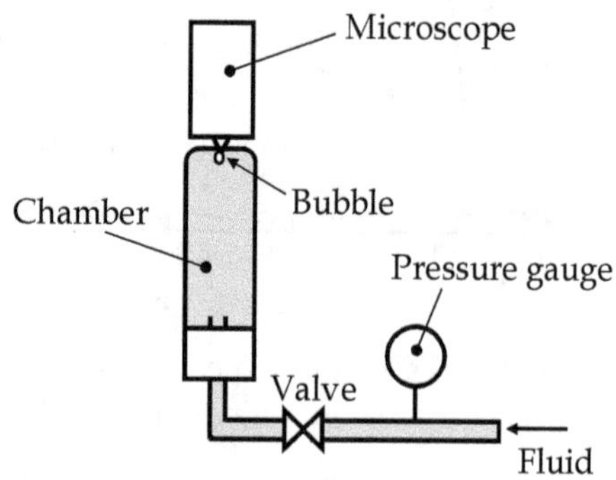

Fig. 18. Schematic diagram of the experiment setup for the bubble lifetime measurement. The size of a bubble is observed using the microscope.

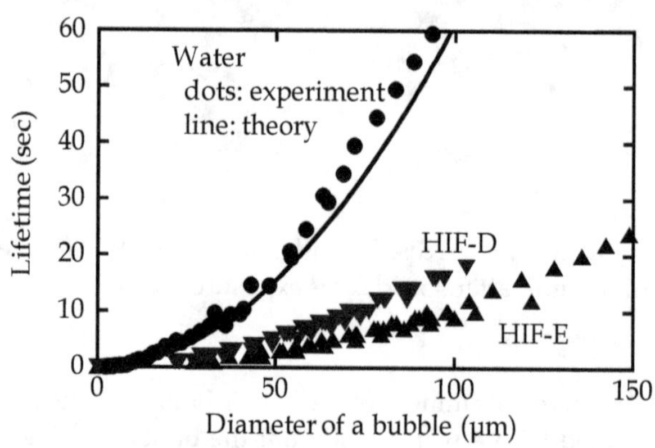

Fig. 19. Bubble lifetimes in various fluids. The lifetimes in G2 fluids (triangles) are shorter than that in water (circles).

3.4 Residual fluid on a wafer

The surface tension and viscosity of water are 72 mN/m and 1 mPa·s, respectively. This extremely high surface tension and low viscosity enable the immersion nozzle to keep water under a final lens (Kubo et al, 2007). On the other hand, it has been reported that a G2 fluid is easy to remain on a wafer (Sewell et al, 2007). After the verification that residual fluid is hard to avoid, some issues arising with residual fluid are discussed. The experimental results for the issues, such as the interaction between a G2 fluid and a resist, metal contamination in a G2 fluid, and fluid darkening due to oxygen-rich residual fluid, are explained.

3.4.1 Scanning test

The surface tension and viscosity of G2 fluids are approximately 30 mN/m and 2~4 mPa·s, respectively. The film pulling velocity (υ_{fp}), which is the minimum scanning velocity to leave fluid behind the nozzle, is proportional to the surface tension (γ) and the inverse of the viscosity (μ) as shown in eq. (6) (Shedd et al., 2006). If a wafer has the same static receding contact angle (SRCA, $\theta_{s,r}$) for water and a G2 fluid, the film pulling velocity for a G2 fluid is approximately one-sixth-fold lower than that for water. Although the higher SRCA can increase the film pulling velocity, the target SRCA over 120° is too high to develop an appropriate topcoat material (Sanders et al., 2008). Therefore, it is difficult to achieve a sufficient scanning speed without residual fluid in an immersion system using a G2 fluid.

$$\upsilon_{fp} = C\frac{\gamma}{\mu}\theta_{s,r}^{3}, \tag{6}$$

where C is the empirical constant.

Figure 20 shows the results of the fundamental scanning test using a prototype topcoat for G2 fluids. HIL-203 (JSR) was used as a G2 fluid for the experiment. Even with the scanning speed of 30 mm/sec, the dynamic receding contact angle became 0° and fluid was left behind the nozzle. As a result, it was confirmed that residual fluid should be allowed to realize a sufficient scanning speed. Since the contact time of a G2 fluid and a resist becomes longer in such a system, the interaction between a G2 fluid and a resist should be discussed.

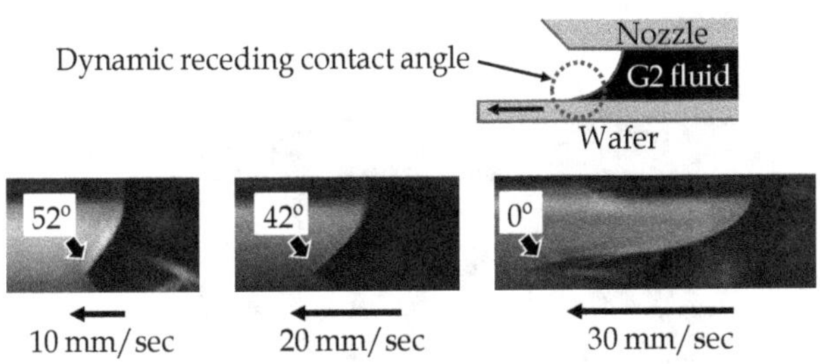

Fig. 20. Side views of the fundamental scanning test. The dynamic receding contact angle on a wafer became 0° with the scanning speed of 30 mm/sec.

3.4.2 Soaking test

To investigate the interaction between a G2 fluid and a resist, the defect inspection of soaked wafers is effective. Figure 21 shows the procedure of a soaking test. Between the exposure and post-exposure bake (PEB), a wafer was soaked in a G2 fluid for 60 seconds and the fluid was removed by spin drying. The defect inspection was carried out using KLA2371 (KLA-Tencor). The defects induced by post-exposure soaking were evaluated with the above procedure.

Figure 22 shows the results of the defect inspection. A large number of defects were found on the resist after soaking in HIF-E. In a microscopic review, there were many stains as shown in Fig. 22. As a result, it was confirmed that it is necessary to use an inert resist for a G2 fluid. On the other hand, it is not apparent that the wafer soaked in HIF-A or HIF-D had more defects than the reference wafer. There were some variations in the number of defects because the experiment was carried out in off-line process. Defect studies should be done in more clean circumstances.

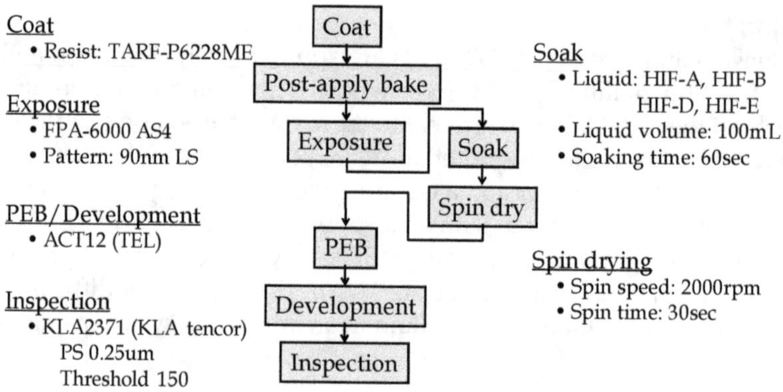

Fig. 21. Experimental procedure of a post-exposure soaking test to investigate the interaction between a G2 fluid and a resist.

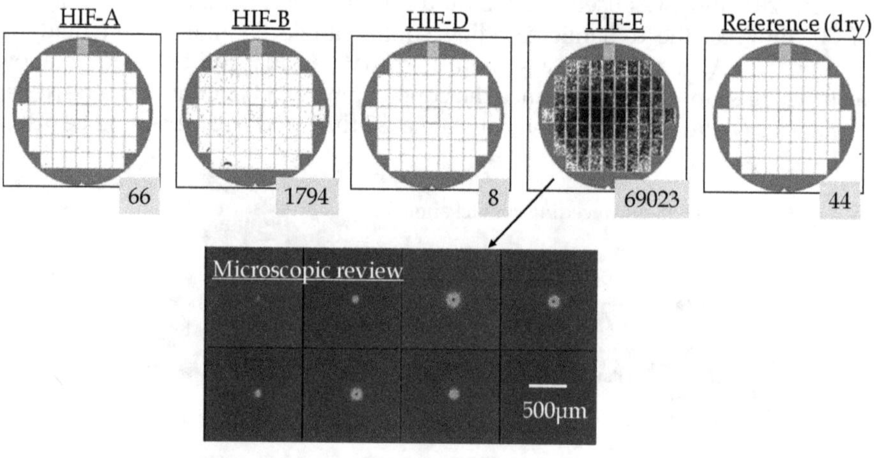

Fig. 22. Results of defect inspection. The number of defects after soaking in HIF-A, B, D, and E are 66, 1794, 8, and 69023, respectively.

3.4.3 Metal contamination

If the fluid contains a lot of metals, a wafer would be polluted by metals in the residual fluid. Thus, the objective of the following experiment is to confirm that G2 fluids are sufficiently clean. The requirement is less than 10^{10} atoms/cm^2 on a silicon wafer.

The conditions for the experiment are shown in Table 4. At first, a silicon wafer was dipped into a fluid for 5 minutes. Then, the wafer was pulled up and was dried by air blowing. Then, the metal concentrations on the wafer were evaluated by vapor phase decomposition inductively coupled plasma mass spectrometry (VPD/ICP-MS). The detection limit of the spectrometry is 0.03×10^{10} atoms/cm^2. The experiment was done twice for each fluid.

Table 5 shows the experimental results. The metal concentrations of the reference mean those on a wafer without dipping into any fluids. In the case of HIF-A, the concentrations of some metals are more than 10^{10} atoms/cm^2 (black-out cells in Table 5). On the other hand, HIF-B and D are sufficiently clean. As a result, the residual fluids of HIF-B and D are acceptable from the viewpoint of metal contamination.

Dipping	Immersion fluid	HIF-A, HIF-B, HIF-D
	Volume of fluid	200 mL
	Dipping time	5 minutes
	Wafer	Bare Si (200 mmϕ)
Analysis	Method	VPD/ICP-MS
	Evaluation area	180 mmϕ
	Detection limit	0.03×10^{10} atoms/cm^2

Table 4. Experimental conditions for metal contamination on a wafer dipped into a G2 fluid.

Fluid		Metal concentration ($\times 10^{10}$atoms/cm^2)													
		Li	Na	Mg	Al	K	Ca	Ti	Cr	Mn	Fe	Ni	Cu	Zn	Pb
Reference	1	<0.03	0.03	<0.03	0.03	<0.03	<0.03	0.43	<0.03	<0.03	<0.03	<0.03	<0.03	<0.03	<0.03
	2	0.06	<0.03	<0.03	0.05	<0.03	<0.03	0.32	<0.03	<0.03	<0.03	<0.03	0.03	<0.03	<0.03
HIF-A	1	0.15	0.30	0.09	1.30	0.28	0.28	0.71	0.29	0.09	1.57	0.87	0.05	0.13	<0.03
	2	0.11	0.92	0.05	0.85	0.97	0.36	0.45	0.13	0.19	0.82	0.40	0.04	0.10	<0.03
HIF-B	1	<0.03	0.45	0.05	0.23	0.13	0.38	0.48	<0.03	<0.03	0.10	<0.03	0.04	0.20	<0.03
	2	<0.03	0.48	0.07	0.29	0.14	0.47	0.58	<0.03	<0.03	0.06	<0.03	0.03	0.40	<0.03
HIF-D	1	<0.03	0.36	0.12	0.57	0.10	0.21	0.51	<0.03	<0.03	0.09	<0.03	0.07	0.06	<0.03
	2	<0.03	0.26	0.09	0.47	0.13	0.17	0.66	<0.03	<0.03	0.10	<0.03	0.04	0.03	<0.03

Table 5. Metal concentrations on a wafer evaluated by VPD/ICP-MS.

3.4.4 Dissolved oxygen

Oxygen in the atmosphere diffuses into the residual fluid on a wafer. When the oxygen-rich residual fluid goes back under the lens, it decreases the transparency of the fluid under the lens. That is the reason why dissolved oxygen is one of the issues arising with a residual fluid. Oxygen concentration in the fluid under the lens was evaluated while a wafer was moving along the sequence as shown in Fig. 23. HIL-203 (JSR) was used as a G2 fluid. A prototype immersion nozzle designed for a G2 fluid was used for the experiment. The processing time for one wafer was 30 seconds and it consisted of a scanning time and a waiting time at the edge of the stage. The fluid under the lens was sucked through the hole opened in the lens and the oxygen concentration in the sucked fluid was measured. The absorbance was calculated by using the relation as shown in Fig. 4.

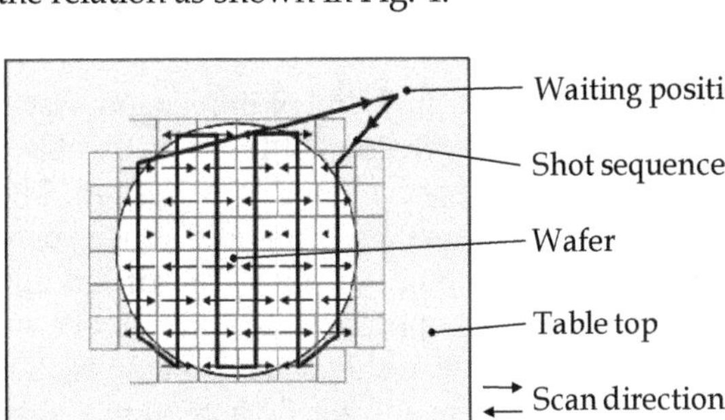

Fig. 23. Sequence of a scanning test on an 8 inches wafer.

Figure 24 shows fluid absorbance induced by dissolved oxygen against the number of scanned wafers. The absorbance rises with the scanning speed but it is low enough even at the scanning speed of 800 mm/sec. This result exhibits that the nozzle can reduce the amount of residual fluid sufficiently and can suppress the fluid darkening. In conclusion, the issue of dissolved oxygen was solved by the immersion nozzle designed for a G2 fluid.

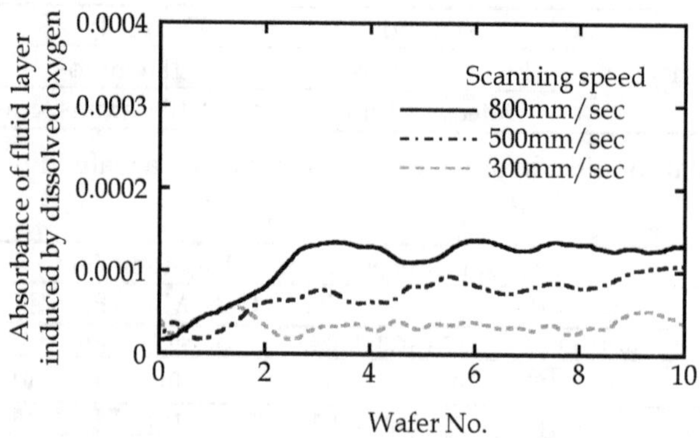

Fig. 24. Fluid absorbance between a lens and a wafer during the scan. The absorbance induced by dissolved oxygen is low enough even at the scanning speed of 800 mm/sec.

4. Remaining challenges

There are some remaining challenges to realize high-index immersion lithography. As described in the second subchapter, the quality of LuAG material does not reach the target specification. Especially, the absorbance of LuAG would be a critical issue. While the intrinsic absorbance of LuAG is lower than the target absorbance, further reduction of impurities is required.

On the other hand, there still remain a few issues to be studied in an immersion system using a G2 fluid. Although fluid absorbance can be kept low enough, dose homogeneity through a fluid layer should be confirmed. Defect study should be done with various resists. It is preferable to examine them with a preproduction tool using a G2 fluid.

For the extendability of high-index immersion lithography, the research activity on new materials such as G3 fluids and high-index resists is needed.

5. Conclusion

It has been discussed the feasibility on a high-index immersion system of 1.55 NA using LuAG and a G2 fluid. Although the IBR correction of LuAG is feasible, the quality of LuAG is not enough and the acceleration of its development is desired. The immersion system using a G2 fluid is being developed without serious issues. It was demonstrated that fluid absorbance can be kept low enough through an in-line purification unit and an oxygen removal unit. Lens contamination can be suppressed by addition of a small amount of water into a G2 fluid. Some issues arising with residual fluid, such as fluid darkening due to reentry of oxygen-rich residual fluid, were solved. By accepting residual fluid on a wafer, the scanning speed and the throughput can be raised.

EUVL is the orthodox candidate of the next generation lithography but still needs real verifications in various items. If EUVL is delayed, high-index immersion lithography will be at the leading edge lithography. It is a common knowledge that strong supports from the industry are indispensable to overcoming the remaining challenges such as LuAG. It is just said, "No market, no tool."

6. Acknowledgment

The author would like to acknowledge DuPont, JSR, and Mitsui Chemicals for supplying G2 fluids and purification units. The author also thanks colleagues for their cooperation.

7. References

Parthier, L.; Wehrhan, G.; Seifert, F.; Ansorg, M.; Aichele, T. & Seitz, C. (2008). High-index lens material LuAG : development status and progress, *presented at SEMATECH Litho Forum*, Bolton Landing, New York, USA, May 12-14, 2008

Parthier, L.; Wehrhan, G.; Seifert, F.; Ansorg, M.; Aichele, T.; Seitz, C. & Letz, M. (2008). Development update of high index lens material LuAG for ArF hyper NA immersion systems, *presented at 5th Int. Symp. on Immersion Lithography*, HI-01, The Hague, Netherlands, September 22-25, 2008

Burnett, J. H.; Kaplan, S. G.; Shirley, E. L.; Horowitz, D.; Josell, D.; Clauss, W.; Grenville, A. & Peski, C. V. (2006). High index materials for 193 nm immersion lithography, *presented at 3rd Int. Symp. on Immersion Lithography*, OO-21, Kyoto, Japan, October 2-5, 2006

Nawata, T.; Inui, Y.; Masada, I.; Nishijima, E.; Mabuchi, T.; Mochizuki, N.; Satoh, H. & Fukuda, T. (2007). High-index fluoride materials for 193 nm immersion lithography, *Proc. SPIE* Vol. 6520, 65201P

Letz, M.; Gottwald, A.; Richter, M.; Liberman, V. & Parthier, L. (2010). Temperature-dependent Urbach tail measurements of lutetium aluminum garnet single crystals, *Phys. Rev. B* 81, 155109

French, R. H.; Liberman, V.; Tran, H. V.; Feldman, J.; Adelman, D. J.; Wheland, R. C.; Qiu, W.; McLain, S. J.; Nagao, O.; Kaku, M.; Mocella, M.; Yang, M. K.; Lemon, M. F.; Brubaker, L.; Shoe, A. L.; Fones, B.; Fischel, B. E.; Krohn, K.; Hardy, D. & Chen, C. Y. (2007). High-index immersion lithography with second-generation immersion fluids to enable numerical aperatures of 1.55 for cost effective 32-nm half pitches, *Proc. SPIE* Vol. 6520, 65201O

Furukawa, T.; Kishida, T.; Miyamatsu, T.; Kawaguchi, K.; Yamada, K.; Tominaga, T.; Slezak, M. & Hieda, K. (2007). High-refractive index materials design for ArF immersion lithography, *Proc. SPIE* Vol. 6519, 65190B

Kagayama, A.; Wachi, H.; Namai, Y. & Fukuda, S. (2007). High-refractive index fluids for second-generation 193-nm immersion lithography, *persented at SPIE Advanced Lithography*, 6519-66, San Jose, California, USA, February 25-March 2, 2007

Zimmerman, P. A.; Byers, J.; Rice, B.; Ober, C. K.; Giannelis, E. P.; Rodriquez, R.; Wang, D.; O'Connor, N.; Lei, X.; Turro, N. J.; Liberman, V.; Palmacci, S.; Rothschild, M.; Lafferty, N. & Smith, B. W. (2008). Development and evaluation of a 193nm immersion generation-three fluid candidates, *Proc. SPIE* Vol. 6923, 69230A

Sakai, K.; Iwasaki, Y.; Mori, S.; Yamada, A.; Ogusu, M.; Yamashita, K.; Nishikawara, T.; Hara, S. & Watanabe, Y. (2008). Feasibility study on immersion system using high-index materials, *Jpn. J. Appl. Phys.* Vol. 47, No. 6, pp. 4853-4861

French, R. H.; Qiu, W.; Yang, M. K.; Wheland, R. C.; Lemon, M. F.; Shoe, A. L.; Adelman, D. J.; Crawford, M. K.; Tran, H. V.; Feldman, J.; McLain, S. J. & Peng, S. (2006). Second generation fluids for 193nm immersion lithography, *Proc. SPIE* Vol. 6154, 615415

Santillan, J.; Otoguro, A.; Itani, T.; Fujii, K.; Kagayama, A.; Nakayama, N.; Tamatani, H. & Fukuda, S. (2006). Novel high refractive index fluids for 193-nm immersion lithography, *Proc. SPIE* Vol. 6154, 61544Q

Sekine, Y.; Kawashima, M.; Sakamoto, E.; Sakai, K.; Yamada, A. & Honda, T. (2007). Feasibility of 37-nm half-pitch with ArF high-index immersion lithography, *Proc. SPIE* Vol. 6520, 65201Q

French, R. H.; Yang, M. K.; Peng, S.; Qiu, W.; Wheland, R. C.; Lemon, M. F.; Crawford, M. K.; Sewell, H.; McCafferty, D. & Markoya, L. (2005). Imaging of 32-nm 1:1 lines and spaces using 193-nm immersion interference lithography with second-generation Immersion fluids to achieve a numerical aperture of 1.5 and a k1 of 0.25, *J. Microlith., Microfab., Microsyst.* Vol. 4, 3, 031103

Wang, Y.; Miyamatsu, T.; Furukawa, T.; Yamada, K.; Tominaga, T.; Makita, Y.; Nakagawa, H.; Nakamura, A.; Shima, M.; Kusumoto, S.; Shimokawa, T. & Hieda, K. (2006). High-refractive-index fluids for the next generation ArF immersion lithography, *Proc. SPIE* Vol. 6153, 61530A

Furukawa, T.; Hieda, K.; Wang, Y.; Miyamatsu, T.; Yamada, K.; Tominaga, T.; Makita, Y.; Nakagawa, H.; Nakamura, A.; Shima, M. & Shimokawa, T. (2006). High refractive index fluid for next generation ArF immersion lithography, *J. Photopolym. Sci. Technol.* Vol. 19, 5, pp. 641-646

Gejo, J. L.; Kunjappu, J. T.; Zhou, J.; Smith, B. W.; Zimmerman, P.; Conley, W. & Turro, N. J. (2007). Outlook for potential third-generation immersion fluids, *Proc. SPIE* Vol. 6519, 651921

Liberman, V.; Rothschild, M.; Palmacci, S. T.; Zimmerman, P. A. & Grenville, A. (2007). Laser durability studies of high-index immersion fluids : fluid degradation and optics contamination effects, *Proc. SPIE* Vol.6520, 652035

Sakai, K.; Mori, S.; Sakamoto, E.; Iwasaki, Y.; Yamashita, K.; Hara, S.; Watanabe, Y. & Suzuki, A. (2008). Progress of high-index immersion exposure system, *presented at 5th Int. Symp. on Immersion Lithography*, HI-06, The Hague, Netherlands, September 22-25, 2008

Honda, T.; Kishikawa, Y.; Tokita, T.; Ohsawa, H.; Kawashima, M.; Ohkubo, A.; Yoshii, M.; Uda, K. & Suzuki, A. (2004). ArF immersion lithography: critical optical issues, *Proc. SPIE* Vol. 5377, pp. 319-328

Kubo, H.; Hata, H.; Sakai, F.; Deguchi, N.; Iwanaga, T. & Ebihara, T. (2007). Immersion exposure tool for the 45-nm HP mass production, *Proc. SPIE* Vol. 6520, 65201X

Sewell, H.; Mulkens, J.; Graeupner, P.; McCafferty, D.; Markoya, L.; Donders, S.; Samarakone, N. & Duesing, R. (2007). Extending immersion lithography with high index materials, *Proc. SPIE* Vol. 6520, 65201M

Shedd, T. A.; Schuetter, S. D.; Nellis, G. F. & Peski, C. V. (2006). Experimental characterization of the receding meniscus under conditions associated with immersion lithography, *Proc. SPIE* Vol. 6154, 61540R

Sanders, D. P.; Sundberg, L. K.; Brock, P. J.; Ito, H.; Truong, H. D.; Allen, R. D.; McIntyre, G. R. & Goldfarb, D. L. (2008). Self-segregating materials for immersion lithography, *Proc. SPIE* Vol. 6923, 692309

10

Scanning Probe Lithography on Organic Monolayers

SunHyung Lee[1], Takahiro Ishizaki[2], Katsuya Teshima[1], Nagahiro Saito[3] and Osamu Takai[3]
[1]*Faculty of Engineering, Shinshu University,*
[2]*National Institute of Advanced Industrial Science and Technology (AIST),*
[3]*EcoTopia Science Institute, Nagoya University,*
Japan

1. Introduction

Lithographic technologies for the surface modification of inorganic and organic surfaces have been developed for various devices such as sensing, data memory, single molecule electronics and biological systems. Nano and micropatterning of organic monolayers have attracted attentions for applications to biological systems in which proteins or DNA are fixed. Photolithography, microcontact printing, and electron beam lithography have usually been used as patterning techniques for organic monolayers (Hayashi et al., 2002; Hong et al., 2003; Saito et al., 2003; Hahn et al., 2004; Kidoaki & Matsuda, 1999). Although the electron-beam lithography can fabricate very small patterns, it requires an ultra-high vacuum system (Harnett et al., 2001). The resolution of photolithography is limited by the light wavelength. Moreover, these methods are based on destructive lithography, i.e., they cause damages to the organic materials.

In particular, nano-lithographic technologies have evolved in order to satisfy persistent demands for miniaturization and high-density integration of semiconductor electric circuits. Scanning probe microscopy (SPM) has been a key tool in achieving this goal. SPM can be used not only as the means that observe surface structure at sub-molecule level by a probe but also as the means that control the atomic and molecular arrangement on a substrate. As a local nano-fabrication means, the lithography technique in nanoscale range by using SPM is called to the scanning probe lithography (SPL) (Kaholek et al., 2004; Blackledge et al., 2000; Tello et al., 2002; Liu et al., 2002). In particular, a variety of lithographic techniques using SPM probe can fabricate nano-scale patterns on an organic monolayer, such as nanoshaving, nanografting, anodization SPL, dip-pen nanolithography (DPN), and electrochemical SPL. The lithography technique is used to break the material surface by using various energy sources. SPM can also be used to break the organic monolayer. For instance, nanoshaving involves mechanical scratching by physical pressure of the probe, and anodization lithography involves anodic oxidation of the substrate surface by an applied bias voltage (above 9 V) between the probe and the substrate (Jang et al., 2002; Kaholek et al., 2004; Sugimura, & Nakagiri, 1995). In the case of the anodization lithography, the oxide layer can be fabricated by anodic oxidation. As another SPL technique, the nanografting procedure combines the fabrication of a nanopattern and the binary SAM using atomic force microscopy (AFM) (Amro et al., 2000; Xu et al., 1999; Liu et al., 2000). As the other process, a probe is scanned on the matrix SAM at a high force. The matrix SAM is removed and simultaneously replaced by another molecule in the scanning area. Fig. 1 shows the nanoshaving, nanografting, and anodization lithography techniques.

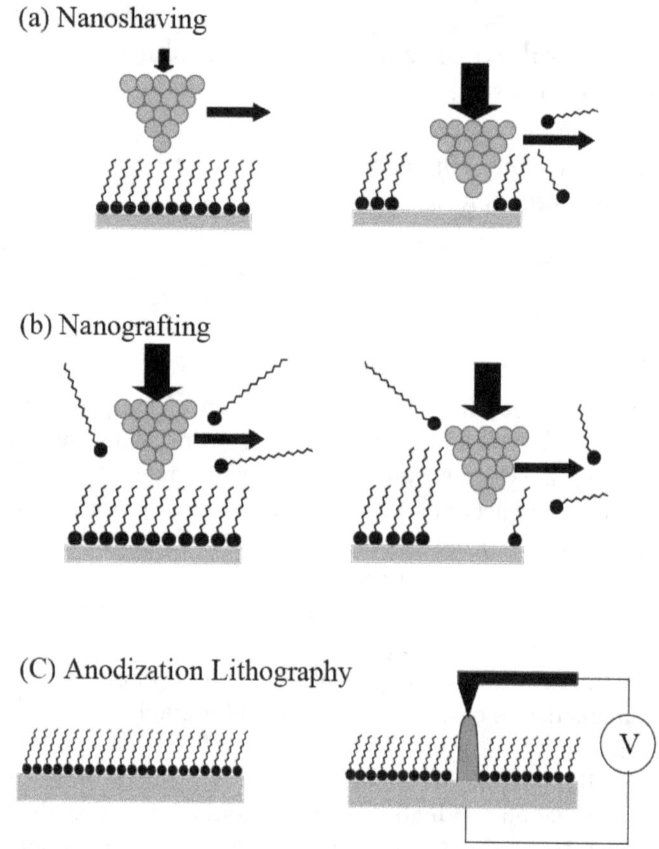

Fig. 1. Various lithographic methods using SPM probe on the organic monolayer. : (a) nanoshaving, (b) nanografting and (c) anodization lithography

In addition, dip-pen nanolithography (DPN) is a nanopatterning technique with a probe which delivers molecules to a surface via a water meniscus in the ambient atmosphere (Pena et al., 2003; Schwartz,2002; Maynor. et al., 2004). This direct-writing technique offers high-resolution patterning capabilities for a number of moleculars and biomolecular materials (ink) on a variety of substrates (paper). Nanografting and DPN are recently developed as "soft" lithographic methods, which do not cause any damage to the organic monolayers. Electrochemical SPL is also one of the soft lithographic methods. This is performed through the water column condensed between the tip of the SPM and the substrate surface. This water column can be used as a minute electrochemical cell. When a bias voltage is applied, a redox reaction occurs on the substrate surface. In the case of SAM, the functional group is converted by this redox reaction and a nanopattern is formed on the SAM surface. Fig. 2 shows the DPN and electrochemical SPL techniques.

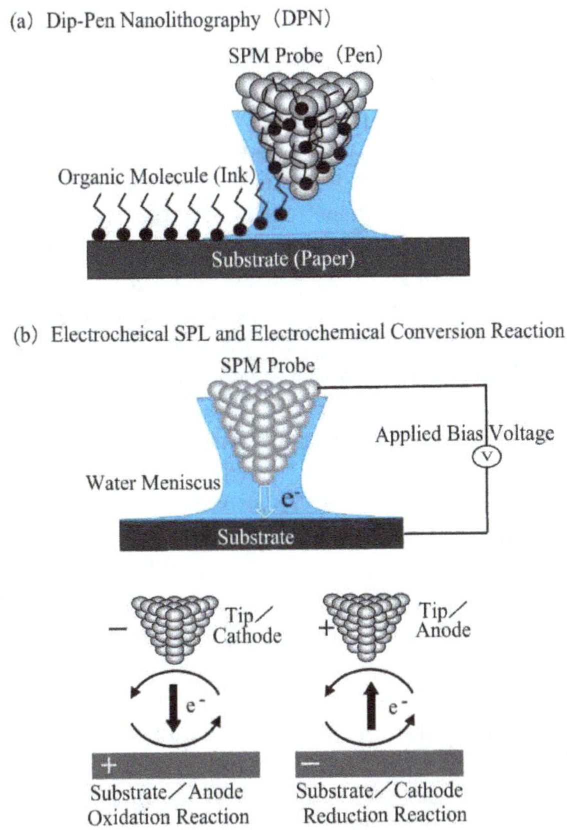

Fig. 2. Schematic illustration of dip-pen nanolithography (DPN) and electrochemical SPL.

In this chapter, we introduce anodization SPL and electrochemical SPL for fabricating the nanostcuture and controlling the function group on the various organic monolayers, such as organosilane self-assembled monolayers (SAMs) and the monolayer covalently attached to silicon through Si-C bond . First, anodization SPL techniques are expected to fabricate nanostructures on surfaces of electronic device. Their technique can remove an organic monolayer and fabricate an oxide structure by applying highly bias voltage between probe and substrate. When a high positive bias voltage is applied to the probe, an oxidation reaction proceeds and an oxide structure forms on the surface. The electrochemical reactions on SPM probe and substrate are shown in Eqs., respectively (Lee et al., 2009).

$$4H_2O + 4e^- \rightarrow 2H_2 + 4OH^- \tag{1}$$

$$M + 2H_2O \rightarrow MO_2 + 4H^+ + 4e^- \ (M: \text{substrate}) \tag{2}$$

In these reactions, electric characteristics of coated materials on the probes are important to formation of an oxide structure. The oxide structure is easily fabricated on the substrate by anodization SPL since the silicon oxide insulator layer is removed. However, the oxide structure is difficult to control and is not attained to single nanometer level though many researchers have reported on anodization SPL. We introduced fabricating technique of a three-dimensional nanostructure (nanoline structure) of silicon oxide on the hydrogen-

terminated Si substrates by anodization SPL, and the effects of coated materials of SPM probe on the sizes of oxidized structures (Lee et al., 2009).
Second, we introduce a novel soft lithography based on electrochemistry through scanning probe electrochemistry for controlling the functional group on the organic monolayers (Lee et al., 2007; Lee & Ishizaki et al., 2007; Saito et al., 2005; Sugimura et al., 2004). Scanning probe electrochemistry in which materials surfaces locally oxidized or reduced by a tip of SPM is a promising technique for constructing nanostructures consisting of organic molecules. This electrochemical SPL is performed through water column which is condensed between the tip of SPM and the substrate surface. This water column can be used as a microscopic electrochemical cell. In the case of electrochemical conversion using a nanoprobe, the electrochemical reactions which proceed at the probe-sample junction are governed by the applied bias voltage and its polarity. When the substrate is polarized positively, anodic reactions, that is, oxidation reactions, proceed on its surface. On the contrary, when the substrate is polarized negatively, cathodic reactions, that is, reductive reactions, proceed. The method is expected as a key technology for future molecular nanodevices. In addition, organosilane self-assembled monolayers (SAMs) have been applied to a resist material for SPL. Here we report the chemical conversion of an organic molecular monolayer in a reversible manner using SPM. The chemical state of the monolayer from its oxidized state to reduced state or vice versa is controlled. First, an amino-terminated SAM was chemically converted into an oxidized SAM by SPL at positive-bias voltages. Moreover, this oxidized SAM was then reconverted into an amino-terminated SAM by SPL at negative bias voltages. We examine the chemical changes undergone in the scanned area from the viewpoint of surface-potential reversibility. Additionally, we introduce a electrochemical SPL to fabricate -COOH groups on an organic monolayer directly attached to silicon, which was synthesized from 1,7-octadien (OD) and hydrogen-terminated silicon. The –COOH groups on the OD-monolayer were also synthesized by the properties of surface produced by SPL.

2. Anodization scanning probe lithography on the organic monolayer on the hydrogen-terminated Si substrate

In this chapter, we introduced fabricating technique of a three-dimensional nanostructure of organic monolayer on the hydrogen-terminated Si substrates by anodization SPL. First, we fabricated the organic monolayer on the hiydrogen-termicated Si(111) wafers with an electrical resistance of 10.0-20.0 Ω-cm. Si substrates were sonicated in acetone and ethanol for 10 min, and then, cleaned by an ultraviolet (UV) light/ozone cleaning method. The substrates were exposed to vacuum UV (VUV) light (172 nm) from an excimer lamp for 30 min under atmospheric pressure and room temperature. Subsequently, the substrates were cleaned in piranha solution (H_2SO_4:H_2O_2 = 3:1) at 100 °C for 10 min and rinsed in ultrapure water. They were then etched for 15 min by immersing in 40 % aqueous ammonium fluoride solution (NH_4F). As a result, the native silicon oxide layer was removed from the substrates, and hydrogen-terminated surfaces were formed. 1-decane monolayer was prepared by a liquid phase method. The hydrogen-terminated substrates were immersed in the solution of 1-decane molecules at 150 °C for 3 h . After the immersion, the organic monolayer coated substrates were cleaned in toluene, acetone, and ethanol, and rinsed in ultrapure water.
Firstly, the 1-decane monolayer was fabricated on the hydrogen-terminated Si substrates through a liquid phase method at 150 °C. Fig. 3(a) shows the chemical structure of 1-decane

molecule. The static water contact angle and film thickness of the monolayer was about 107° and 1.12 nm, respectively. A smooth surface coated with densely packed CH_3 groups (alkyl monolayer) shows a water contact angle of approximately 110°. In addition, its thickness was considered reasonable since the chain length of 1-decane molecule was estimated to be 1.32 nm. It is indicated that the 1-decane molecules formed an organic monolayer on the silicon substrate. Fig. 3(b) shows the XPS Si2p spectrum of the Si surface coated with the 1-decane monolayer. As clearly shown in Fig. 3(b), the peak of native silicon oxide was not observed at all. These results indicate that the 1-decane molecules reacted directly to the hydrogen-terminated silicon surfaces and formed the organic monolayer there. Fig. 3(c) and 3(d) show the AFM topographic and friction images of the Si surfaces after immersing in 1-decene solution. The flat terraces with steps for silicon one-atom and the changes of high friction on steps were, respectively, observed on the topographic and friction images. In the topographic image [Fig. 3(c)], the intervals of the steps and the terraces were 180 nm and 3.2 ± 0.3 Å, respectively. In addition, the difference of friction force between the flat terraces and steps was about 10 mV. These results indicated to the surface profiles of hydrogen-terminated silicon. It was therefore found that the 1-decane molecules were vertically and densely assembled, and their monolayer was formed through doubly-boned terminated group attached to the hydrogen-terminated silicon surface.

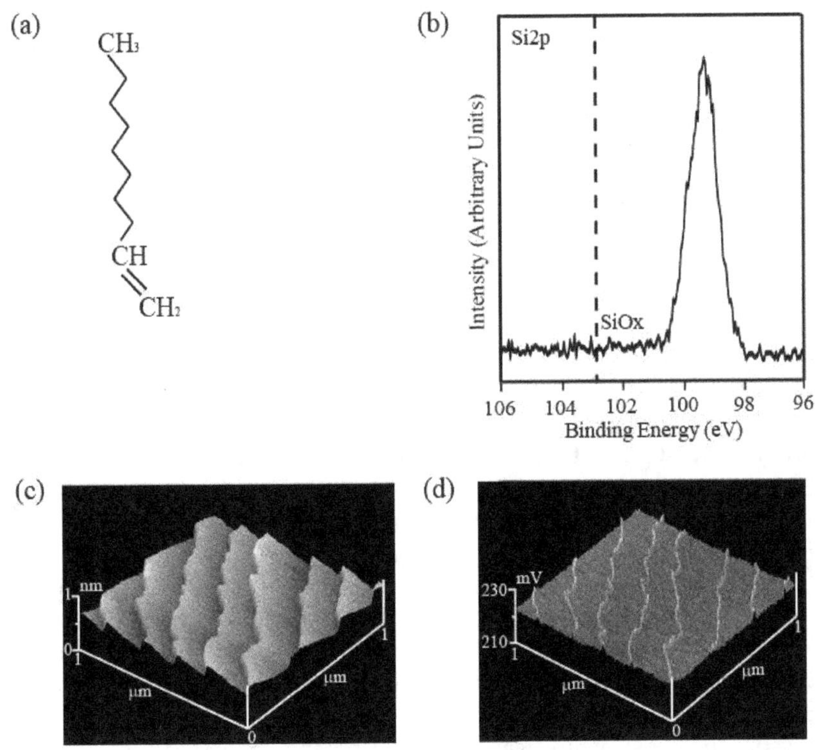

Fig. 3. (a) The chemical structure of 1-decane (b) XPS Si2p spectra (c) topographic image (d) friction image of sample surfaces after prepared 1-decane monolayer.

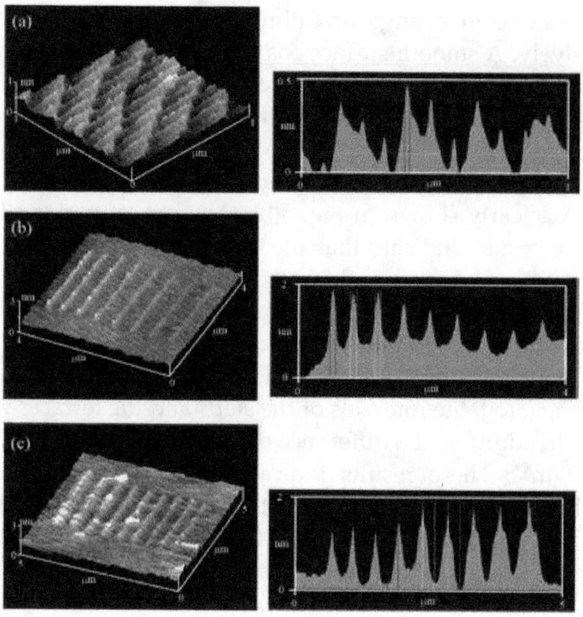

Fig. 4. The topographic images of SiOx line structure fabricated on the 1-decane monolayer and profiles for each topographic image by using various tip: (a) diamond coated Si tip, (b) Si tip and (c) Au coated Si tip.

Next, we investigated that the effect of scanning rate on nanoline width of silicon oxide fabricated by anodization SPL. An anodization SPL was carried out on the 1-decane monolayer in air (humidity in ranging from 30 to 40 %) at the applied bias voltage of 9 V. In the anodization SPL, 1-decane monolayer was removed and the SiOx nanoline structures were formed by scanning the probes. Fig. 4(a)-4(c) show the topographic images and profiles of SiOx nanoline structure fabricated by the anodization SPL. In these experiments, various probes were used to fabricate oxide nanostructures [Fig. 4(a); diamond-coated Si probe, Fig. 4(b); Si probe (i.e., uncoated Si probe), Fig. 4(c); Au-coated Si probe]. Additionally, we investigated the effect of coated materials on the formation of the oxide nanostructure. The ranging of scanning rates for SPL is 0.1 to 5 μm/s. In these topographic images, the nanoline structure and flat terraces with steps were obviously observed under all scanning conditions. However, the widths of nanoline structures were changed with the scanning rates. Fig. 5 shows the variation in the width of nanoline structures with scanning rates and surface compositions of probes. When the diamond coated probe is used for anodization SPL, the nanoline widths were found to remain constant under all scanning conditions, and the scanning rates had no effect on the line width. The width was approximately 15 nm, which is one of the finest nanostructures in the field of SPL technique. The highly reproducible structure was fabricated by anodization SPL using the diamond-coated Si probe, and this technique is thought to be able to apply various industrial fields. On the other hand, when the Si probe (uncoated-Si) and Au-coated Si probe were used for anodization SPL, the line widths were markedly changed with the scanning rates. In the case of Au-coated Si probe, as the scanning rate increased from 0.1 to 5 μm/s, the width of line structures drastically decreased from 375 to 125 nm. For Si probe (uncoated Si probe), the line width gradually decreased with increasing scanning rate, reaching about 100 nm at 5 μm/s. The width variations were Au-coated Si probe > Si probe (uncoated) > diamond-coated Si probe. These variations of the line width could be explained the band-gap energy of the coated materials. The band-gap energies of Au, Si and diamond are respectively about 0, 1.2 and 5.47 eV. That is to say, their conductive property is thought to be greatly dependent on the line width. These results indicate that high reproducibility of oxide nanoline structures is attainable by means of anodization SPL using the diamond-coated probe or a probe which has relatively low conductivity.

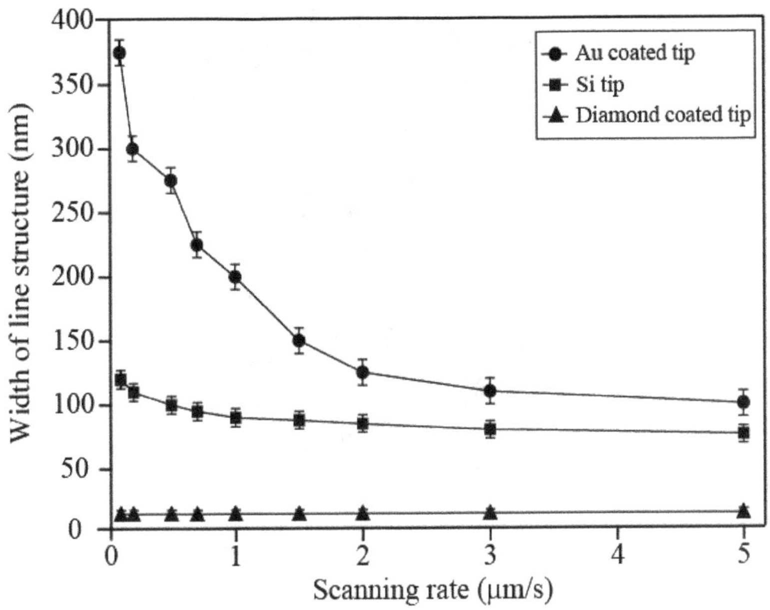

Fig. 5. The change of the line width with increasing scanning rate.

Finally, the effect of applied bias voltage on nanoline width of silicon oxide was investigated by use of various SPM probes. Fig. 6(a)-(c) show the topographic images of SiOx nanoline structures fabricated by diamond-coated Si, Si (uncoated) and Au-coated Si probes, respectively. The applied bias voltage and scanning rate were, respectively, fixed at 7 V and 1 μm/s. The obtained widths were 15, 60 and 100 nm in Fig. 6(a)-(c), respectively. The line width fabricated at the bias voltage of 7 V decreased compared to that at 9 V when Au-coated Si and uncoated Si probes were used. For the diamond-coated probe, the nanoline structure maintained the width of 15nm, that is, the applied bias voltages had no effect on the nanoline width. These results are also attributed to the band-gap energy of the coated materials. In particular, three-dimensional oxide nanostructures fabricated by the diamond-coated probe showed highly reproducibility even though various scanning rates and applied bias voltages are used in aondization SPL. Therefore, this technique is expectable to be applied to fabricate a wide variety of nanodevices, that have three-dimensional structures, in various industrial fields.

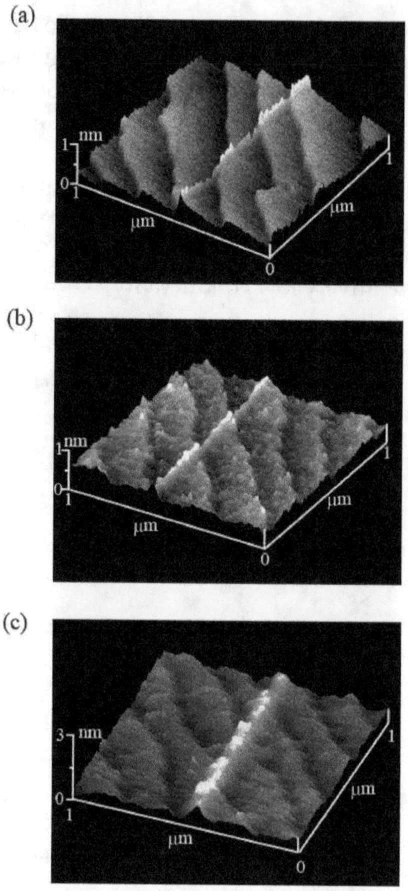

Fig. 6. Topographic images of line structure fabricated by (a) diamond coated Si tip, (b) Si tip and (c) Au coated Si tip, respectively. The applied bias voltage and scanning rate are 7 V and 1µm/s, respectively.

3. Electrochemical scanning probe lithography on organic monolayer

A future tool of nanolithography may be nanopattern drawing based on SPM. Several reports describe SPL in which a sample surface is chemically converted or mechanically deformed at the probe-sample junction. The characteristic feature of all these methods is that they allow permanent nanopatterns to be drawn on the substrate. Scanning probe electrochemistry in which materials surfaces locally oxidized or reduced by a tip of SPM is a promising technique for constructing nanostructures consisting of organic molecules. This electrochemical SPL is performed through water column which is condensed between the tip of SPM and the substrate surface. This water column can be used as a microscopic electrochemical cell. When the bias voltage is applied, redox reaction proceeds on the substrate surface. The method is expected as a key technology for future molecular nanodevices. In addition, organosilane SAMs have been applied to a resist material for SPL.

4. Nano-probe electrochemistry on amino-terminated self-assembled monolayers toward nano memory

In this chapter, we investigated that an amino-terminated SAM was electrochemically converted into an oxidized SAM by SPL at positive bias voltages. Moreover, this oxidized SAM was then reconverted into an amino-terminated SAM by SPL at negative bias voltages. The chemical conversions of amino groups were confirmed by Kelvin probe force microscopy (KPFM), atomic force microscopy (AFM) and the site-selective adsorption of carboxyl-modified fluorescent spheres. We examined the chemical changes undergone in the scanned area from the viewpoint of surface-potential reversibility. Fig. 7 schematically illustrates of the experimental procedure. An amino-terminated SAM was prepared by chemical vapor deposition (CVD) from *p*-aminophenyltrimethoxysilane (APhS) on n-type silicon (111) wafers with electrical resistivity of 4-6 Ω/cm. First, the silicon substrate was cleaned in acetone, ethanol, and deionized water, in that order. After cleaning, the silicon substrate was irradiated by 172 nm vacuum ultraviolet light in air for 20 min. This removed organic contaminants and introduced silanol groups onto the substrate surface. Next, each cleaned silicon substrate was placed together with a glass vessel filled with APhS liquid in a Teflon container. The Teflon container was sealed and placed in an oven with the temperature kept at 100 °C. The reaction time was 1 h. The heated APhS liquid vaporized and hydrolyzed. The hydrolyzed APhS reacted with the silanol groups on the silicon substrate resulting in the fabrication of an amino-terminated SAM.

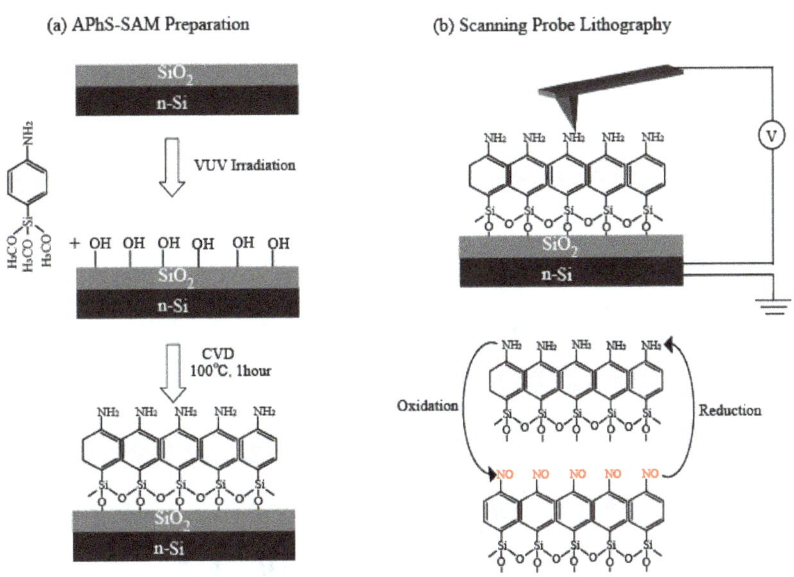

Fig. 7. Preparation and electrochemical scanning probe lithography of APhS-SAM.

We fabricated the APhS-SAM through the CVD method. The formation of APhS SAM was confirmed by the ellipsometry, water contact angle and X-ray photoelectron spectroscopy (XPS) measurement. In our SPL system, electrons were transferred between a gold-coated probe and the silicon substrate through the APhS SAM and an adsorbed water layer. The adsorbed water played the same role as water in a typical, macroscopic electrochemical cell system. The electrochemical conversion was conducted with a gold-coated silicon nanoprobe with a force constant of 1.8 N/m and a resonance frequency of 23.2 kHz. The probe was scanned in air at a relative humidity of 35% to 40%.

Firstly, the formation of APhS SAM was confirmed by the ellipsometry, water contact angle and XPS measurement. The static water contact angle of the sample surface after preparation was about 60°. Its film thickness obtained by ellipsometry was ca. 0.58 nm. This thickness was considered reasonable since the chain length of the APhS molecule was estimated to be 0.6 nm by semiempirical molecular orbital calculation using the AM1 Hamiltonian. The XPS N1s spectrum of the sample seen in Fig. 8 shows that the N1s binding energy is 399.6 eV. This binding energy agrees approximately with a previously reported value (400.0 eV). These results indicated the formation of APhS SAM on the SiO_2 substrate.

To show the effect of adsorbed water, the nano-probe was scanned at the pressure of 10^{-6} Pa. Fig. 9 shows topographic and surface-potential images scanned at 2 and 5 V in both air and vacuum. On the basis of these images, it is obvious that the chemical reaction does not proceed in vacuum. Thus, adsorbed water is considered necessary for the chemical conversion.

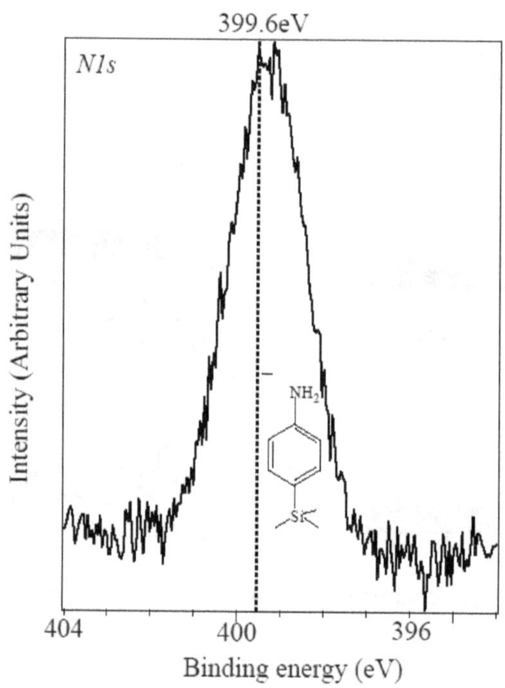

Fig. 8. XPS *N1s* spectrum of the Si surface covered with APhS-SAM.

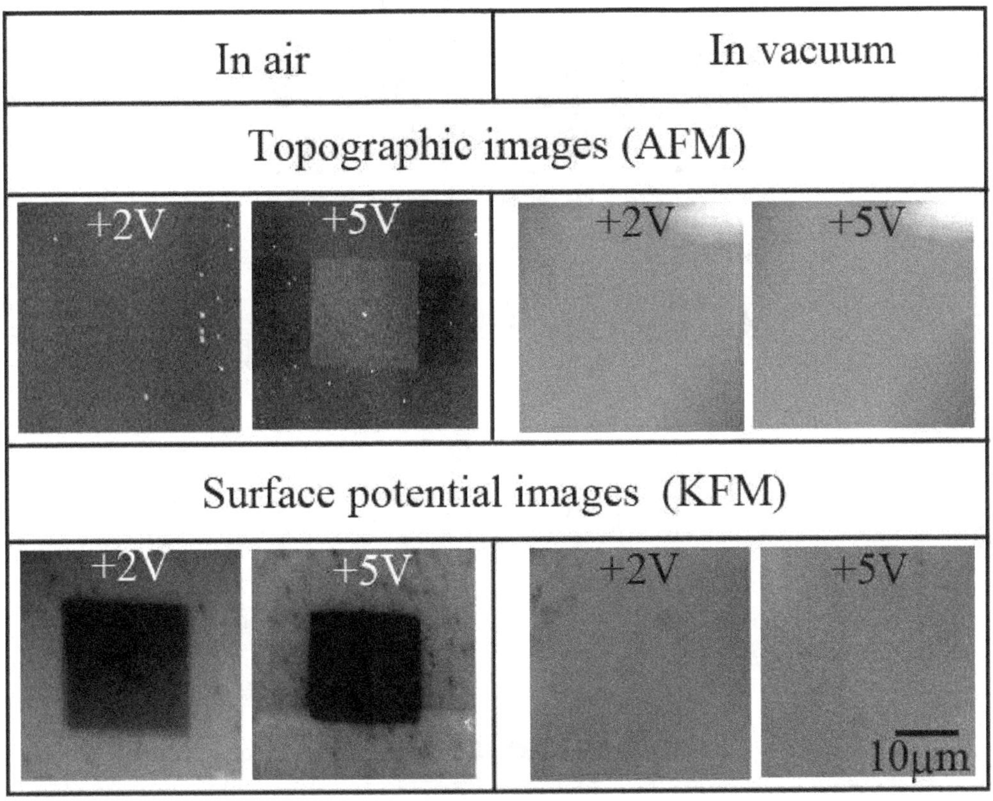

Fig. 9. Topographic and surface-potential images scanned at +2 and +5 V in air and vacuum.

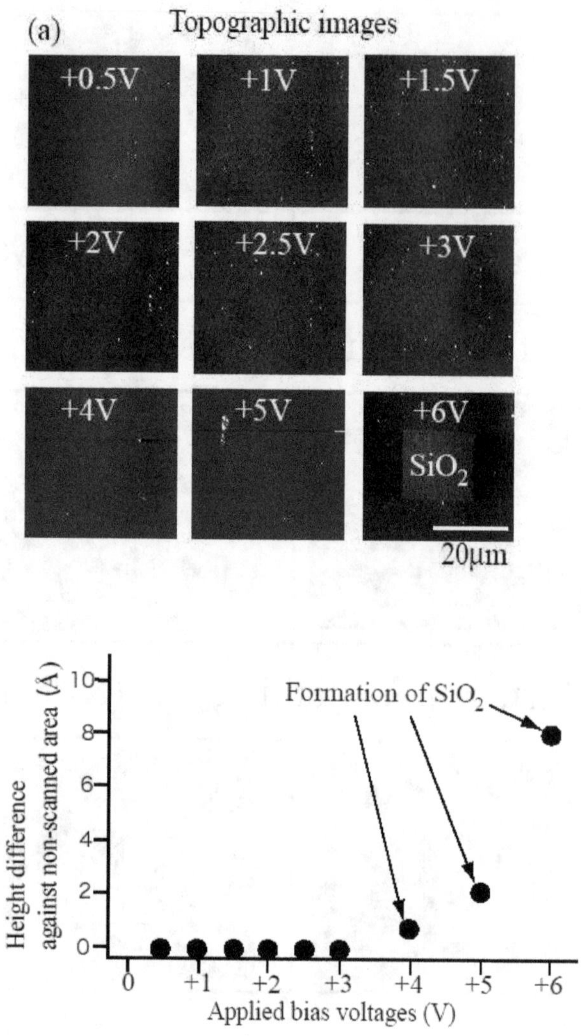

Fig. 10. (a) AFM topographic images and (b) the height difference against the non-scanned area after scanning at positive bias voltages.

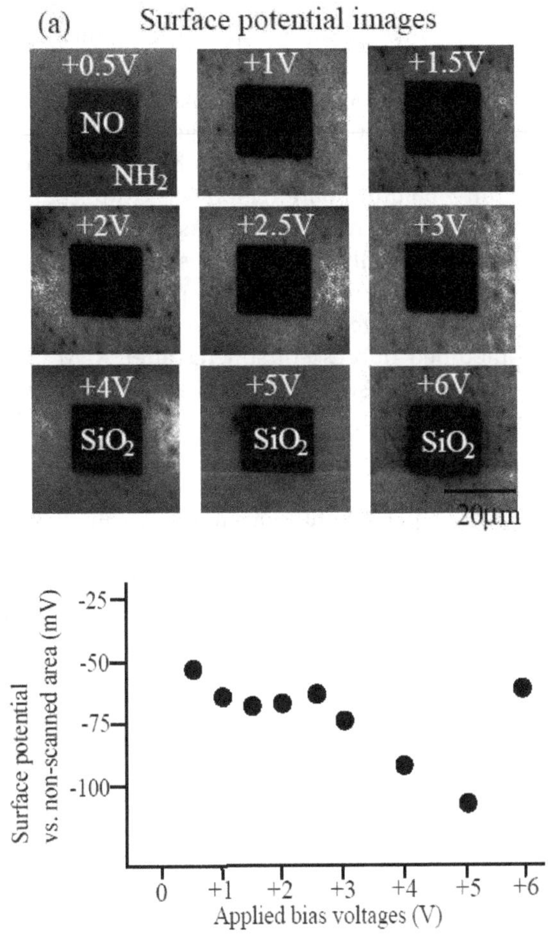

Fig. 11. (a) Surface potential images obtained by KFM and (b) the change in surface potential on the area scanned at positive bias voltages.

The nanoprobe was scanned across the sample surface over an area of 20μm X 20μm at bias voltages of 0.5-6 V. Fig. 10 shows both AFM topographic images and the height difference against the nonlithographic area after scanning. A slight protuberance can be observed in the topographic images of samples scanned at the bias voltages of 4 to 6 V. This protuberance became much higher at the bias voltages of 4 to 6V. These protuberances are due to the production of SiO_2, which resulted from the decomposition of the as-deposited APhS SAM and the oxidation of silicon. This demonstrates that there is no possibility of chemical reversibility at these bias voltages since the SAM has been damaged. Specifically, there is no possibility of surface-potential reversibility. On the other hand, no change can be observed in the topographic images of samples scanned at the bias voltages of 0.5-3 V. At these voltages, it is possible that the framework of the SAM molecules remained intact. Fig. 11 shows both surface-potential images obtained by KFM and the change in surface potential on the scanned region. The surface potential shifted negatively, which can be roughly explained by the apparent dipole moment of the SAM. The apparent dipole moment of the untreated APhS SAM is in the direction from the sample surface to the substrate. This can explain the negative shifts of the surface potential in the scanned area. In addition, amino surfaces were converted into nitroso surface at the bias voltages of 1 V to 3 V because surface potential contrast was nearly constant. In surface potential image, the surface potential reversed with the applied bias voltage.

To confirm surface-potential reversibility, a nanoprobe scanning series was performed as follows. At first, (a) a 60 μm × 60 μm square region was oxidized, and (b) a 20 μm × 20 μm square region in the 60 μm × 60 μm square region was reduced. The scan rates were 0.5 and 1.0 Hz, respectively. Fig. 12 shows illustrates of these (a) experimental processes and (b) surface potential image of scanned area. In surface potential image, the surface potential reversed with the applied bias voltage. Fig. 13 shows (a) schematic illustration of selective adsorption of carboxyl- modified fluorescent spheres after AFM lithography and (b) dark field image acquired by optical microscope after immersion of the sample in Fig. 12 in a pH 4 solution containing carboxyl-modified fluorescent spheres. The -COOH and $-NH_2$ groups in the pH 4 solution were converted into $-COO^-$ and $-NH_3^+$ ion groups. Thus, the selective adsorption of fluorescent spheres onto the NH_2 regions proceeded due to attractive electrostatic interaction. In the pH 4 solution, amino-modified fluorescent spheres were repulsed in regions with –NO terminated groups. Therefore, the bright and dark areas correspond to the NH_2 and NO surface, respectively. These indicated that the nitroso terminated surfaces were reconverted into amino terminated surfaces with a negative bias voltage.

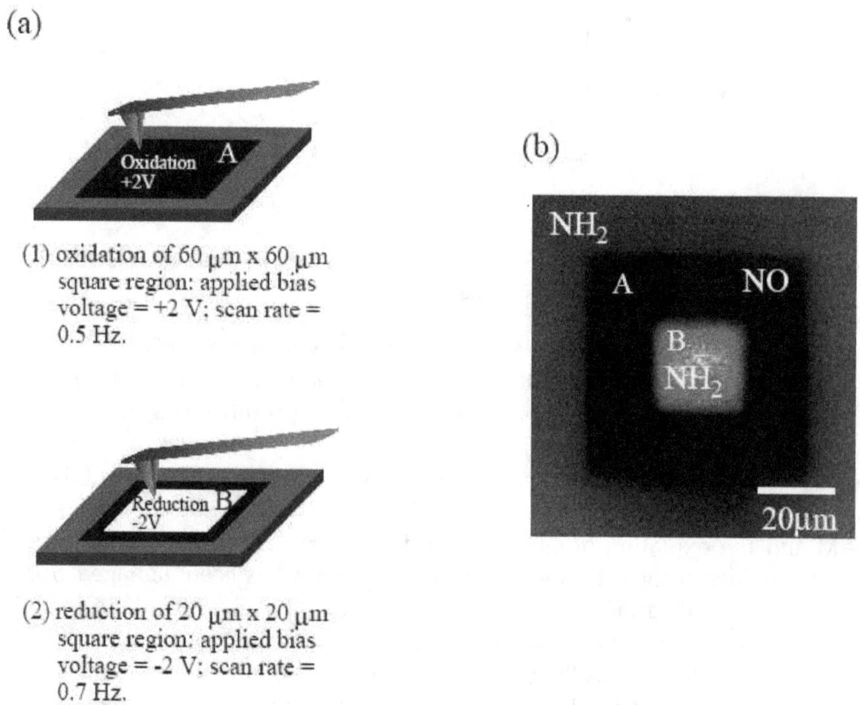

Fig. 12. (a) Schematic illustrations for a series of the lithography and (b) the obtained surface potential image.

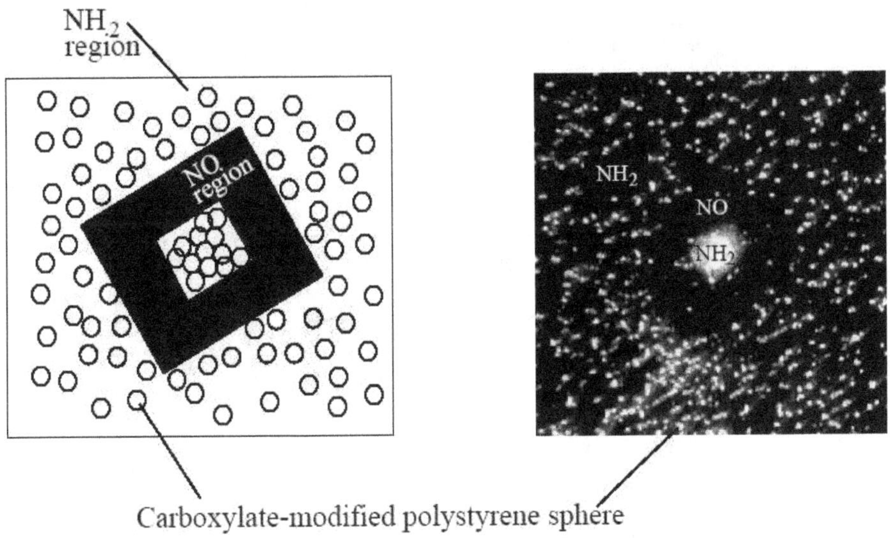

Fig. 13. (a) Schematic illustration of selective adsorption of carboxylate-modified polystyrene spheres after AFM lithography and (b) dark field image acquired by optical microscope after immersion in a pH 4 solution containing carboxylate-modified polystyrene spheres, followed by successive scanning probe lithography.

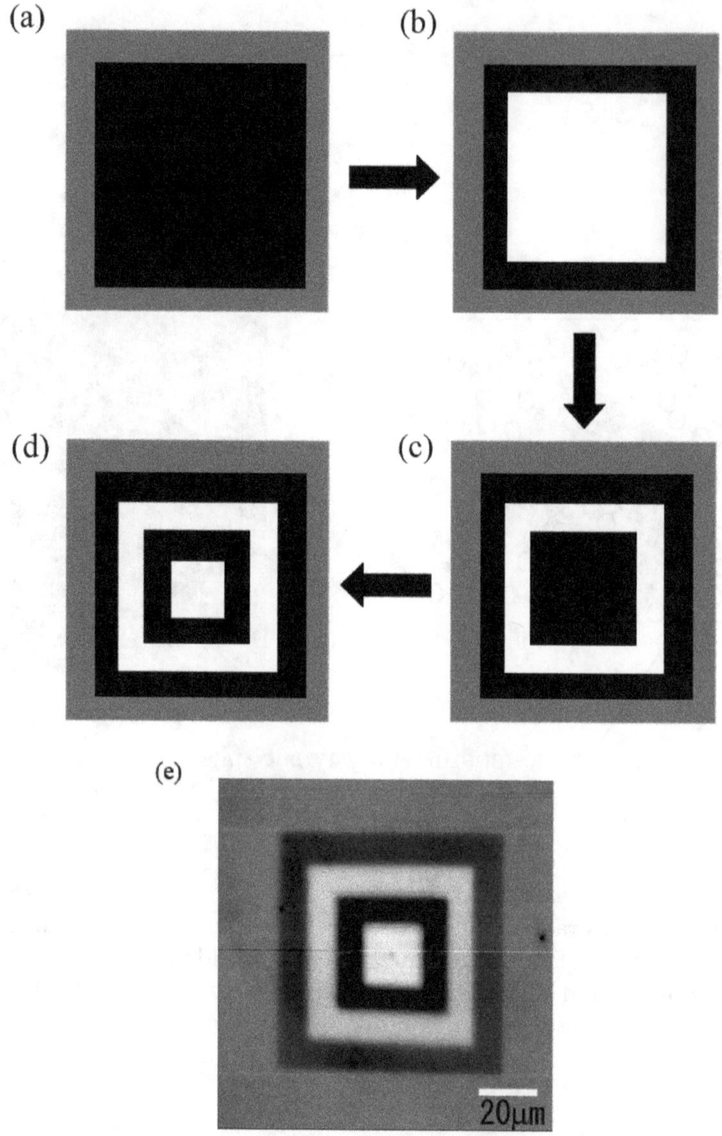

Fig. 14. Illustrations of reversibility processes ((a)-(d)) and the surface potential image of multu-scanned area (e).

In addition, we investigated the multi-reversible conversion of the APhS surface. Firstly, (a) a 80 μm × 80 μm square region was oxidized, and (b) a 60 μm × 60 μm square region in the 80 μm × 80 μm square region was reduced. Moreover, (c) the 40 μm × 40 μm square region in the 60 μm × 60 μm square region was oxidized, and then (d) the 20 μm × 20 μm square region the 40 μm × 40 μm square region was reduced. The scan rates were 0.5, 0.7, 1.0, and 2.0 Hz, respectively. Fig. 14 (a) and (b) show illustrations of these experimental processes and the surface potential image of the scanned area, respectively. In the surface potential image, the surface potential reversed with the applied bias voltage. These indicated the multi-reversible conversion of amino terminated surfaces. Thus, we can control the surface-potential reversibility on the amino-terminated SAM by controlling the applied bias voltage. This reaction formula is as follows .

$$-NH_2 + H_2O \leftrightarrow -NO + 4H^+ + 4e^- \quad (3)$$

Finally, by making use of this phenomenon of surface-potential reversibility, we demonstrated surface-potential memory. First, a square region 10 μm × 10 μm was oxidized at the bias voltage of 2 V. Next, dotted areas in the oxidized region were selectively reduced at the bias voltage of -2 V. Finally, the 10 μm × 10 μm square region was again oxidized at the bias voltage of 2 V. Fig. 15 shows the surface potential changes corresponding to (a) "writing" and (b) "erasing" with the experimental process. In Fig. 15 (a), sixteen bright areas corresponding to the chemically converted region can be observed. These sixteen areas disappeared after "erasing," as shown in Fig. 15 (b). Although this "surface potential memory" has not yet been highly integrated, it has the potential to perform as ultra-integrated memory.

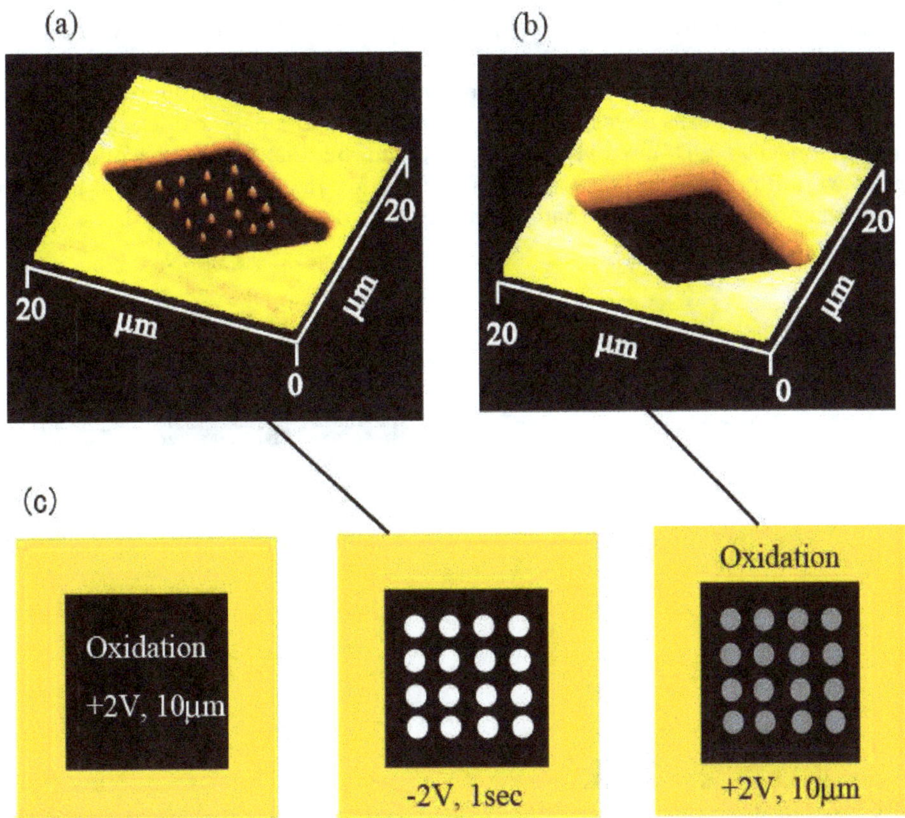

Fig. 15. Surface-potential memory: (a) "writing" state, (b) "erasing" state (c) and the experimental process.

4.1 Electrochemical lithography of 1,7-octadiene monolayers covalently linked to hydrogen-terminated silicon using scanning probe microsoopy

The organic monolayer covalently attached to silicon through Si-C bond has been expected to have better chemical resistivity compared to organosilane monolayers. The Si-C interface provides a good electronic property for molecular devices constructed on the silicon. In particular, the construction of hybrid organic- molecule/silicon devices is a promising approach for the future molecular devices. To realize such devices, it is vital to establish the fabrication technology for microstructure of the organic monolayer. The electrochemical SPL is performed through the water column condensed between the tip of SPM and the substrate surface. This water column can be used as a minute electrochemical cell. When the bias voltage is applied, a redox reaction proceeds on the substrate surface. Through reversible chemical SPL, we successfully controlled this redox reaction so that an NH_2-terminated organosilane monolayer surface was converted into an NO-terminated surface. However, this organosilane monolayer suffered from electrical defects due to the presence of SiOx. A 1,7-octadiene (OD) monolayer was directly formed on a hydrogen-terminated silicon surface by radical reaction. In this section, we report the electrochemical conversion of the vinyl-terminated groups on an OD monolayer induced by a nanoprobe.

Si (111) with electrical resistivity of 10.0-20.0 Ω/cm was used as for the substrates. Fig. 16 shows schematic illustrations of the experimental process. The substrates were cleaned in piranha solution ($H_2SO_4 : H_2O_2 = 3 : 1$) at 100 °C for 10 min and rinsed in ultrapure water. They were then etched for 15 min by immersing in 40 % aqueous ammonium fluoride solution (NH_4F). The surface was hydrogen-terminated and silicon oxide was removed by this immersion. The OD monolayer was prepared by the liquid phase method. The substrates were immersed in OD solution at 120 °C for 1h. After immersion, the samples were cleaned in toluene, acetone, ethanol and rinsed in ultrapure water.

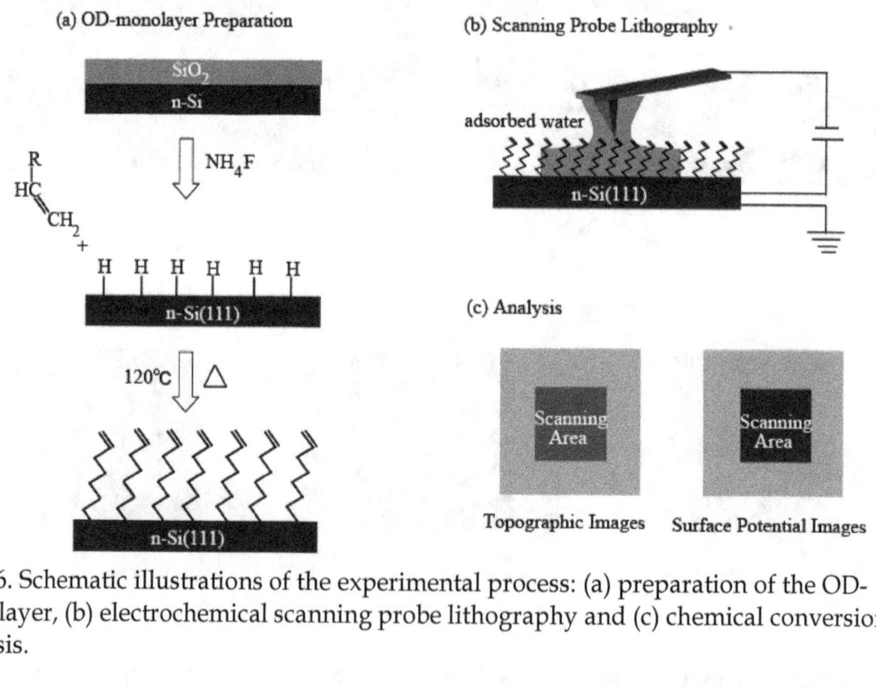

Fig. 16. Schematic illustrations of the experimental process: (a) preparation of the OD-monolayer, (b) electrochemical scanning probe lithography and (c) chemical conversion analysis.

Fig. 17 (a) shows the XPS Si *2p* spectra of silicon substrate surfaces before and after the immersion in 40 % aqueous ammonium fluoride solutions (NH_4F). In the spectrum of the sample surface before the immersion, the peak at 104 eV attributed to SiO_2 was observed, while no appreciable peak related to the oxide was observed in the spectrum after the immersion. This indicates that the native oxide layer on the silicon substrate is completely removed after the immersion. Fig. 17 (b) show the AFM topographic image of the silicon substrate surface. The AFM image has flat terraces with the steps for silicon one-atom. The distance and the height difference between steps were evaluated to be 180 nm and 0.32±0.03 nm, respectively, as shown in Fig. 17 (b). These results reveal that the silicon surfaces are terminated with hydrogen. In order to deposit the OD monolayers, the substrate was immersed in the OD solution heated at 120 °C for 1 h. After the immersion, the water contact angle of the OD monolayers became saturated at approximately 80° at the reaction time of 1 hour. Furthermore, the film thickness of 1.2 nm corresponded approximately to the distance from Si to -CH_2 end groups in the precursor, as determined by ellipsometry. Fig. 17 (c) shows the topographic image of the OD monolayer surface. The distance and the height difference between steps, as shown in Fig. 17 (c), were 120 nm and 0.28±0.03 nm, respectively. These values are well in agreement with that of hydrogen-terminated silicon surface. The OD monolayers were stably attached to the hydrogen-terminated Si surfaces, since the parallel monoatomic steps were observed on the OD molecule. This means that the OD molecule was deposited at a monolayer on the substrate. These surfaces are very stable and can be stored for several weeks without any change in the topographic properties.

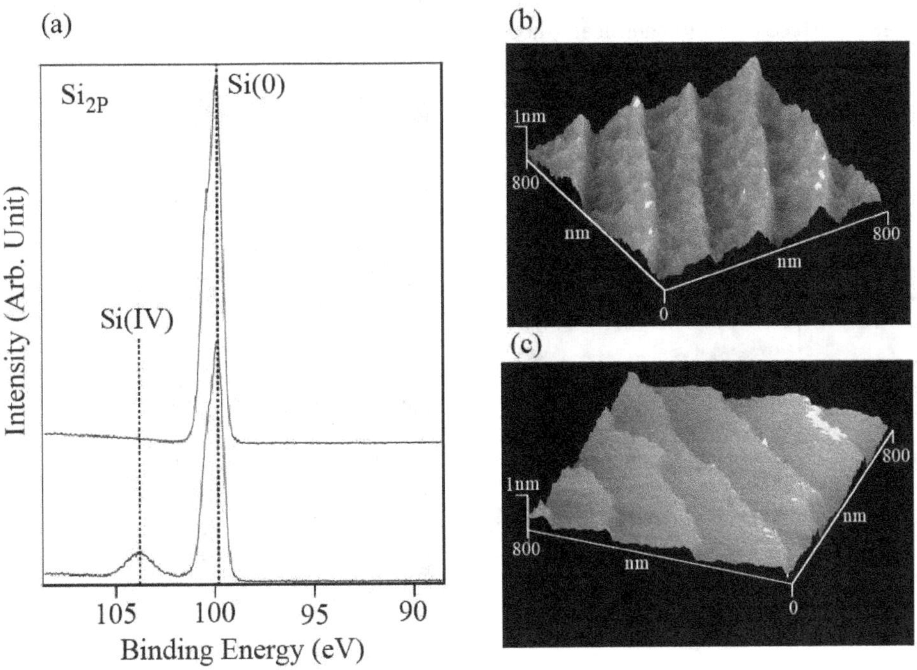

Fig. 17. (a) Si2p XPS spectra of the Si(111) surface after and before etching in 40% NH4, (b)AFM image (800nm×800nm)of a H-terminated Si(111) surface, (c) AFM image (800nm×800nm) of a OD-monolayer surface

The OD monolayers were selectively oxidized in air through the photomask by vacuum ultraviolet (VUV) light irradiation for 10 min. The areas irradiated were converted into -COOH groups due to photochemical oxidation, thus dividing the small surface into distinct $-CH_2$ and -COOH end groups regions. Fig. 18 (a) shows the XPS *C1s* spectra of OD monolayer surfaces before and after the irradiation. Its peak of 289.6 eV is assigned to the carboxyl group. Fig. 18 (b) shows the surface potential image (KFM) of the OD monolayer. The dark and bright regions correspond to the CPD images of low and high surface potential, respectively. In this figure, the surface potential for the irradiated OD monolayer surfaces was 20 mV lower than that of the unirradiated surfaces. The change of the surface potential indicates that $-CH_2$ end groups on the OD monolayer were chemically converted into -COOH end groups. The end groups of the OD monolayers were confirmed by the selective adsorption of amino-modified fluorescence spheres in a pH 4 solution. The -COOH and $-NH_2$ groups in the pH 4 solution were converted into -COO- and $-NH_3^+$ ion groups, so that the selective adsorption of fluorescence spheres on to the substrate proceeded due to their attractive interaction to the surface. Under this pH condition, the regions of $-CH_2$ end groups on the surface were not negatively charged and the amino-modified polystyrene fluorescence spheres did not adsorb onto it. Fig. 18 (c) shows an image acquired by dark field microscopy of the micropatterned CH_2 / COOH sample after immersion. The lighter areas between the dark rectangular regions correspond to the COOH terminated regions. This dark-field image indicates that amino-modified polystyrene fluorescence spheres selectively adsorbed on the -COOH end group regions since scattered light due to surface roughness can be observed. From these results, we determined that an $-CH_2$ end group had been successfully converted into the COOH terminated surface through chemical lithography, i.e., photolithography.

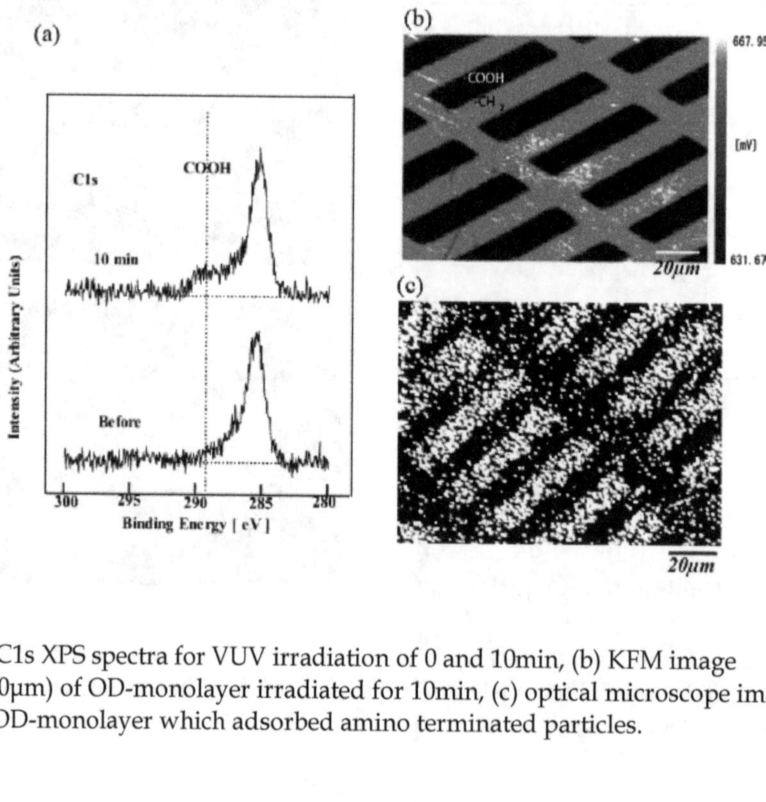

Fig. 18. (a) C1s XPS spectra for VUV irradiation of 0 and 10min, (b) KFM image (150μm×150μm) of OD-monolayer irradiated for 10min, (c) optical microscope image of the irradiated OD-monolayer which adsorbed amino terminated particles.

A gold-coated probe was scanned in air on OD monolayers at the bias voltage of 1 V and the scanning rate of 2 Hz. The areas scanned were a square, 20 μm on a side. Fig. 19 (a) and (b) shows the topographic and the surface potential images, respectively. The surface potential for the area scanned was 20 mV more negative than that of the area non-scanned and no change was observed in the topographic images. The value of surface potential difference between the scanned and non-scanned areas by SPL is well in agreement with that by photo oxidation, i.e., VUV light. These results indicate that the scanned OD monolayer surfaces are completely oxidized and chemically converted into COOH terminated surfaces without decomposition of the OD monolayer and siloxane networks. In order to verify that the chemical conversion with SPL was electrochemically proceeded on the areas scanned through the water column, the SPL was carried out in vacuum. The conditions were as follows: the bias voltage : 1 V, the pressure : 10^{-4} Pa, the scanned rate : 2 Hz, and the scanned areas : a square of 20 μm on a side. Fig. 20 shows the topographic and the surface potential images on the scanned area. In both images, no change was observed. This means that the chemical conversion on the scanned areas does not proceed under such a condition. Therefore, we can conclude that the chemical conversions with SPL are based on electrochemical reactions through the water column.

Electrochemical SPL was performed on the OD monolayer at the scanning rate of 2 Hz and at bias voltages of -3 V to 3 V. Fig. 21 shows representative surface potential images, topographic images, the changes in surface potential, and the height difference against the non- lithographic regions. In topographic images, no change in height difference was observed at any of the bias voltages. This indicates that the scanning caused no change in microscopic morphology under these conditions. On the other hand, the surface potential changed remarkably. The dark and bright regions in the surface potential images correspond to low and high surface potential regions, respectively. When the bias voltage was positively applied, oxidation proceeded on the substrate. At the positive bias voltage, the scanned regions were oxidized and showed lower surface potential than the unscanned areas. The surface potential contrast was nearly constant at approximately -18 mV at bias voltages of 1 V to 2 V. However, the surface potential contrast at bias voltages of 2 V to 3 V gradually decreased in proportion to voltage evolution.

On the other hand, the surface potential contrast at negative bias voltages gradually increased in proportion to voltage evolution. The surface potential contrast at bias voltages of 0 V to -1 V was negative, indicating that some oxidation proceeded under these conditions. This was due to the difference in contact potential between the Au-coated probe and the Si substrate. The surface potential constant was positive at bias voltages from -1.5 V to -3 V. This change in surface potential indicates that reduction reactions occurred on the substrate surface due to the applied negative bias voltage. Considering these AFM and KPFM results, the electrochemical conversion of the vinyl-terminated groups is believed to have been governed by the applied bias voltage.

Using XPS, we investigated the conversion of vinyl terminated groups at each bias voltage. Fig. 22 (a) and (b) show XPS Si*2P* and C*1s* spectra for sample surfaces after probe-scanning at bias voltages of 1 V and 3 V. The C*1s* spectrum in Fig. 22 (b) shows an additional peak from -COOH groups at 288.5 eV at the bias voltage of 1 V, but not at 3 V. In addition, a silicon oxide peak at 103.4 eV can be seen in Fig. 22 (a) at the bias voltage of 3 V, but not at 1 V. The intensity of the alkyl chain peak in Fig. 22 (b) at the bias voltage of 3 V was observed to be lower than that at 1 V. These XPS results indicate that the vinyl functional groups were oxidized and converted into carboxyl groups at the bias voltage of 1 V. The surface potential of the carboxyl surface was lower than that of the vinyl-terminated surface because the carboxyl groups had more negative dipole moment. This agrees with the KFM results. In addition, OD molecules on the sample surface were decomposed and silicon oxide was formed at the bias voltage of 3 V. However, the fact that there was no change of AFM morphology at the bias voltage of 3 V was probably due to the formation of "depthless" silicon oxide, that is, the partial decomposition of OD molecules, and to the effect of absorbed water on the sample surface.

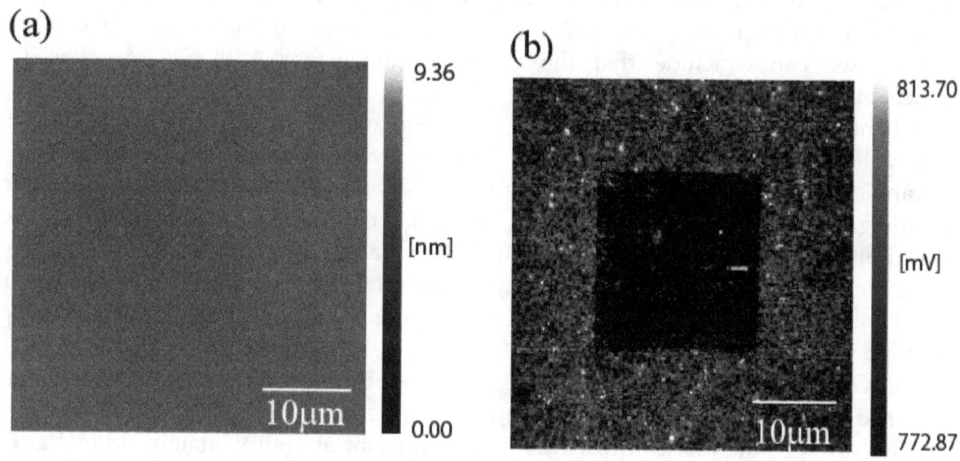

Fig. 19. (a) Topographic images and (b) surface potential image after the scanning at the rate of 2Hz.

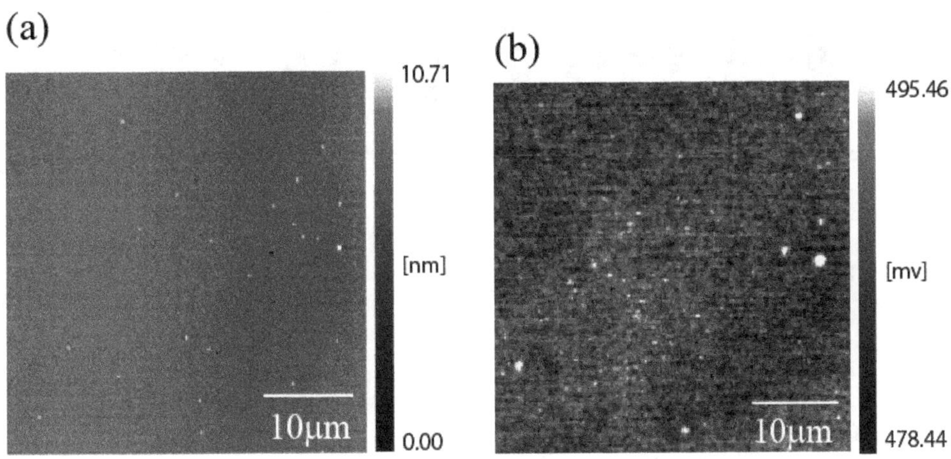

Fig. 20. (a) Topographic image and (b) surface potential image after the scanning in vacuum.

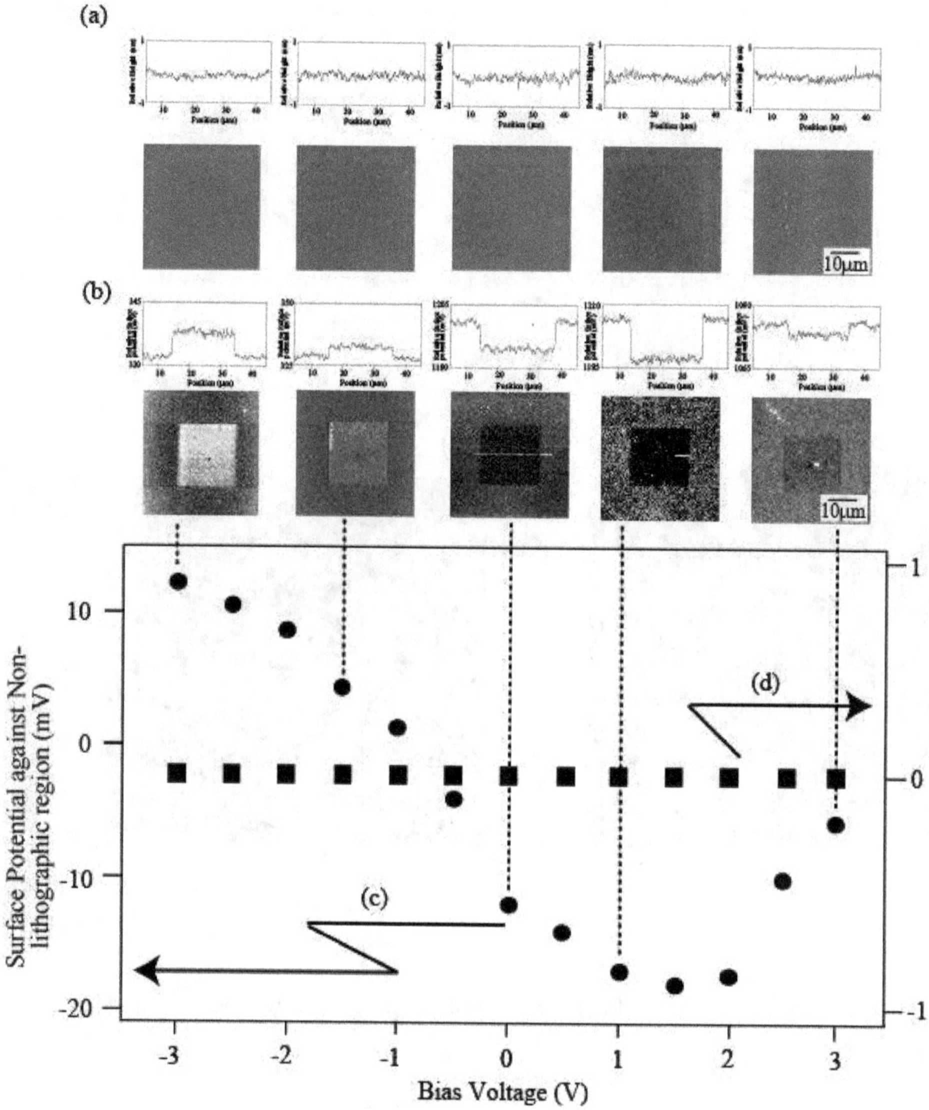

Fig. 21. (a) Representative topographic images, (b) representative surface potential images, (c) change of surface potential against non-lithographic regions; and (d) height difference against non-lithographic regions. Electrochemical SPL was performed at bias voltages of -3V to 3V.

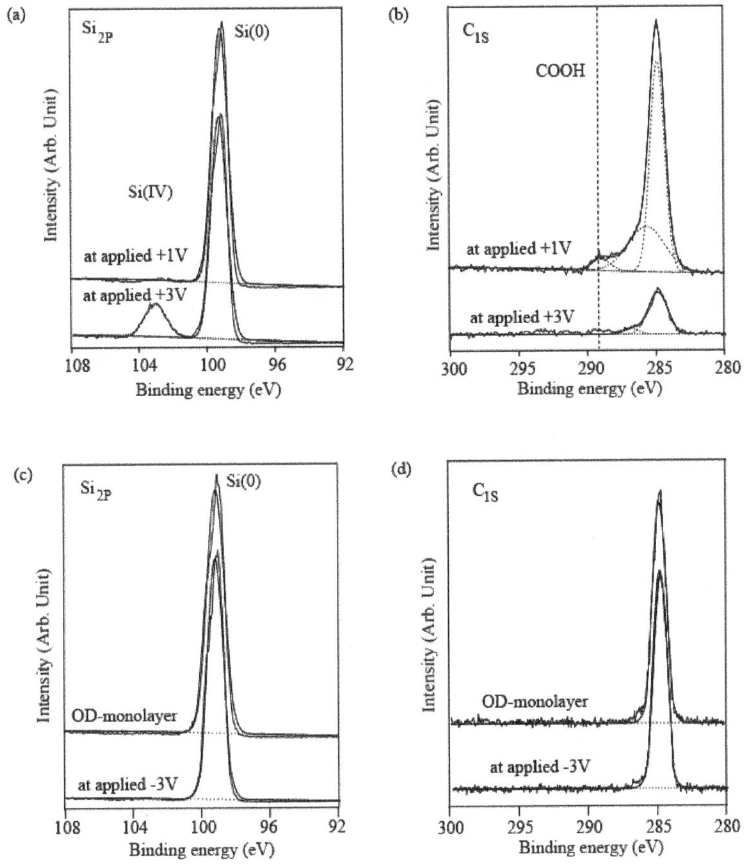

Fig. 22. (a) XPS Si2*P* spectra and (b) XPS C*1s* spectra of sample surfaces after probe-scanning at bias voltages of 1V and 3V. (c) XPS Si2*P* spectra and (d) XPS C*1s* spectra of an OD-monolayer surface and a sample surface after probe-scanning at the bias voltage of -3V.

Fig. 22 (c) and (d) show XPS Si2*P* and C*1s* spectra of an unscanned OD-monolayer surface and a sample surface scanned at the bias voltage of -3 V. In the C*1s* XPS spectrum in Fig. 22 (d), the intensity of the alkyl chain peak for the surface scanned at the bias voltage of -3 V was the same as that for the OD monolayer. The silicon oxide peaks in the Si2*p* spectra were not observed in all cases [Fig. 22 (a) and (c)]. These XPS results indicate that OD molecules were not decomposed at negative bias voltage. We believe that the vinyl functional groups were reduced and converted into cyclobutane rings. No peak for these cyclobutane rings was observed in the C*1s* spectra since such a peak is generally weak. However, in view of the AFM and KFM results, we consider that cyclobutane rings form in the same manner as they are known to in photochemical and thermal reactions.

Fig. 23 shows a schematic illustration of the mechanism of electrochemical SPL on the OD monolayer. In this Section, two factors were considered: the alkyl radical reaction from frictional heat due to the probe scanning, and the redox reaction on the sample surface caused by polarization due to the applied bias voltage. First, alkyl radicals were formed by frictional heat. Next, the redox reaction occurred on the sample surface in the radical atmosphere. With positive bias voltages, the oxidation reaction easily occurred on the sample surface due to polarization in adsorbed water. Thus, the conversions on the sample surface were governed by the oxidation reaction. The vinyl-terminated groups of the OD monolayer were converted into carboxyl groups at positive bias voltage. However, the reduction reaction on the sample surface rarely occurred at negative bias voltages because the dissolved oxygen was preferentially reduced in adsorbed water. Thus, in this case, the surface reaction was governed by alkyl radicals. The formation of cyclobutane rings was considered to have occurred due to alkyl radical combinations at the negative bias voltage.

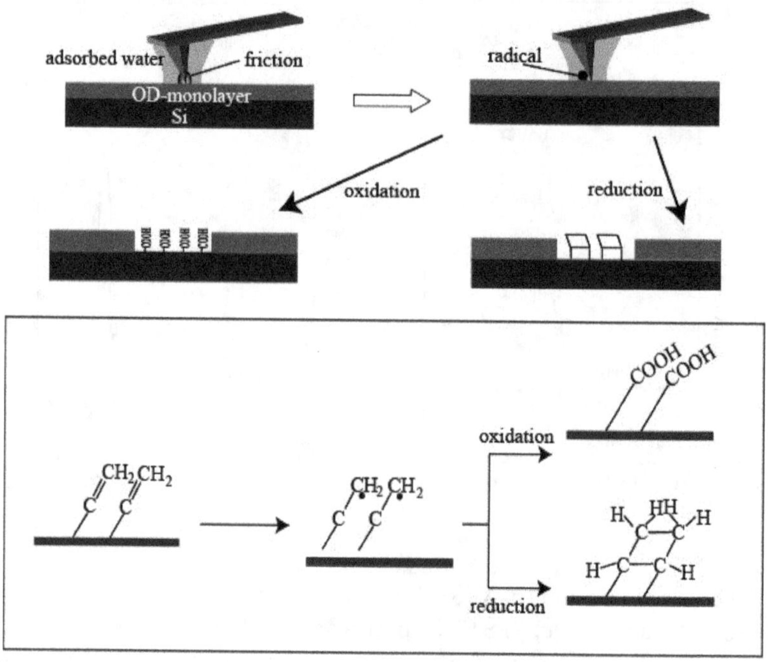

Fig. 23. Schematic illustration of the redox reaction induced by electrochemical SPL.

In support of the SPM and XPS results, the oxidized groups of the OD monolayer were confirmed by the selective adsorption of amino-modified fluorescent spheres. The -COOH and $-NH_2$ groups in the pH 4 solution were converted into $-COO^-$and $-NH_3^+$ ion groups. Thus, the selective adsorption of fluorescent spheres onto the COOH regions proceeded due to attractive electrostatic interaction. In the pH 4 solution, regions with vinyl terminated groups were not negatively charged, and the amino-modified polystyrene fluorescent spheres were repulsed. Fig. 24 (a) shows the mechanism of this selective adsorption of the amino-modified fluorescence spheres. Fig. 24 (b) shows a dark field image of samples scanned at the bias voltage of 1 V after immersion in the solution of amino-modified fluorescent spheres. The bright areas correspond to the areas scanned at the bias voltage of 1 V which site-selectively adsorbed the fluorescent spheres. This confirms that vinyl terminated groups of the OD monolayer were converted into COOH terminated groups by scanning at the applied bias voltage of 1 V.

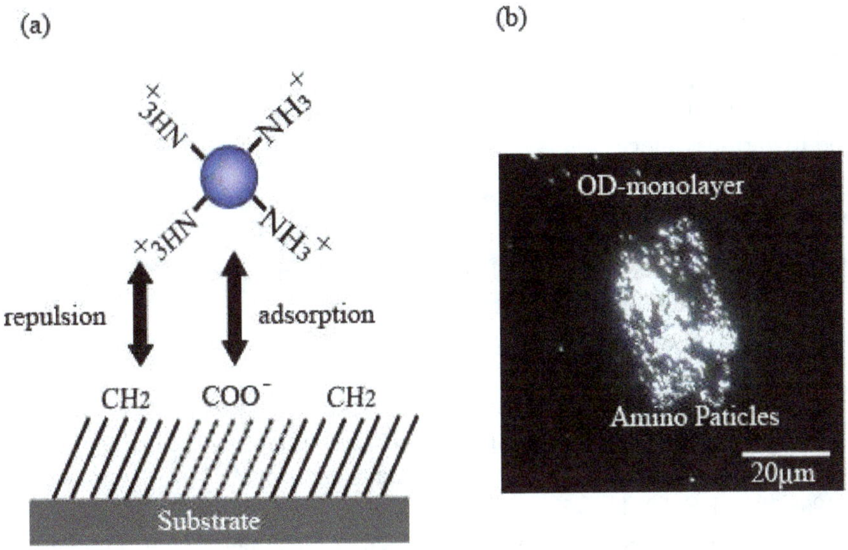

Fig. 24. (a) Selective adsorption of amino-modified fluorescent spheres on a surface scanned at the bias voltage of 1V; (b) dark field image of a patterned surface after immersion in a pH 4 solution containing amino-modified fluorescent spheres.

5. Conclusion

In this chapter, we introduced the chemical conversion of functional groups on the organic monolayer by electrochemical SPL. The three-dimensional nanostructures of silicon oxide were successfully fabricated by decomposing the 1-decane monolayer and subsequent oxidizing the hydrogen-terminated Si surfaces via anodization SPL. The size and reproducibility of oxide nanoline structures were greatly dependent on the sorts of probes for anodization SPL. In the case of Au-coated Si and uncoated Si probes, the obtained nanoline structures were changed with the scanning rates and the applied bias voltages. On the other hand, the nanotexture fabrication using the diamond-coated probe showed one of the finest structures (15 nm nanoline) and highly reproducibility even though any fabrication conditions such as scanning rate and applied bias voltage are used in the anodization SPL.The amino surface on SAM was oxidized and converted into a nitroso surface at bias voltages of 0.5 to 3 V. The functional groups on APhS SAM were reversibly converted by controlling the applied bias voltage. It was also demonstrated that the surface-potential memory was based on surface potential reversibility. In addition, the vinyl-terminated groups of the OD monolayer were site-selectively oxidized and chemically converted into carboxyl groups at bias voltages of 1 to 2 V. OD molecules on the sample surface were decomposed and silicon oxide was formed at bias voltages greater than 3 V. On the other hand, CH_2-terminal groups were converted into cyclobutane rings at bias voltages of less than −1.5 V. Recently, the research into applications of SAMs has progressed rapidly because of their ability to modify surfaces functionally and provide hydrophobicity, hydrophilicity, or biocompatibility. However, the reproducibility of the formation of SAMs is difficult. In this chapter, the formation mechanism of SAMs has been studied for high reproducibility. The control of surface properties by fabrication of micro/nanosized domains composed of SAMs is expected to be applied in the field of biomaterials. In addition, electrochemical SPL is also expected to be applied to various devices.

6. References

Hayashi, K.; Saito, N.; Sugimura, H.; Takai, O. & Nagagiro, N. (2002). Regulation of the Surface Potential of Silicon Substrates in Micrometer Scale with Organosilane Self-Assembled Monolayers. *Langmuir,* Vol.8, No.20 (September 2002), pp. 7469-7472, ISSN 0743-7463.

Hong, L.; Sugimura, H.; Furukawa, T. & Takai, O. (2003). Photoreactivity of Alkylsilane Self-Assembled Monolayers on Silicon Surfaces and Its Application to Preparing Micropatterned Ternary Monolayers. *Langmuir,* Vol.19, No.6 (February 2003), pp. 1966-1069, ISSN 0743-7463.

Saito, N.; Wu, Y.; Hayashi, K.; Sugimura, H. & Takai,O. (2003). Principle in Imaging Contrast in Scanning Electron Microscopy for Binary Microstructures Composed of Organosilane Self-Assembled Monolayers. *The Journal of Physical Chemistry B,* Vol.107, No.3 (December 2002), pp. 664–667, ISSN 1089-5647.

Hahn, J. & Webber, S. E. (2004). Graphoepitaxial Deposition of Cationic Polymer Micelles on Patterned SiO_2 Surfaces. *Langmuir,* Vol.20, No.4 (January 2004), pp. 1489-1494, ISSN 0743-7463.

Kidoaki, S. & Matsuda, T. (1999). Adhesion Forces of the Blood Plasma Proteins on Self-Assembled Monolayer Surfaces of Alkanethiolates with Different Functional Groups Measured by an Atomic Force Microscope. *Langmuir,* Vol.15, No.22 (September 1999), pp. 7639-7646, ISSN 0743-7463.

Harnett, C. K.; Satyalakshmi, K. M. & Craighead, H. G. (2001). Bioactive Templates Fabricated by Low-Energy Electron Beam Lithography of Self-Assembled Monolayers. *Langmuir,* Vol.17, No.1 (December 2000), pp. 178-182, ISSN 0743-7463.

Kaholek, M.; Lee, W. K.; LaMattina, B.; Caster, K. C. & Zauscher, S. (2004). Fabrication of Stimulus-Responsive Nanopatterned Polymer Brushes by Scanning-Probe Lithography. *Nano Letters,* Vol.4, No.2 (January 2004), pp. 373–376, ISSN 1530-6984.

Blackledge, C.; Egebretson, D. A. & McDonald, J. D. (2000). Nanoscale Site-Selective Catalysis of Surface Assemblies by Palladium-Coated Atomic Force Microscopy Tips: Chemical Lithography without Electrical Current. *Langmuir,* Vol.16, No.22 (September 2000), pp. 8317–8323, ISSN 0743-7463.

Tello, M.; García, F. & García, R. (2002). Linewidth Determination in Local Oxidation Nanolithography of Silicon Surfaces. *Journal of Applied Physics,* Vol.92, No.7 (September 2002), pp. 4075-4080, ISSN 1089-7550.

Liu, S.; Maoz, R.; Schmid, G. & Sagiv, J. (2002). Template Guided Self-Assembly of [Au_{55}] Clusters on Nanolithographically Defined Monolayer Patterns. *Nano Letters,* Vol.2, No.10 (August 2002), pp. 1055-1060, ISSN 1530-6984.

Jang, C. H.; Stevens, B. D.; Carlier, P. R.; Calter, M. A. & Ducker, W. A. (2002). Immobilized Enzymes as Catalytically-Active Tools for Nanofabrication. *Jounal of the American Chemical Society,* Vol.124, No.41 (September 2002), pp. 12114-12115, ISSN 0002-7863.

Kaholek, M.; Lee, W. K.; Ahn, S. J.; Ma, H.; Caster, K. C.; LaMattina, B. & Zauscher, S. (2004). Stimulus-Responsive Poly(N-isopropylacrylamide) Brushes and Nanopatterns Prepared by Surface-Initiated Polymerization. *Chemical of Materials,* Vol.16, No.19 (August 2004), pp. 3688-3696, ISSN 1520-5002.

Sugimura, H. & Nakagiri, N. (1995). Degradation of a Trimethylsilyl Monolayer on Silicon Substrates Induced by Scanning Probe Anodization. *Langmuir,* Vol.11, No.10 (October 1995), pp. 3623-3625, ISSN 0743-7463.

Amro, N. A.; Xu, S. & Liu, G. Y. (2000). Patterning Surfaces Using Tip-Directed Displacement and Self-Assembly. *Langmuir,* Vol.16, No.7 (March 2000), pp. 3006-3009, ISSN 0743-7463.

Xu, S.; Miller, S.; Laibinis, P. E. & Liu, G. Y. (1999). Fabrication of Nanometer Scale Patterns within Self-Assembled Monolayers by Nanografting. *Langmuir,* Vol.15, No.21 (August 1999), pp. 7244-7251, ISSN 0743-7463.

Liu, G. Y.; Xu, S. & Qian, Y. (2000). Nanofabrication of Self-Assembled Monolayers Using Scanning Probe Lithography. *Accounts of Chemical Research,* Vol.33, No.7 (March 2000), pp. 457-466, ISSN 1520-4898.

Pena, D. J.; Raphael, M. P. & Byers, J. M. (2003). "Dip-Pen" Nanolithography in Registry with Photolithography for Biosensor Development. *Langmuir,* Vol.19, No.21 (September 2003), pp. 9028-9032, ISSN 0743-7463.

Schwartz, P. V. (2002). Surface Functionalization and Stabilization of Mesoporous Silica Spheres by Silanization and Their Adsorption Characteristics. *Langmuir,* Vol.18, No.10 (April 2002), pp. 4014-4019, ISSN 0743-7463.

Maynor, B. W.; Li, J.; Lu, C. & Liu, J. (2004). Site-Specific Fabrication of Nanoscale Heterostructures: Local Chemical Modification of GaN Nanowires Using Electrochemical Dip-Pen Nanolithography. *Jounal of the American Chemical Society,* Vol.126, No.20 (May 2004), pp. 6409-6413, ISSN 0002-7863.

Lee, SH.; Saito, N. & Takai, O. (2009). Highly reproducible technique for three-dimensional nanostructure fabrication via anodization scanning probe lithography. *Applied Surface Science,* Vol.255, No.16 (May 2009), PP. 7302-7306, ISSN 0169-4332.

Lee, SH.; Ishizaki, T.; Saito, N. & Takai, O. (2007). Electrochemical soft lithography of an 1,7-octadiene monolayer covalently linked to hydrogen-terminated silicon using scanning probe Microscopy. *Surface Science,* Vol.601, No.18 (September 2007), pp. 4206-4211, ISSN 0039-6028.

Lee, SH.; Ishizaki, T.; Saito, N. & Takai, O. (2007). Local generation of carboxyl groups on an organic monolayer through chemical conversion using scanning probe anodization.

Materials Science and Engineering C, Vol.27, No.5-8 (September 2007), pp. 1241-1246, ISSN 0928-4931.

Saito, N.; Lee, SH.; Ishizaki, T.; Hieda, J.; Sugimura, H. & Takai, O. (2005). Surface-Potential Reversibility of an Amino-Terminated Self-Assembled Monolayer Based on Nanoprobe Chemistry. *The Journal of Physical Chemistry B,* Vol.109, No.23 (May 2005), pp. 11602-11605, ISSN 1089-5647.

Sugimura, H.; Saito, N.; Lee, SH. & Takai, O. (2004). Reversible nanochemical conversion. *Journal of Vacuum Science & Technology B*, Vol.22, No.6 (November 2004), pp. L44-46, ISSN 1520-8567.

11

Extremely Wetting Pattern by Photocatalytic Lithography and its Application

Yuekun Lai[1,2], Changjian Lin[1*] and Zhong Chen[2*]
[1]*State Key Laboratory for Physical Chemistry of Solid Surfaces and College of Chemistry and Chemical Engineering, Xiamen University*
[2]*School of Materials Science and Engineering, Nanyang Technological University*
[1]*China*
[2]*Singapore*

1. Introduction

Wettability is an important property governed by not only chemical composition, but also geometrical structure as well (Ichimura et al., 2000; Lai et al., 2009a; Wang et al., 1997). Super-wetting and antiwetting interfaces, such as superhydrophilic and superhydrophobic surfaces with special liquid-solid adhesion have recently attracted worldwide attention. Superhydrophilicity and superhydrophobicity are defined based on the conventional water contact angle experiment. If the contact angle is smaller than 5°, the surface is said to be superhydrophilic. Superhydrophobic refers to surface with contact angle greater than 150°. Such two extremely cases have attracted much interest due to their importance in both theoretical research and practical application (Lafuma & Quéré, 2003; Liu et al., 2010; Gao & Jiang, 2004).

In recent years, patterned thin films have received considerable attentions due to their interesting properties for a range of applications, such as optoelectronic devices, magnetic storage media, gas sensors, and fluidic systems. Compared to the conventional thin film technology, such as physical vapor deposition (Li et al., 2006; Zhang & Kalyanaraman, 2004), chemical vapor deposition (Jeon et al., 1996; Slocik et al., 2006) and sputtering (Rusponi et al., 1999), solution-based deposition method is becoming popular for the fabrication of patterning films due to the low temperature process under ambient environment, less energy and time consumption, and easier control of the experimental parameters (Lai et al., 2010a; Liu et al., 2007; Yoshimura & Gallage, 2008). Although traditional photolithographic technique is excellent for preparing sub-micrometer or even only sub-100-nanometer pattern (Cui & Veres, 2007; Li et al., 2009), it is a complex multi-step process (wafer cleaning; barrier layer formation; photoresist coating; soft-baking; mask alignment; exposure and development; and hard-baking) and needs to remove part of the film and all the photoresist used. Direct and selective assembly of nanostructured materials from precursors paves a new avenue for the fabrication of electronic optical microdevices.

Wetting micropatterns with different physical or chemical properties, without the need for ultra-precise positioning, have frequently been acted as templates for fabricating various functional materials in a large scale. The great difference in contact angle of the two extreme cases provides a potentially powerful and economical platform to directly and precisely construct patterned nanostructures in aqueous solution. In general, wetting micropatterns with low contact angle contrast (≤120°) on smooth substrates can be formed by photolithography (Falconnet et al., 2004; Kobayashi et al., 2011), microcontact lithography (Csucs et al., 2003; Kumar et al., 1992), colloidal patterning (Michel et al., 2002; Bhawalkar et al., 2010), electron beam lithography (Wang & Lieberman, 2003; Zhang et al., 2007), nanoimprint lithography (Jiao et al., 2005; Zhang et al., 2006), dip-pen nanolithography (Huang et al., 2010a; Lee et al., 2006; Xu & Liu, 1997), and so on. Among these methods, photocatalytic lithography employing semiconductors to photocatalytic decompose of organic monolayer is one of the most practical techniques because it able to accurately transfer an entire photomask pattern to a target substrate at a single exposure time under environmental condition (Bearinger et al., 2009; Lee & Sung, 2004; Nakata et al., 2010; Tatsuma et al., 2002; Wang et al., 2011). Moreover, it can greatly reduce the photoresist waste. The resolution of the patterning is greatly dependant on the mask alignment and light source exposure. Under optimal condition, a resolution of micrometer- or submicrometer-scale pattern of alkylsiloxane self-assembled monolayers can be achieved with UV light projection irradiation. Once patterned on the surface, organic monolayer has been applied in various ways to restrict corrosion or induce nanostructures growth. Firstly, patterned layer itself may serve as etching mask to protect the substrate to generate pattern with certain thickness/aspect ratio. Secondly, patterned organic layer may be employed as barrier to inhibit the liquid phase deposition of nanostructures to generate functional composite pattern with diverse shape and density. So far, only a few reports have been available on the fabrication and application of superhydrophilic–superhydrophobic patterning by photocatalytic lithography under ambient conditions (Lai et al., 2008a; Nishimoto et al., 2009; Zhang et al., 2007).

Upon UV irradiation, the electron-hole pairs in semiconductor TiO_2 can be generated and migrated to its surface, where the hole reacts with OH- or adsorbed water to produce highly reactive hydroxyl radicals (Zhao et al., 1998). These hydroxyl radicals can further oxidize and decompose most organic compounds. Recently, we found that the pollutant solution can be rapidly decomposed on a nanotube array TiO_2 film with UV irradiation (Lai et al., 2006, 2010b; Zhuang et al., 2007). Considering its effectiveness for the photocatalytic decomposition of organic compounds, the photocatalysis of such TiO_2 nanotube film can be a promising way to decompose the low energy hydrophobic fluoroalkyl chains. So it is possible to achieve a conversion from superhydrophobicity to superhydrophilicity due to the amplification effect of the rough aligned nanotube structure. By using a patterned photomask to control the site-selective decomposition by UV light, that is photocatalytic lithography, superhydrophilic cells can be accurately transferred to a target substrate at a single exposure time under environmental condition. Therefore, these two types of extreme wettability coexist on the surface directly to make up of superhydrophilic-superhydrophobic pattern.

In this chapter, we firstly discuss the wettability on TiO_2 nanostructure film by electrochemical anodization. Secondly, we demonstrate using a novel synthetic process to prepare wetting pattern with a high contrast (superhydrophilic–superhydrophobic) on TiO_2 nanotube structured film by a combination of SAM technique and photocatalytic lithography. The resultant micropattern has been characterized with scanning electron microscopy, optical microscopy, electron probe microanalyzer and X-ray photoelectron spectroscopy. Finally, we focus on the technological details and potential future application of wetting template to induce and direct the assembly of functional nanostructure to form uniform micropatterns. For example, the patterning, biomedical and sensing application of the wetting template and functional composite nanostructure pattern (TiO_2, ZnO, OCP and CdS) (Lai et al., 2009b, 2010a-d).

2. Wettability on TiO_2 nanostructures by electrochemical anodization

Wettability of solid surfaces is a very important property of solid surface. Surfaces with extreme wetting properties, e.g. superhydrophilic and superhydrophobic, can be prepared by introducing certain rough structures on the originally "common" hydrophilic and hydrophobic surfaces. Various ways of preparing TiO_2 semiconductor films on the different solid substrates have been developed, including sol-gel technique (Shen et al., 2005), sputtering (Takeda et al., 2001), chemical vapor deposition (Rausch & Burte, 1993), liquid phase deposition (Katagiri et al., 2007), hydrothermal (Yun et al., 2008), and electrochemical anodizing. Among them, the electrochemical anodizing is verified to be a convenient technique for fabricating nanostructured TiO_2 films on titanium substrates (Lai et al., 2004, 2008b, 2009c; Gong et al., 2003). Moreover, the conductive titanium support substrate can be an advantage for fabricating functional material composites through electrochemical depositions to further improve their photoelectrochemical activities.

Figure 1a shows a typical FESEM image of the titanium substrate before electrochemical anodization. The surface of the substrate was relatively smooth, with features of parallel polished ridges and grooves at the micron scale (Lai et al., 2010a). Figure 1b shows the top view SEM image of the typical TiO_2 nanotube array film by anodizing under 20 V for 20 min. After anodization, shallow cavities as large as several micrometers in diameter were present on the surface of the sample. This is probably due to the anisotropic oxidation of the underlying Ti grains (Crawford & Chawla, 2009; Yasuda et al., 2007). From the high magnification image (Fig. 1c), it can be seen that vertically aligned TiO_2 nanotubes with inner diameter of approximately 80 nm covered the entire surface including the shallow polygonal micropits. The side view image shows that the self-assembled layers of the TiO_2 nanotubes were open at the top and closed at the bottom with thickness about 350 nm (inset of Fig. 1c). Water droplet can quickly spread and wet the as-grown vertically aligned TiO_2 nanostructure film due to capillary effect caused by the rough porous structure, indicating such TiO_2 structure film by electrochemical anodizing is superhydrophilic. A more hydrophobic behaviour, on the other hand, was obtained after coating the TiO_2 film with fluoroalkyl silane. The inset of Fig. 1b shows the intrinsic contact angle (CA) on the as-prepared vertically aligned TiO_2 nanotube surface and its corresponding 1*H*,1*H*,2*H*,2*H*-perfluorooctyltriethoxysilane (PTES, Degussa Co., Ltd.) modified surface is nearly 0° (superhydrophilic) and 156° (superhydrophobic), respectively. However, the CA for the "flat" TiO_2 surface and its corresponding PTES modified sample is about 46° (hydrophilic) and 115° (hydrophobic), respectively. From these results, we know the top surface of the vertically aligned nanotubes has an amplification effect to make hydrophilic and hydrophobic surfaces become superhydrophilic and superhydrophobic, respectively. After UV irradiation for 30 min, the water CA on the TiO_2 nanotube film and "flat" TiO_2 film decreased to 0° and 26°, as a consequence of the photocatalytic activity of TiO_2 films (Balaur et al., 2005; Lai et al., 2010a). Moreover, the sample showed hydrophobic character once again when it was treated with PTES. Therefore the surface can be reversibly switched between superhydrophobic and superhydrophilic by alternating SAM and UV photocatalysis on the rough TiO_2 nanotube arrays (shown in Fig. 1d). Compared with the large wettability contrast on this type of rough surface (larger than 150°), the wettability of a "flat" TiO_2 film can only be reversibly changed within the small range between 26° and 115°.

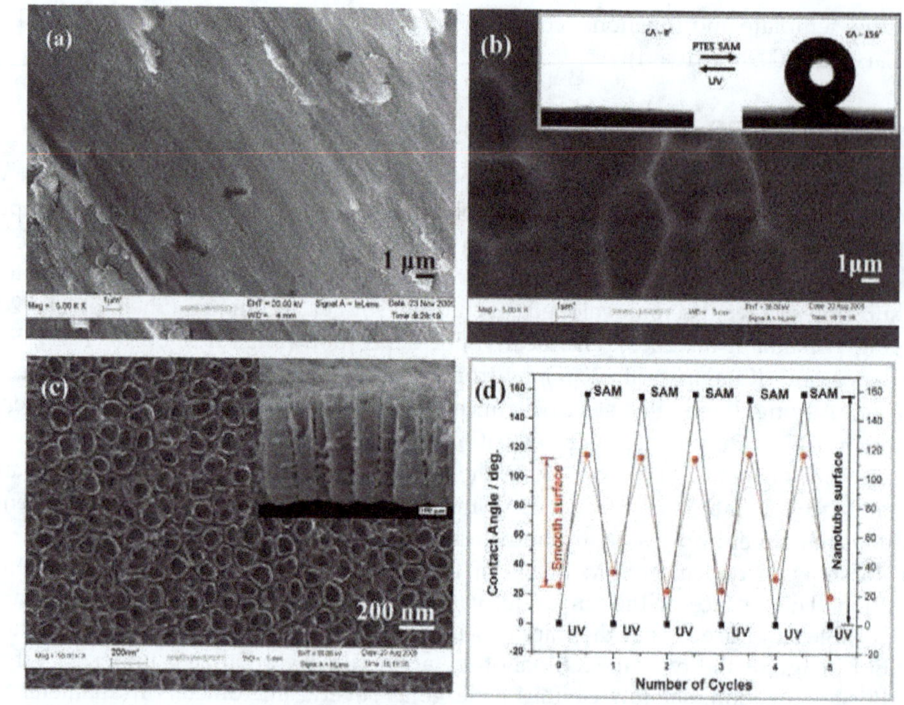

Fig. 1. SEM images of the mechanically polished and cleaned titanium substrate (a), low magnification of a nanotube structured TiO_2 film (b), and a higher magnification of the TiO_2 nanotube array film (c). Reversible surface wettability on a ''flat'' TiO_2 film and rough nanotube TiO_2 film by alternating SAM and UV photocatalysis (d). The inset of (b) shows the shape of a water drop on the PTES-modified and UV-irradiated TiO_2 nanotube array film. The inset of (c) shows the side view of a TiO_2 nanotube array film.

Recently, we designed three types of superhydrophobic nanostructure models consisting of a nanopore array (NPA), a nanotube array (NTA), and a nanovesuvianite structure (NVS) apply a facile electrochemical process (Figure 2) (Lai et al., 2009a). Based on basic principles of roughness-enhanced hydrophobicity and capillary-induced adhesion, these different porous structures were expected to create interfaces with decreasing adhesive forces. The surface adhesive forces could be effectively tuned by solid-liquid contact ways at the nanoscale and air-pocket ratio in open and sealed systems. The magnitude of the adhesive force of a droplet for a superhydrophobic surface descends in the order "area-contact" > "line-contact" > "point-contact". A continuous three-phase (solid-air-liquid) contact line (TCL) generates serious CA hysteresis and surface adhesion, while a discrete TCL is energetically advantageous to drive a droplet off a superhydrophobic surface, showing lower surface adhesion. Therefore, the water droplet behavior on these superhydrophobic surfaces could be greatly changed from pinning to sliding by adjusting the solid-liquid contact way.

Capillary adhesive force plays a dominant role in imparting adhesive behavior on NPA and NTA nanostructures will sealed cells, while the open NVS nanostructure, which had extremely low adhesion capacity for water, acted solely by van der Waals attraction between water and PTES molecules. A possible explanation is as follows. As the droplet gradually retracted from the sample surface, the meniscus on each nanotube nozzle would be changed from concave to convex (Figure 3a). This could result in an increased volume of air sealed in each nanotube by the liquid/air interface. According to Boyle's law (West, 1999), there is an inverse relationship between the volume (V) and pressure (P) for an ideal gas under the conditions of constant temperature and quality. Therefore, this expansion of air would result in the formation of a negative pressure (ΔP). In this case, the volume of air sealed in the nanotubes was varied by their depths, so longer tubes would be expected to have lower air-expansion ratios (ΔV/V), thus lesser negative pressures. For a fixed nanotube diameter, a longer nanotube would therefore require a smaller pulling-off force, and the total surface adhesive force would be smaller.

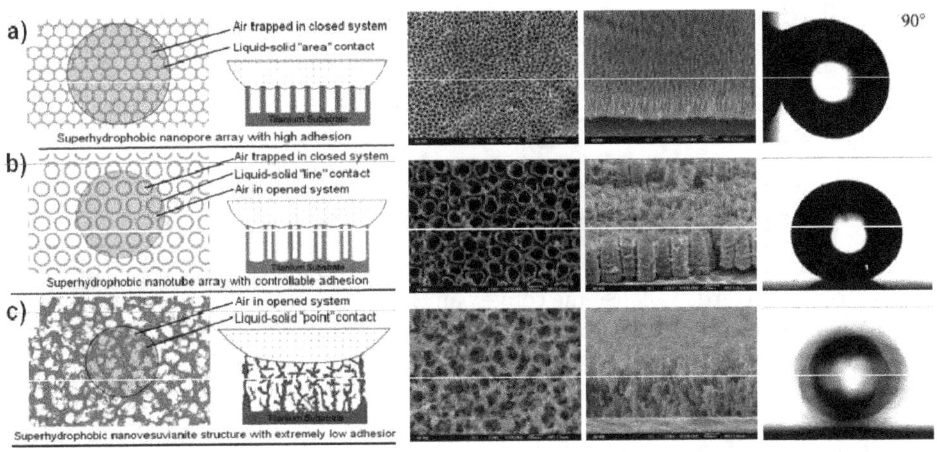

Fig. 2. Schematic models, SEM images and corresponding water behavior on three types of superhydrophobic porous-nanostructure with water adhesive forces ranging from high to low. (a) Superhydrophobic NPA with high adhesion. (b) Superhydrophobic NTA with controllable adhesion. (c) Superhydrophobic NVS with extremely low adhesion.

Figure 3b shows the curves of the water CAs and adhesive forces with respect to the diameter of individual nanotube. When the nanotube diameter decreases, the force drastically increased, while the CA slightly decreased. When the diameter was tuned from 78 nm down to 38 nm, the surface adhesive force of the superhydrophobic NTA film increased 2.06 times, while the decline in magnitude of the water CA was not more than 2%, showing that the negative pressure caused by the volume change of air sealed in the nanotubes could effectively tune the surface adhesive force. The NTA structures in this study had variable length, with values of (0.35 ± 0.04) μm, (0.76 ± 0.05) μm, (1.18 ± 0.07) μm, while their diameter (~80 nm) was fixed. Figure 3c shows the curves for water CAs and adhesive forces obtained with PTES-modified NTA-nanostructure surfaces differing in nanotube length. With lengths extending from 0.35 μm to 0.76 μm and 1.18 μm, the CA change was very small, not more than 2%, which could be due to the minor variation in nanotube diameter. However, the adhesive force linearly decreased from 21.5 μN down to 16.7 μN and 12.2 μN for the above increases in length, respectively. It was evident that the water adhesive force of the superhydrophobic NTA-nanostructure surfaces could be tuned by varying the diameters and also lengths of the nanotubes. These findings are valuable to deepen insight into the roles of nanostructures in tailoring surface water-repellent and adhesive properties for exploring new applications.

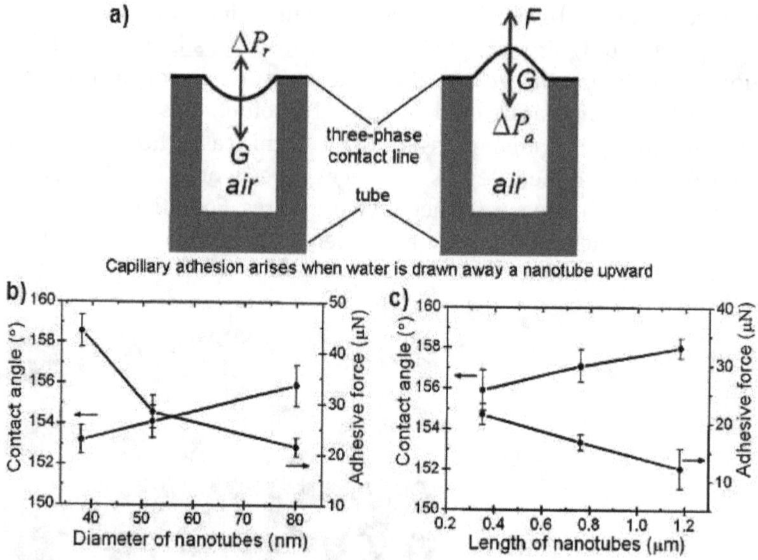

Fig. 3. (a) Capillary adhesion arises when a water droplet sitting on the tube nozzle is gradually drawn upwards because the convex air/liquid interface produces an inward pressure ΔP. (b,c) The curves of water contact angles and adhesive force on the superhydrophobic NTA nanostructures with respect to the diameter and length of nanotubes.

3. Wetting pattern by photocatalytic lithography

A novel approach for constructing superhydrophilic-superhydrophobic micropattern on the nanotube structured TiO_2 films has developed by using photocatalytic lithography (Figure 4a) (Lai et al., 2008a). At the first step, the as-prepared amorphous TiO_2 nanotubes by electrochemical anodizing of titanium sheets were calcinated at 450°C to form anatase phase, then treated with a methanolic solution of hydrolyzed 1 wt% PTES for 1 h and subsequently heated at 140°C for 1 h, and at the second step, the superhydrophobic film is selectively exposed to UV light through a copper grid (photomask) to photocatalytically cleave the fluoroalkyl chain. It is noteworthy, from the characterization of chemical composition before and after UV irradiation by X-ray photoelectron spectroscopy, that the intensities of the F1*s* and F*KLL* are decreased greatly and those of the Ti2*p* and O1*s* are increased after exposing the PTES modified surface to UV light for 20 min (Fig. 4b). From the inset high-resolution spectra (Fig. 4c), the peaks of $-CF_2$ (at 291.8 eV) and $-CF_3$ (at 294.1 eV) are obviously vanished after UV light irradiation, while the strength of silicon peaks in the XPS spectra remains unchanged but shifts from 102.8 to 103.3 eV, suggesting that Si-O-Si networks have already formed due to UV irradiation. According to these results, we believe that the hydrophobic fluoroalkyl chains have been completely decomposed and removed by the photocatalytic reactions at TiO_2 nanotube films. Similarly, a serial of fluoroalkyl silane monolayer pattern (e.g. heptadecafluorodecyltrimethoxysilane, octadecyltriethoxysilane, and methyltriethoxysilane) can be successfully fabricated in our case. Although various monolayer patterns can be prepared with a resolution about micro-scale or submicro-scale under optimal condition, we will focus on the application of the PTES micro-pattern with a general TEM copper grid as a photomask by photocatalytic lighography.

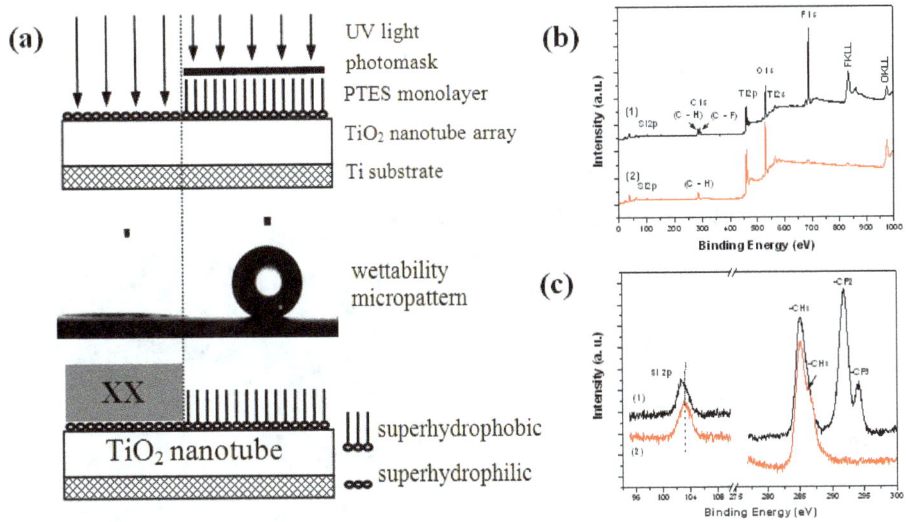

Fig. 4. Schematic outline of the procedures to fabricate nanostructured patterning film by electrochemical deposition based on superhydrophilic-superhydrophobic micropattern (a). Survey-scan X-ray photoelectron spectra of the PTES modified nanotube TiO_2 films before (1) and after (2) 20 min UV irradiation (b). The high-resolution spectra of Si2*p* and C1*s* regions (c).

Figure 5 shows the optical micrograph of the as-obtained superhydrophilic-superhydrophobic pattern by focusing on the droplet within the superhydrophilic regions. A uniform pattern is formed due to the site-selective wetting by water droplets within the superhydrophilic regions (Fig. 5a). A light dot array (inset of Fig. 5a) is seen when focusing on the top of the droplets, indicating that the confined droplet has a hemispherical dome. To further verify the resulting micropatterns with an extreme wettability contrast, fluorescein sodium was used as a probe to label the surface of the films. Figure 5b shows the fluorescent micrograph of the resultant superhydrophilic–superhydrophobic micropatterns on the TiO_2 nanotube array films. As shown, geometrically identical square superhydrophilic regions and dark superhydrophobic regions transferred well from the photomask to form a well-defined pattern. The UV-irradiated regions become superhydrophilic owing to the photocatalytic cleavage of the PTES molecule and the enhanced roughness of the nanotube structure, while the non-irradiated parts remain superhydrophobic without any change. Because the difference in the water CA between the irradiated and non-irradiated regions is larger than 150°, the liquid containing the fluorescent probe selectively appears only on the uniform superhydrophilic grids and not on the neighboring superhydrophobic regions. Therefore, a clear, well-defined fluorescent pattern in line with the dimensions of the Cu grid can be obtained. These results indicate that the micropatterned template composed of superhydrophilic and superhydrophobic regions was fabricated successfully.

The UV irradiation times had a great effect on the quality of the resulting pattern. For example, it cannot exhibit a sufficient wettability contrast between the irradiated and non-irradiated regions to form a well-defined pattern within 5 min. This is attributed to the hydrophobic fluoroalkyl chain in the PTES molecule that was not efficiently cleaved under a short-time UV irradiation. However, with a long-time UV irradiation (i.e., 60 min), the adjacent PTES molecule covered by the Cu grid can be remotely oxidized by a TiO_2 nanotube photocatalyst or the diffusion, scattering, and diffraction of the incident light (Haick & Paz, 2001; Kubo et al., 2004). Therefore, to obtain a higher pattern resolution, the optimized UV irradiated time in our case was controlled in the range of 10-30 min.

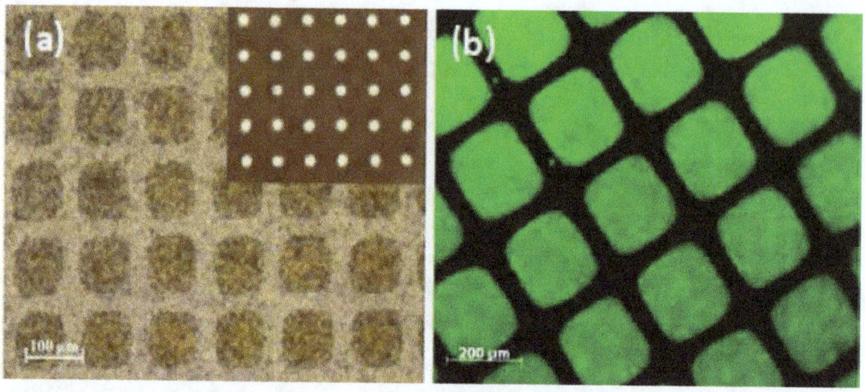

Fig. 5. (a) Optical micrograph of the as-obtained superhydrophilic-superhydrophobic pattern by focusing on the water droplet within the superhydrophilic regions. (b) Fluorescence microscopy image of the fluorescein probes on the as-prepared superhydrophilic-superhydrophobic micropattern.

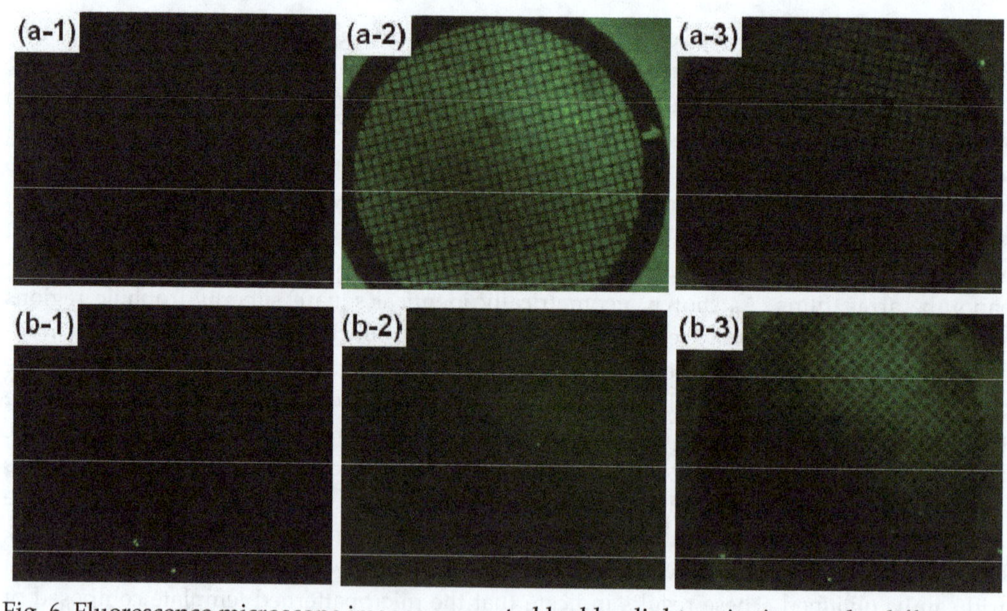

Fig. 6. Fluorescence microscope images generated by blue light excitation on the different micropatterned templates with the adsorption of FITC-BSA, (a-1, a-2, a-3) the PTES template with a pH value of 7.5, 4.5 and 2.5, respectively; (b-1, b-2, b-3) the PTES-APTS template with a pH value of 7.5, 4.5 and 2.5, respectively.

Based on the molecular self-assembly and photocatalytic lithography techniques, micropatterned templates of PTES or PTES-APTS(3-aminopropyltriethoxysilanes) with different wettabilities were fabricated on the titania film. The adsorption behavior of bovine serum albumin (BSA) on the above two templates was investigated using fluorescent labeling (FITC) in buffer solution with different pH values. The results showed that, for the PTES template with great wettability differences, BSA would preferentially adsorb on the superhydrophilic regions. For the APTS-PTES template with smaller differences in wettability, competitive adsorption phenomenon on the super-hydrophobic regions was found due to the hydrophobic interaction force between the albumin and the surface. As the pH value decreased to 2.5, the phenomenon of competitive adsorption was prominent with the albumin adsorbed in the super-hydrophobic areas. The adsorption feature of the albumin may be closely related to the wettability and surface energy of the materials. This technique has promising applications in bio-compatible coatings where drugs could be encapsulated in specific areas of the coating using simple microfabrication methods.

4. Application of wetting pattern

Uniform self-assembly of functional inorganic nanomaterials is a fundamental challenge. Nature adopts a superior approach in biomineralization, where "matrix" macromolecules induce nucleation of inorganic crystals at specific locations with controlled size and morphology, and sometimes even with defined growth orientation. We apply the biomimetic principles derived from liquid phase processes to the assembly of nanoscale functional materials into microscale systems. We carefully control surface wettability to promote etching or heterogeneous nucleation at designated superhydrophilic regions while completely suppress these processes elsewhere (superhydrophobic regions), therefore enable the controlled top-down or bottom-up assembly of inorganic nanomaterials directly from solution. Following this principle, arrays of crystalline TiO_2 nanotube, ZnO nanorods, CdS semiconductor materials and octacalcium phosphate (OCP) biomaterials were nucleated and assembled directly from solution onto Ti substrates at the desired precise locations and then fabricated into arrays of photodetector or matrix devices for large-area microelectronic applications. This strategy of micropatterned nanocomposites will be helpful to develop various micropatterned functional nanostructured materials.

4.1 Template for preparing functional pattern

Figure 7 shows an optical microscopy image of the TiO_2 nanotube micropattern produced using a grid micropattern with different wet etching times (Lai et al., 2009b). A patterning with a clear outline was formed in a short time for 30 s (Fig. 7a). With an increase in the wet etching time (Fig. 7b and c), identical micropatterns with higher aspect ratios can be fabricated. When the etching was prolonged to 240 s (Fig. 7d), the size of the grids increased slightly, indicating that the PTES-SAM layer at the edge of the superhydrophobic lines is more easily etched as compared to the inner superhydrophobic area. This is due to the loose and disordered SAMs resulting from the scattered UV light photocatalytic degradation and the transfer of the active hydroxyl radicals at the edge of the grids. Moreover, the isotropic etching of the Ti substrate underneath the boundary leads to the collapse of the upper nanotube array structures. Therefore, the pattern can be obtained with a clear boundary in a short time after the wet etching in the aqueous solution, and the depth of the etching can be controlled simply by adjusting the etching time.

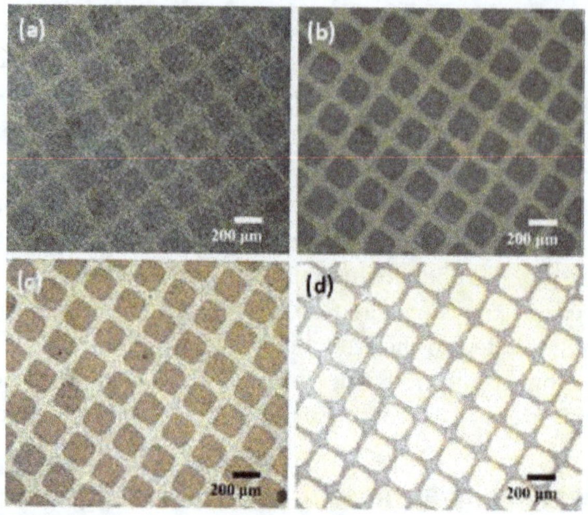

Fig. 7. Optical micrograph images of the time-resolved evolution process of the resulting vertical aligned TiO_2 nanotube micropattern on a superhydrophilic-superhydrophobic template by wet chemical etching in 0.1 wt% HF solution: (a) 30, (b) 60, (c) 120, and (d) 240 s.

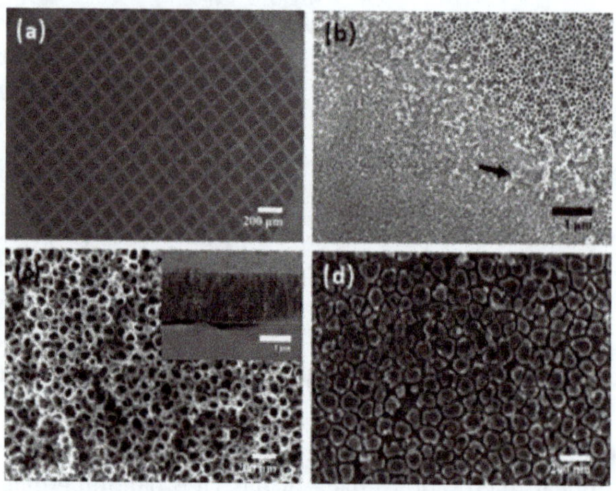

Fig. 8. (a) The SEM micrograph of the ordered vertical aligned TiO_2 nanotube array pattern by vapor-condense etching for 10 min. Magnified images of the corresponding (b) superhydrophilic-superhydrophobic boundary, (c) superhydrophobic area, and (d) superhydrophilic area. The inset figure shows the corresponding cross-sectional image.

The micropattern with a higher aspect-ratio can also be fabricated by a developed vapor etching technique. As water evaporated, the vapor containing HF dewetted the superhydrophobic lines while condensing on the superhydrophilic grids to etch the vertical aligned TiO_2 nanotube layer. Figure 8a and b shows the as-prepared TiO_2 micropattern with a clear boundary by water vapor etching for 10 min. This resulted from the selective condensing of the vapor from the water solution containing 5 wt% HF in the superhydrophilic regions. Some collapsed residue (indicated by the black arrow) covers the boundary due to the rapid etching of the bottom nanotubes and the resultant bubbles;

however, they can be easily removed by sonication. The vertical aligned TiO_2 nanotube film with a length of about 1.53 μm shows no obvious change within the superhydrophobic regions (Fig. 8c), while the TiO_2 nanotubes in the superhydrophilic regions are etched completely for 20 min (Fig. 8d), that is to say, a TiO_2 micropattern with a high aspect ratio of ~20 can be obtained. This technique is particularly attractive in generating large-area functional nanostructure patterns in a high throughput fashion with a high aspect-ratio.

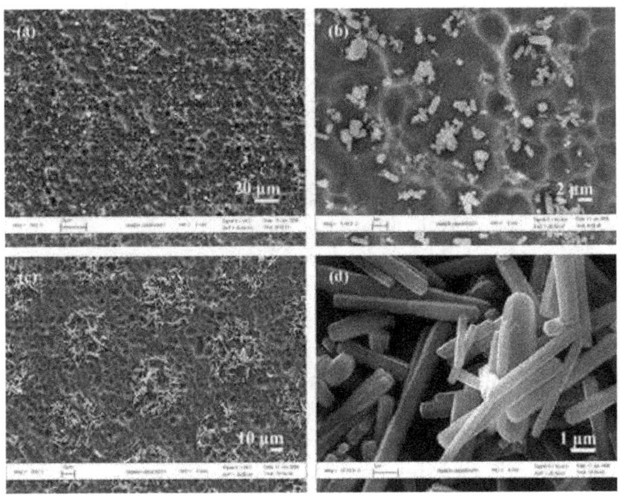

Fig. 9. SEM images of the patterned ZnO nanostructured micropatterns by liquid phase deposition at 90°C for different times: (a, b) 30 min; (c, d) 90 min.

Figure 9 shows representative top-view FESEM images of the ZnO/TiO_2 micropatterns by liquid-phase deposition after different times. After growth for 30 min (Fig. 9a,b), the nucleation and growth of ZnO crystals with various morphologies and sizes (nanoparticles and nanorods) are sparsely dispersed within the predefined superhydrophilic regions. In the superhydrophobic regions, the nanotube structure is retained with its original morphology due to the indirect contact with the solution and the effective protection by the PTES monolayer. Upon further increase in the deposition time to 90 min (Fig. 9c,d), it was observed that ZnO nanorods were the predominant structural features. The average diameter of the grown ZnO nanorods increases greatly to about 400–800 nm in diameter and 3-5 μm in length, which may be due to lower nucleation rate and higher growing space for ZnO nanorods. The superhydrophilic microdots are almost covered with randomly lying ZnO nanorods. Moreover, there are different growth orientations and some connections into adjacent nanorods. It is of interest to note that a two dimensional (2D) pattern with smaller density of randomly packed ZnO nanorods, instead of the 3D pattern consisted of well-aligned vertical nanorods, which were confined and grown within the superhydrophilic regions.

For practical application of thin film devices, the position and orientation-control of ZnO nanorods are very important because they directly relate to their physical and chemical performances (Koumoto et al., 2008; Masuda et al., 2006). In order to precisely control the spatial orientation of the ZnO nanostructures, we developed a new technique which is able to make the ZnO nanorods grow along the vertically aligned titania nanotubes rather by disordered deposition on the titania nanotube array surface. In this technique, resistance discrepancy was adopted to make the entrance of the tubes more conductive than the bottom of the tubes, to induce the epitaxial growth with spatial organization of uniform

ZnO nanorods along the direction of the nanotubes. The quasi-perpendicular ZnO nanorods nucleate and grow uniformly and selectively throughout the superhydrophilic regions of the TiO_2 nanotube surface by electric field assisted deposition at 90°C for 3 min, while no nanorods are observed in the superhydrophobic regions (Fig. 10a,b) (Lai et al., 2010a). The EDS spectra also reveal that the presence of Zn, Ti and O elements on the superhydrophilic regions, while the elemental components in the superhydrophobic areas are only Ti and O. The inset of Fig. 10b shows the hexagonal end facet of a vertically aligned ZnO nanorod with a diameter about 100–150 nm growing on top of the TiO_2 nanotube array surface. Therefore, the density, size and orientation of ZnO nanorods are very sensitive to the presence of electric fields. A 3D AFM profile image (Fig. 10c) shows that the microscopic structure of the ZnO crystal deposition consisted in dense column arrays, which are induced and directed by the wettability template. The thickness of vertical ZnO nanorod film is in the range of 800–900 nm. Furthermore, the three dimensional confocal microscopy image (Fig. 10d) also shows that the growth of the ZnO nanorod pattern is identical with the superhydrophilic/superhydrophobic template.

On the basis of the versatile superhydrophilic-superhydrophobic template, we can successfully control the growth of ZnO nanostructures in the superhydrophilic regions under mild reaction conditions and in the absence of seed and noble metal catalyst. In the superhydrophobic regions, the growth is suppressed. This special template can be utilized to generate different nanostructured ZnO patterns with clearly defined edges. Hence, it is expected that this novel micropatterned technique based on the superhydrophilic-superhydrophobic template will become a powerful tool for fabricating various types of micropatterned nanomaterials and devices.

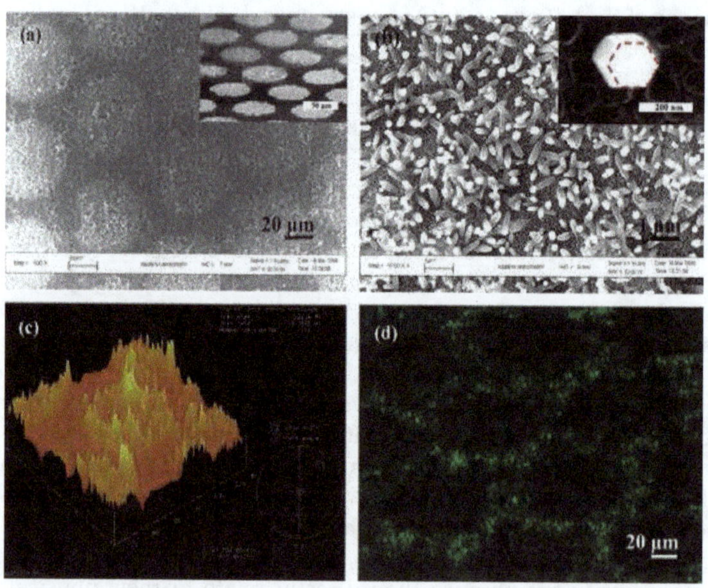

Fig. 10. (a,b) Typical SEM images of the vertically aligned ZnO nanorods selectively grown on superhydrophilic patterning regions by the developed electric field assisted deposition technique at 90°C for 3 min. The inset in (a) shows the side view SEM image of the corresponding ZnO nanorod micropattern. The inset in (b) shows the higher magnified SEM image of a ZnO nanorod with hexagonal end facet. (c) 3D AFM image of the ZnO/TiO_2 micropattern. (d) Confocal microscopy image of the perpendicular ZnO nanorod array.

4.2 Biomedical arrays

The typical SEM image of TiO_2 nanotube array surface before and after the deposition of OCP film by electrochemical technique for 5 min is shown in Fig. 11a,b. The superhydrophobic-superhydrophilic micropatterned TiO_2 was used as a micro-template to selectively deposit nano-OCP crystals on the superhydrophilic regions by an electrochemical deposition to form a special micropatterned nano-OCP. The deposition electrolyte was consisted of 0.042 mol/L $Ca(NO_3)_2$ and 0.025 mol/L $NH_4H_2PO_4$. The pH value was adjusted to approximately 4.2 with 0.05 mol/L NaOH solution. The precipitation was carried out galvanostatically at a cathodic current of 0.5 mA cm^{-2} under 67.5°C for a certain time (Wang et al., 2008). It can see that quasi-perpendicular ribbon-like crystals of several hundred nanometers in width are uniformly grown on the TiO_2 nanotube array surface. Fig. 11c shows a typical fluorescence microscope of the superhydrophilic-superhydrophobic micropattern on TiO_2 nanotube surface. As can be seen, the green dot patterns are clearly imaged through the fluorescence contrast between the UV-irradiated superhydrophilic and photomasked superhydrophobic regions. The photoirradiated dot exhibiting a uniformly stronger fluorescence against the surrounding dark background is due to the highly affinity to solution resulting in the absorption of fluorescent probes into the irradiated nanotube array films. Therefore, a clear well-defined fluorescence pattern based on the superhydrophobic-superhydrophilic pattern is obtained. Fig. 11d displays the identical patterning of OCP biomaterials deposited on the superhydrophilic-superhydrophobic patterns on TiO_2 nanotube array surface. It is obvious that the size of the white OCP dots is equal to that of the superhydrophilic area on template, indicating the deposited regions were only located within the superhydrophilic dots where photocatalytic degradation of PTES-SAMs was performed.

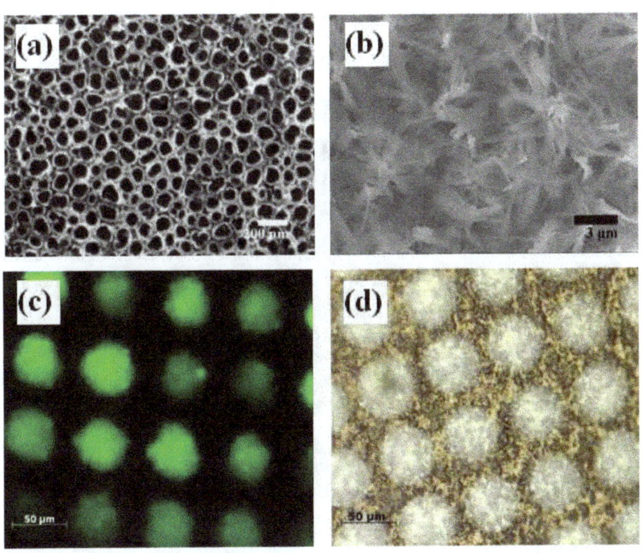

Fig. 11. SEM images of the (a) TiO_2 nanotube array film fabricated by electrochemical anodization; (b) OCP nanostructure layer on TiO_2 nanotube array film by electrochemical deposition for 5 min. Optical fluorescence pattern of the superhydrophilic-superhydrophobic template (c) and patterned OCP thin films selectively deposited in pre-defined superhydrophilic regions by electrochemical deposition for 5 min (d).

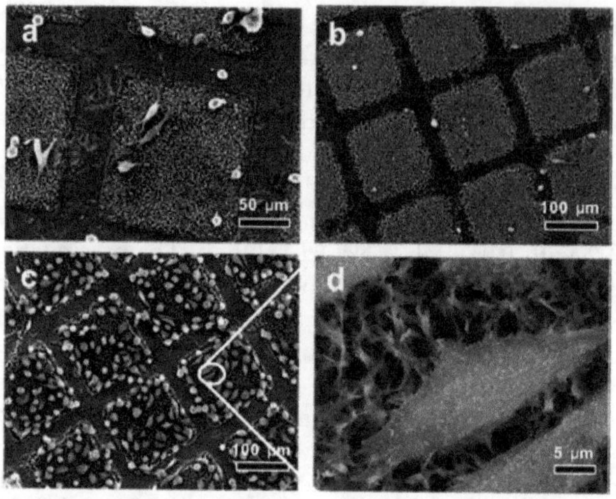

Fig. 12. SEM micrographs of MG-63 cells cultured on the patterned OCP coatings with different deposition time for 6 h, (a) 1 min; (b) 3 min; (c) 5 min; (d) higher magnification.

The *in-vitro* MG-63 cell tests were used to study the biological performance of the as-obtained OCP micropatterns (Huang et al., 2010b). The results showed that MG-63 cells were found preferentially attached on the superhydrophilic regions with OCP thin films, while the superhydrophobic regions with the PTES monolayers can effectively prevented the adhesion of cells on the surface, indicating that the cells had the selective adhesion action on the tiny units of OCP films. Moreover, the cells adhered on the OCP film deposited for a longer period (5 min) are more active to spread on the OCP nanobelt covering surface. It is promising for developing a new cell chip for high throughput evaluation of the cell behaviors.

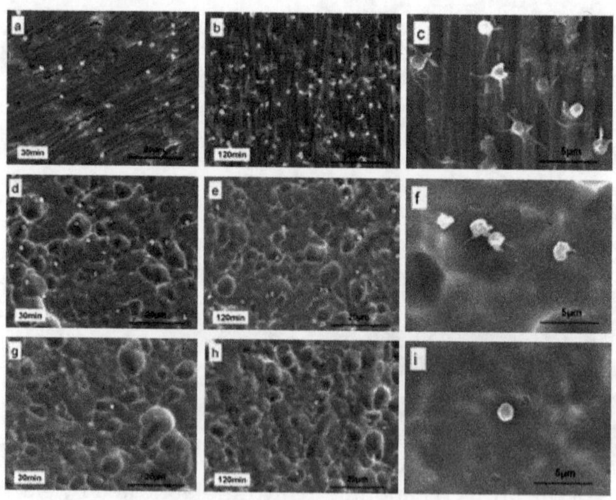

Fig. 13. SEM images of adhered platelets on various kinds of surfaces at 37°C for different periods. (a-c) mechanically polished and cleaned Ti substrate; (d-f) superhydrophilic surface; (g-i) superhydrophobic surface; (a,d,g) 30 min; (b,e,h) 120 min; (c,f,i) magnified images of the corresponding images of (b,e,h).

The *in vitro* experimental indicated that the superhydrophobic nanotube TiO_2 layers exhibit a remarkable resistance to platelets attachment (Fig. 13) (Yang et al., 2010). It is indicated that the superhydrophobic nanotube TiO_2 layers exhibit a remarkable resistance to platelets attachment. As shown in Fig. 13a,d,g, abundant platelets adhere to both the plain Ti surface and the superhydrophilic TiO_2 nanotube layers after 30 min incubation. Comparatively, after 120 min incubation, a large number of platelets adhered and spread out on both the plain Ti surface (77 ± 7.4 per 5000 μm^2, Fig. 13b,c) and the superhydrophilic surface which was obtained by exposing the TiO_2 nanotubes under a UV irradiation (22 ± 1.5 per 5000 μm^2, Fig. 13e,f), only very few of platelets (1 ± 0.8 per 5000 μm^2) was found to adhere on the superhydrophobic TiO_2 nanotube layers (Fig. 13h). Moreover, even though some platelets were occasionally seen attached on the superhydrophobic surface, they looked smooth without any growth of pseudopods (Fig. 13i), implying that the platelets adhered on the superhydrophobic TiO_2 nanotube surface remain inactive and hardly grow and spread out for a long period. The quantities and morphologies of adhered platelets and their corresponding interactions on the different samples are illustrated in Figure 14. Therefore, the construction of superhydrophobic surface on biomedical implants could pave a way to improve the blood compatibility of the biomedical devices and implants.

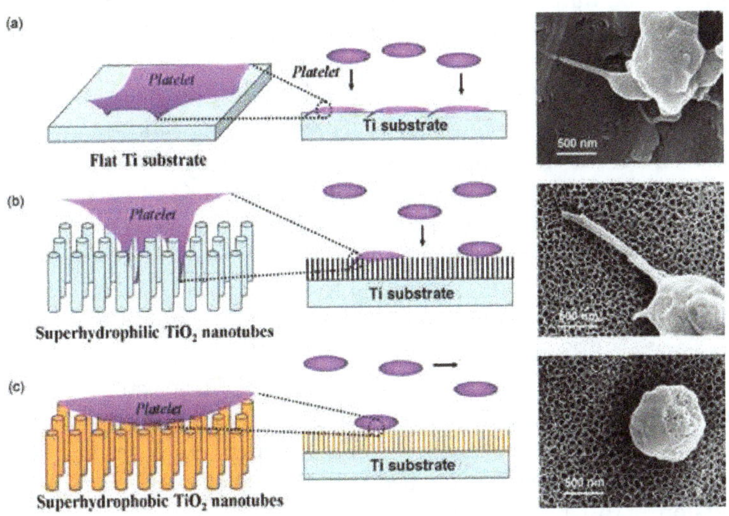

Fig. 14. Schematic illustration of the quantity and morphology of platelet and corresponding interactions on the three kinds of surfaces. (a) Plain Ti substrate; (b) Superhydrophilic TiO_2 nanotubes; and (c) Superhydrophobic TiO_2 nanotubes.

4.3 Sensing devices

Based on photocatalytic lithography, we demonstrate a facile, rapid and practical approach to fabricate Ag nanoparticle (NP) patterns on TiO_2 films by means of pulse-current electrodeposition technique (Huang et al., 2011). The size and density of as-deposited Ag NPs can be controlled by changing deposition charge density. Moreover, the resultant patterned Ag NP films exhibited particle size-as well as density-dependent UV-vis absorption and SERS enhancement effect. It was found that the patterned Ag NP films produced under the deposition charge density of 2.0 C cm^{-2} exhibited the intense UV-vis and Raman peaks. Two dimensional surface enhanced Raman scattering (SERS) mapping of

Rhodamine 6G (R6G) on the patterned Ag NP films demonstrated a high throughput localized molecular adsorption and micropatterned SERS effect.

Furthermore, the elemental distributions of the as-prepared Ag NPs arrays were also observed by electron probe microanalyzer, which are shown in Fig. 15a-c. Figure 15a shows the Ag element distribution map. As shown in the map, the green dot patterns are clearly images obtained through element concentration contrast between the UV-irradiated superhydrophilic and photomasked superhydrophobic regions. The green dots exhibiting a uniform Ag concentration against the surrounding black regions indicate that Ag NPs are uniformly deposited and confined to the superhydrophilic regions. Figure 15b,c shows the element distribution maps of Ti and O, respectively, which are also in line with the dimensions of the photomask. The blue superhydrophilic regions (dot patterns) show lower Ti and O concentrations due to the preferential deposition of Ag NPs in the superhydrophilic areas, while the yellow and red superhydrophobic regions have higher Ti and O concentration. The corresponding line-scan signal intensity profiles of Ag, O, and Ti elements across the dot pattern (red line direction). The intense Ag signals in the superhydrophilic regions as well as the Ti and O signals in the superhydrophobic regions suggest that Ag NPs are deposited only in the dot areas and that the other regions are the exposed TiO_2 nanotube films. The consistent intensity of the Ag signals indicates that Ag NPs are uniformly deposited on the superhydrophilic regions.

In addition, a two-dimensional point-by-point SERS mapping of the patterned Ag NP film whose deposited charge density is 2.0 C cm^{-2} was obtained using R6G as the probe molecule. Figure 15d,e show the optical image and the corresponding SERS mapping image of the patterned Ag NP film. The mapping area was approximately 140 × 100 μm^2 and the data acquisition time was 1 s. A signal to baseline from 594.0 to 623.4 cm^{-1} was chosen for the acquisition of the SERS mapping. The bright and dark areas respectively represent higher and lower intensity of the SERS signal. It is clear that the geometrically identical gray superhydrophilic areas (circle) with a strong SERS activity and the dark superhydrophobic areas without any SERS activity form a high-resolution SERS intensity distribution map. As can be seen from the SERS mapping results, most SERS peak area is in a very narrow intensity window as shown by the contrast in color codes. Furthermore, the SERS peak area is uniform over the superhydrophilic region with several high intensity spots represented by white color codes. On comparing the SERS mapping with the SEM detection, it is reasonable to conclude that the homogeneous SERS signal in the circle areas reflects the uniform dispersion of Ag NPs on the superhydrophilic areas. The high-resolution SERS intensity distribution and micropatterned SERS effect of the Ag NP film might make it potentially useful in high-throughput molecule detection and bio-recognition.

To gain insight into the dependence of SERS enhancement on the size and density of Ag NPs, the SERS spectra of R6G absorbed on the different patterned Ag NP films were detected, which are shown in Fig. 15f. Because of the small particle size and low density, which are not the optimum size and distribution for SERS, the signal enhancement is rather weak below the charge density of 0.5 C cm^{-2}. The enhancement behavior of the substrate, however, is obviously improved under the charge density of 1.0 C cm^{-2}. In particular, the patterned Ag NP film prepared under a charge density of 2.0 C cm^{-2} exhibits the highest intensity, which is attributed to large size and high density of Ag NPs, as shown by the SEM results. On increasing the charge density to 2.5 C cm^{-2}, the signal becomes weaker. The size-correlated enhancement may be explained by the EM mechanism (Zeng et al., 2008). The intensity of the SERS signal might also be controlled by the NPs density, which changes the interparticle spacing as well as the hot spots among the NPs (Felidj et al., 1999; Lu et al., 2005). The average surface enhancement factor for pyridine on the Ag NP film with a charge density of 2.0 C cm^{-2} was calculated to be 1.3×10^5.

Fig. 16a shows the typical SEM micrographs of the CdS nanosphere micropatterns after 3 min deposition on the superhydrophobic-superhydrophilic template of TiO_2 nanotube films (Lai et al., 2010d). The bright rectangular areas corresponded to the deposition of CdS nanosphere crystals on the superhydrophilic regions. The boundary between the CdS pattern and the surrounding superhydrophobic regions is clearly visible at a higher magnification (inset). The dispersed CdS nanosphere crystals grew on the top of TiO_2 nanotube arrays within the rectangular superhydrophilic region (Fig. 16c). Most of the crystals were less than 90 nm in diameter due to the confinement by the inner diameter of nanotube, though a few larger spheres (~120 nm) were seen across neighboring tube openings. While on the superhydrophobic areas (Fig. 16d), there was almost no CdS crystal. The high growth selectivity was also confirmed by the EDS analysis, revealing that CdS spheres easily nucleate and grow on the hydroxyl groups (-OH) terminated regions (Fig. 16e), but not on the $-CF_3$ terminated areas (Fig. 16f). Since the difference of the water contact angle between the superhydrophilic and superhydrophobic regions is larger than 150°, electrolyte solution is preferentially presented on the uniform superhydrophilic dots. No water droplets go to the neighboring superhydrophobic regions. Although a few CdS particles resulted from homogeneous precipitation attached onto superhydrophobic surface due to van der Waals interactions and gravity, they can be easily removed by ultrasonication. Therefore, a clear and well-defined CdS pattern in line with the dimensions of the superhydrophobic-superhydrophilic template has been obtained.

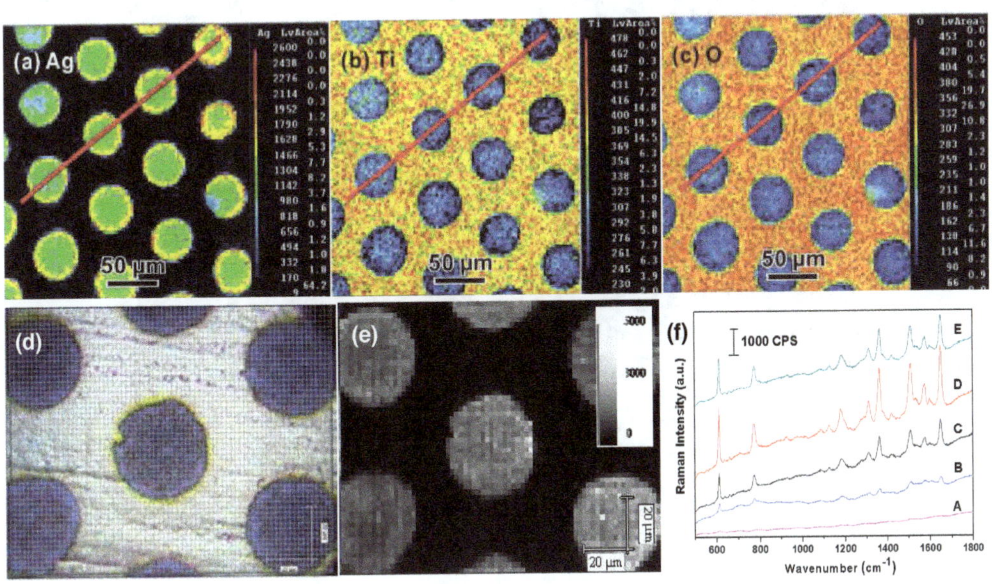

Fig. 15. Typical element distribution maps of Ag (a),Ti (b), O (c), and optical microscopy(d) and the corresponding R6G SERS mapping (e) of the patterned Ag NP films with an area of 140 × 100 μm^2 using the peak area at 614 cm^{-1} as the reference. (f) Raman spectra of the patterned Ag NP films from different charge density: curve A, bare subsrate; curve B, 0.5 C/cm^2; C, 1.0 C/cm^2; D, 2.0 C/cm^2; and E, 2.5 C/cm^2.

Fig. 16b shows the photocurrent spectra of the couple CdS/TiO_2 nanotube array electrode prepared under different electrodeposition times. It is apparent that the pure TiO_2 nanotube array samples have a photo-response wavelength lower than 400 nm due to its band-gap of 3.2 eV (curve a). The decoration of CdS nanospheres with a smaller energy band-gap (2.4 eV) can significantly extend the photo-response range from 380 nm to about 500 nm.Moreover, the CdS modified TiO_2 nanotube array electrodes can also greatly increase the photocurrent response under UV light, especially for the samples obtained under 2 min electrodeposition (curve b), which thus would be the optimal deposition time. This is attributed to the uniform dispersed CdS nanospheres with suitable size decorated onto the TiO_2 nanotubes. This allows for more efficient electron transfer and lower electron-hole recombination rate which leads to enhanced light harvesting at the directly grown CdS/TiO_2 heterojunctions. With the increase of time (curve c and d), more CdS particles with bigger size started to randomly distribute on top of TiO_2 nanotube arrays. Such composite nanostructures would weaken the light absorption of the uniform CdS/TiO_2 heterojunction underlayer, which has resulted in a lower photocurrent in both UV and visible light region.

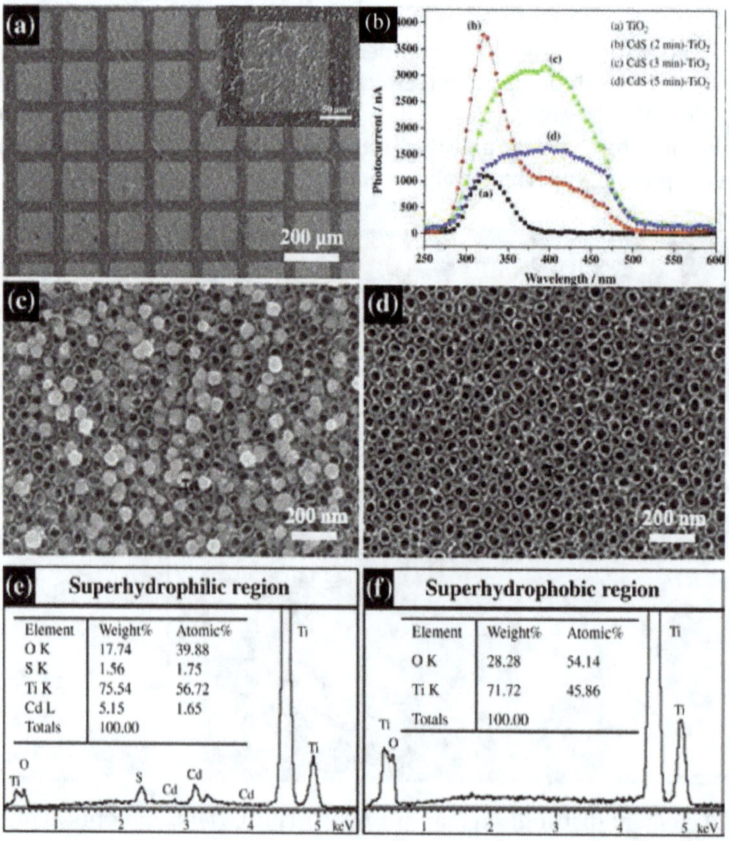

Fig. 16. (a) Typical SEM images of the CdS micropattern; (b) Photocurrent spectra of micropatterned CdS film on TiO_2 nanotube array electrode. (curve a-d): pure TiO_2; 2 min; 3 min; and 5 min. (c) Superhydrophilic region; (d) superhydrophobic region. EDX spectrum of the corresponding superhydrophilic (e) and superhydrophobic regions (f).

5. Summary and outlook

Extremely wetting micropattern (superhydrophilic/superhydrophobic) on TiO_2 nanostructure surface by using SAM technique and photocatalytic lithography has been studied intensely as it provides a cost effective template to construct well defined functional composited pattern. Numerous potential applications have also been proposed and investigated in biomedical, sensors and micro-nano devices. We believe that the photocatatlytic lithography patterning technique presented in this chapter should be general to create micro-scale wetting pattern on other semiconductor substrates and these developments will open the door for more widespread application of the wetting pattern in practical fields.

6. Acknowledgments

The authors thank the National Natural Science Foundation of China (grants 51072170, 21021002), the National Basic Research Program of China (grant 2007CB935603) and the National High Technology Research and Development Program of China (grant 2009AA03Z327), and the Environment and Water Industry Programme Office (EWI) under the National Research Foundation of Singapore (grant MEWR651/06/160) for their financial supports.

7. References

Balaur, E., Macak, J.M., Taveira, L. & Schmuki, P. (2005). Tailoring the wettability of TiO_2 nanotube layers. *Electrochem. Commun.*, Vol. 7, No. 10, 1066-1070, 1388-2481

Bearinger, J.P., Stone, G, Hiddessen, A.L., Dugan, L.C., Wu, L.G., Hailey, P., Conway, J.W., Kuenzler, T., Feller, L., Cerritelli, S. & Hubbell, J.A. (2009). Phototocatalytic lithography of poly(propylene sulfide) block copolymers: Toward high-throughput nanolithography for biomolecular arraying applications. *Langmuir,* Vol. 25, No. 2, 1238-1244, 0743-7463

Bhawalkar, S.P., Qian, J., Heiber, M.C. & Jia, L. (2010). Development of a colloidal lithography method for patterning nonplanar surfaces. *Langmuir,* Vol. 26, No. 22, 16662-16666, 0743-7463

Crawford, G.A. & Chawla, N. (2009). Porous hierarchical TiO_2 nanostructures: Processing and microstructure relationships. *Acta Mater.*, Vol. 57, No. 3, 854-867, 1359-6454

Csucs, G., Kunzler, T., Feldman, K., Robin, F. & Spencer, N.D. (2003). Microcontact printing of macromolecules with submicrometer resolution by means of polyolefin stamps. *Langmuir,* Vol. 19, No. 15, 6104-6109, 0743-7463

Cui, B. & Veres, T. (2007). Fabrication of metal nanoring array by nanoimprint lithography (NIL) and reactive ion etching. *Microelectron. Eng.*, Vol. 84, No. 5-8, 1544-1547, 0167-9317

Falconnet, D., Koenig, A., Assi, T. & Textor M. (2004). A combined photolithographic and molecular-assembly approach to produce functional micropatterns for applications in the biosciences. *Adv. Funct. Mater.*, Vol. 14, No. 8, 749-756, 1616-301X

Felidj, N., Aubard, J. & Levi, G. (1999). Discrete dipole approximation for ultraviolet-visible extinction spectra simulation of silver and gold colloids. *J. Chem. Phys.*, Vol. 111, No. 3, 1195-1208, 0021-9606

Gao, X.F. & Jiang, L. (2004). Water-repellent legs of water striders. *Nature,* Vol. 432, No. 7013, 36-36, 0028-0836

Gong, D.W., Grimes, C.A. & Varghese O.K. (2001). Titanium oxide nanotube arrays prepared by anodic oxidation. *J. Mater. Res.,* Vol. 16, No. 12, 3331-3334, 0884-2914

Haick, H. & Paz, Y. (2001). Remote photocatalytic activity as probed by measuring the degradation of self-assembled monolayers anchored near microdomains of titanium dioxide. *J. Phys. Chem. B,* Vol. 105, No. 15, 3045-3051, 1089-5647

Huang, L., Braunschweig, A.B., Shim, W., Qin, L.D., Lim, J.K., Hurst, S.J., Huo, F.W., Xue, C., Jong, J.W. & Mirkin, C.A. (2010). Matrix-assisted dip-pen nanolithography and polymer pen lithography. *Small,* Vol. 6, No. 10, 1077-1081, 1613-6810

Huang, Y.X., Lai, Y.K., Lin, L.X., Sun, L. & Lin, C.J. (2010). Electrochemical construction and biological performance of micropatterned CaP films. *Acta Phys. -Chim. Sin.,* Vol. 26, No. 8, 2057-2060, 1000-6818

Huang, Y.X., Sun, L., Xie, K.P., Lai, Y.K., Liu, B.J., Ren, B. & Lin, C.J. (2011). A SERS study of Ag nanoparticles electrodeposited on patterned TiO_2 nanotube films, *J. Raman Spectrosc.,* DOI: 10.1002/jrs.2830, 0377-0486, Vol. 42, No. 5, 986-991

Ichimura, K., Oh, S.K. & Nakagawa, M. (2000). Light-driven motion of liquids on a photoresponsive surface. *Science,* Vol. 288, No. 5471, 1624-1626, 0036-8075

Jeon, N.L., Clem, P.G., Payne, D.A. & Nuzzo, R.G. (1996). A monolayer-based lift-off process for patterning chemical vapor deposition copper thin films. *Langmuir,* Vol. 12, No. 22, 5350-5355, 0743-7463

Jiao, L.Y., Gao, H.J., Zhang, G.M., Xie, G.Y., Zhou, X., Zhang, Y.Y., Zhang, Y.Y., Gao, B., Luo, G., Wu, Z.Y., Zhu, T., Zhang, J., Liu, Z.F., Mu, S.C., Yang, H.F. & Gu, C.Z. (2005). Fabrication of metallic nanostructures by negative nanoimprint lithography. *Nanotechnology,* Vol. 16, No. 12, 2779-2784, 0957-4484

Katagiri, K., Ohno, K., Masuda, Y. & Koumoto, K. (2007). Growth behavior of TiO_2 particles via the liquid phase deposition process. *J. Ceram. Soc. Jpn.,* Vol. 115, No. 1348, 831-834, 1882-0743

Kobayashi, T., Shimizu, K., Kaizuma, Y. & Konishi, S. (2011). Formation of superhydrophobic/superhydrophilic patterns by combination of nanostructure-imprinted perfluoropolymer and nanostructured silicon oxide for biological droplet generation. *Appl. Phys. Lett.,* Vol. 98, No. 12, 123706, 0003-6951

Koumoto, K., Saito, N., Gao, Y.F., Masuda, Y. & Zhu, P.X. (2008). Nano/micro patterning of inorganic thin films. *Bull. Chem. Soc. Jpn.,* Vol. 81, No. 11, 1337-1376, 0009-2673

Kubo, W., Tatsuma, T., Fujishima, A. & Kobayashi, H. (2004). Mechanisms and resolution of photocatalytic lithography. *J. Phys. Chem. B,* Vol. 108, No. 9, 3005-3009, 1520-6106

Kumar, A., Biebuyck, H.A., Abbott, N.L. & Whitesides, G.M. (1992). The use of self-assembled monolayers and a selective etch to generate patterned gold features. *J. Am. Chem. Soc.,* Vol. 114, No. 23, 9188-9189, 0002-7863

Lafuma, A. & Quéré, D. (2003). Superhydrophobic states. *Nat. Mater.,* Vol. 2, No. 7, 457-460, 1476-1122

Li, L.Y., Yang, Y., Yang, G.L., Chen, X.M., Hsiao, B.S., Chu, B., Spanier, J.E. & Li, C.Y. (2006). Patterning polyethylene oligomers on carbon nanotubes using physical vapor deposition. *Nano Lett.,* Vol. 6, No. 5, 1007-1012, 1530-6984

Lai, Y.K., Sun, L., Zuo, J. & Lin, C.J. (2004). Electrochemical fabrication and formation mechanism of TiO_2 nanotube array on metallic titanium surface. *Acta Phys. -Chim. Sin.*, Vol. 20, No. 9, 1063-1066, 1000-6818

Lai, Y.K., Sun, L., Chen, Y.C., Zhuang, H.F., Lin, C.J. & Chin, J.W. (2006). Effects of the structure of TiO_2 nanotube array on Ti substrate on its photocatalytic activity. *J. Electrochem. Soc.*, Vol. 153, No. 7, D123-127, 0013-4651

Lai, Y.K., Lin, C.J., Wang, H., Huang, J.Y., Zhuang, H.F. & Sun, L. (2008). Superhydrophilic-superhydrophobic micropattern on TiO_2 nanotube films by photocatalytic lithography. *Electrochem. Commun.*, Vol. 10, No. 3, 387-391, 1388-2481

Lai, Y.K., Lin, C.J., Huang, J.Y., Zhuang, H.F., Sun, L. & Nguyen, T. (2008). Markedly controllable adhesion of superhydrophobic spongelike nanostructure TiO_2 films. *Langmuir,* Vol. 24, No. 8, 3867-3873, 0743-7463

Lai, Y.K., Gao, X.F., Zhuang, H.F., Huang, J.Y., Lin, C.J. & Jiang, L. (2009). Designing superhydrophobic porous nanostructures with tunable water adhesion. *Adv. Mater.,* Vol. 21, No. 37, 3799-3803, 0935-9648

Lai, Y.K., Huang, J.Y., Gong, J.J., Huang, Y.X., Wang, C.L., Chen, Z. & Lin, C.J. (2009). Superhydrophilic-superhydrophobic Template: A simple approach to micro- and nanostructure patterning of TiO_2 films. *J. Electrochem. Soc.,* Vol. 156, No. 11, D480-484, 0013-4651

Lai, Y.K., Zhuang, H.F., Sun, L., Chen, Z. & Lin, C.J. (2009). Self-organized TiO_2 nanotubes in mixed organic-inorganic electrolytes and their photoelectrochemical performance. *Electrochim. Acta,* Vol. 54, No. 26, 6536-6542, 0013-4686

Lai, Y.K., Lin, Z.Q., Huang, J.Y., Sun, L., Chen, Z. & Lin, C.J. (2010). Controllable construction of ZnO/TiO_2 patterning nanostructures by superhydrophilic/superhydrophobic templates. *New J. Chem.*, Vol. 34, No. 1, 44-51, 1144-0546

Lai, Y.K., Huang, J.Y., Zhang, H.F., Subramaniam, V.P., Tang, Y.X., Gong, D.G., Sundar, L., Sun, L., Chen, Z. & Lin, C.J. (2010). Nitrogen-doped TiO_2 nanotube array films with enhanced photocatalytic activity under various light sources. *J. Hazard. Mater.*, Vol. 184, No. 1-3, 855-863, 0304-3894

Lai, Y.K., Huang, Y.X., Huang, J.Y., Wang, H., Chen, Z. & Lin, C.J. (2010). Selective formation of ordered arrays of octacalcium phosphate ribbons on TiO_2 nanotube surface by template-assisted electrodeposition. *Colliods Surf. B,* Vol. 76, No. 1, 117-122, 0927-7765

Lai, Y.K., Lin, Z.Q., Chen, Z., Huang, J.Y. & Lin, C.J. (2010). Fabrication of patterned CdS/TiO_2 heterojunction by wettability template-assisted electrodeposition. *Mater. Lett.*, Vol. 64, No. 11, 1309-1312, 0167-577X

Lee, J.P. & Sung, M.M. (2004). A new patterning method using photocatalytic lithography and selective atomic layer deposition. *J. Am. Chem. Soc.*, Vol. 126, No. 1, 28-29, 0002-7863

Lee, S.W., Oh, B.K., Sanedrin, R.G., Salaita, K., Fujigaya, T. & Mirkin, C.A. (2006). Biologically active protein nanoarrays generated using parallel dip-pen nanolithography. *Adv. Mater.*, Vol. 18, No. 9, 1133-1136, 0935-9648

Li, Z.W., Gu, Y.N., Wang, L., Ge, H.X., Wu, W., Xia, Q.F., Yuan, C.S., Chen, Y., Cui, B. & Williams, R.S. (2009). Hybrid nanoimprint-soft lithography with sub-15 nm resolution. *Nano Lett.,* Vol. 9, No. 6, 2306-2310, 1530-6984

Liu, S.H., Wang, W.C.M., Mannsfeld, S.C.B., Locklin, J., Erk, P., Gomez, M., Richter, F. & Bao, Z.N. (2007). Solution-assisted assembly of organic semiconducting single crystals on surfaces with patterned wettability. *Langmuir,* Vol. 23, No. 14, 7428-7432, 0743-7463

Liu, M.J., Zheng, Y.M., Zhai, J. & Jiang, L. (2010). Bioinspired super-antiwetting interfaces with special liquid-solid adhesion. *Accounts Chem. Res.,* Vol. 43, No. 3, 368-377, 0001-4842

Liu, X.J., Ye, Q.A., Yu, B., Liang, Y.M., Liu, W.M. & Zhou, F. (2010). Switching water droplet adhesion using responsive polymer brushes. *Langmuir,* Vol. 26, No. 14, 12377-12382, 0743-7463

Lu, Y., Liu, G.L. & Lee, L.P. (2005). High-density silver nanoparticle film with temperature-controllable interparticle spacing for a tunable surface enhanced Raman scattering substrate. *Nano Lett.,* Vol. 5, No. 1, 5-9, 1530-6984

Masuda, Y., Kinoshita, N., Sato, F. & Koumoto, K. (2006). Site-selective deposition and morphology control of UV- and visible-light-emitting ZnO crystals. *Cryst. Growth Des.,* Vol. 6, No. 1, 75-78, 1528-7483

Michel, R., Reviakine, I., Sutherland, D., Fokas, C., Csucs, G., Danuser, G., Spencer, N.D. & Textor, M. (2002). A novel approach to produce biologically relevant chemical patterns at the nanometer scale: Selective molecular assembly patterning combined with colloidal lithography. *Langmuir,* Vol. 18, No. 22, 8580-8586, 0743-7463

Nakata, K., Nishimoto, S., Yuda, Y., Ochiai, T., Murakami, T. & Fujishima, A. (2010). Rewritable superhydrophilic-superhydrophobic patterns on a sintered titanium dioxide substrate. *Langmuir,* Vol. 26, No. 14, 11628-11630, 0743-7463

Nishimoto, S., Sekine, H., Zhang, X.T., Liu, Z.Y., Nakata, K., Murakami, T., Koide, Y. & Fujishima, A. (2009). Assembly of self-assembled monolayer-coated Al_2O_3 on TiO_2 thin films for the fabrication of renewable superhydrophobic-superhydrophilic structures. *Langmuir,* Vol. 25, No. 13, 7226-7228, 0743-7463

Rausch, N. & Burte, E.P. (1993). Thin TiO_2 films prepared by low-pressure chemical vapor-deposition. *J. Electrochem. Soc.,* Vol. 140, No. 1, 145-149, 0013-4651

Rusponi, S., Costantini, G., de Mongeot, F.B., Boragno, C. & Valbusa, U. (1999). Patterning a surface on the nanometric scale by ion sputtering. *Appl. Phys. Lett.,* Vol. 75, No. 21, 3318-3320, 0003-6951

Shen, G.X., Chen, Y.C. & Lin, C.J. (2005). Corrosion protection of 316 L stainless steel by a TiO_2 nanoparticle coating prepared by sol-gel method. *Thin Solid Films,* Vol. 489, No. 1-2, 130-136, 0040-6090

Slocik, J.M., Beckel, E.R., Jiang, H., Enlow, J.O., Zabinski, J.S., Bunning, T.J. & Naik, R.R. (2006). Site-speciric patterning of biomolecules and quantum dots on functionalized surfaces generated by plasma-enhanced chemical vapor deposition. *Adv. Mater.,* Vol. 18, No. 16, 2095-2100, 0935-9648

Takeda, S., Suzuki, S., Odaka, H. & Hosono, H. (2001). Photocatalytic TiO_2 thin film deposited onto glass by DC magnetron sputtering. *Thin Solid Films,* Vol. 392, No. 2, 338-344, 0040-6090

Tatsuma, T., Kubo, W. & Fujishima, A. (2002). Patterning of solid surfaces by photocatalytic lithography based on the remote oxidation effect of TiO_2. *Langmuir,* Vol. 18, No. 25, 9632-9634, 0743-7463

Wang, H., Lin, C.J., Hu, R., Zhang, F. & Lin, L.W. (2008). A novel nano-micro structured octacalcium phosphate/protein composite coating on titanium by using an electrochemically induced deposition. *J. Biomed. Mater. Res. A,* Vol. 87A, No. 3, 698-705, 1549-3296

Wang, H., Duan, J.C. & Cheng, Q. (2011). Photocatalytically patterned TiO_2 arrays for on-plate selective enrichment of phosphopeptides and direct MALDI MS analysis. *Anal. Chem.,* Vol. 83, No. 5, 1624-1631, 0003-2700

Wang, R., Hashimoto, K., Fujishima, A., Chikuni, M., Kojima, E., Kitamura, A., Shimohigoshi, M. & Watanabe, T. (1997). Light-induced amphiphilic surfaces. *Nature,* Vol. 388, No. 6641, 431-432, 0028-0836

Wang, Y.L. & Lieberman, M. (2003). Growth of ultrasmooth octadecyltrichlorosilane self-assembled monolayers on SiO_2. *Langmuir,* Vol. 19, No. 4, 1159-1167, 0743-7463

West, J.B. (1999). The original presentation of Boyle's law. *J. Appl. Physiol.,* Vol. 87, No. 4, 1543-1545, 8750-7587

Xu, S. & Liu, G.Y. (1997). Nanometer-scale fabrication by simultaneous nanoshaving and molecular self-assembly. *Langmuir,* Vol. 13, No. 2, 127-129, 0743-7463

Yang, Y., Lai, Y.K., Zhang, Q.Q., Wu, K., Zhang, L.H., Lin, C.J. & Tang, P.F. (2010). A novel electrochemical strategy for improving blood compatibility of titanium-based biomaterials. *Colloids Surf. B,* Vol. 79, No. 1, 309-313, 0927-7765

Yasuda, K., Macak, J.M., Berger, S., Ghicov, A. & Schmuki, P. (2007). Mechanistic aspects of the self-organization process for oxide nanotube formation on valve metals. *J. Electrochem. Soc.,* Vol. 154, No. 9, C472-478, 0013-4651

Yoshimura, M. & Gallage, R. (2008). Direct patterning of nanostructured ceramics from solution - differences from conventional printing and lithographic methods. *J. Solid State Electrochem.,* Vol. 12, No. 7-8, 775-782, 1432-8488

Yun, H., Lin, C.J., Li, J. Wang, J.R. & Chen, H.B. Low-temperature hydrothermal formation of a net-like structured TiO_2 film and its performance of photogenerated cathode protection. *Appl. Surf. Sci.,* Vol. 255, No. 5, 2113-2117, 0169-4332

Zeng, J.B., Jia, H.Y., An, J., Han, X.X., Xu, W.Q., Zhao, B. & Ozaki, Y. (2008). Preparation and SERS study of triangular silver nanoparticle self-assembled films. *J. Raman Spectrosc.,* Vol. *39*, No. 11, 1673-1678, 0377-0486

Zhang, C. & Kalyanaraman, R. (2004). In situ lateral patterning of thin films of various materials deposited by physical vapor deposition. *J. Mater. Res.,* Vol. 19, No. 2, 595-599, 0884-2914

Zhang, G.M., Zhang, J., Xie, G.Y., Liu, Z.F. & Shao H.B. (2006). Cicada wings: A stamp from nature for nanoimprint lithography. *Small,* Vol. 2, No. 12, 1440-1443, 1613-6810

Zhang, G.J., Tanii, T., Kanari, Y. & Ohdomari, I. (2007). Production of nanopatterns by a combination of electron beam lithography and a self-assembled monolayer for an antibody nanoarray. *J. Nanosci. Nanotechnol.,* Vol. 7, No. 2, 410-417, 1533-4880

Zhang, X.T., Jin, M., Liu, Z.Y., Tryk, D., Nishimoto, S., Murakami, T. & Fujishima, A. (2007). Superhydrophobic TiO_2 surfaces: Preparation, photocatalytic wettability conversion, and superhydrophobic-superhydrophilic patterning. *J. Phys. Chem. C,* Vol. 111, No. 39, 14521-14529, 1932-7447

Zhao, J.C., Wu, T.X., Wu, K.Q., Oikawa, K., Hidaka, H. & Serpone, N. (1998). Photoassisted degradation of dye pollutants. 3. Degradation of the cationic dye rhodamine B in aqueous anionic surfactant/TiO_2 dispersions under visible light irradiation: Evidence for the need of substrate adsorption on TiO_2 particles. *Environ. Sci. Technol.*, Vol. 32, No. 16, 2394-2400, 0013-936X

Zhuang, H.F., Lin, C.J., Lai, Y.K., Sun, L. & Li, J. (2007). Some critical structure factors of titanium oxide manotube array in its photocatalytic activity. *Environ. Sci. Technol.*, Vol. 41, No. 13, 4735-4740, 0013-936X

Permissions

All chapters in this book were first published by InTech Open; hereby published with permission under the Creative Commons Attribution License or equivalent. Every chapter published in this book has been scrutinized by our experts. Their significance has been extensively debated. The topics covered herein carry significant findings which will fuel the growth of the discipline. They may even be implemented as practical applications or may be referred to as a beginning point for another development.

The contributors of this book come from diverse backgrounds, making this book a truly international effort. This book will bring forth new frontiers with its revolutionizing research information and detailed analysis of the nascent developments around the world.

We would like to thank all the contributing authors for lending their expertise to make the book truly unique. They have played a crucial role in the development of this book. Without their invaluable contributions this book wouldn't have been possible. They have made vital efforts to compile up to date information on the varied aspects of this subject to make this book a valuable addition to the collection of many professionals and students.

This book was conceptualized with the vision of imparting up-to-date information and advanced data in this field. To ensure the same, a matchless editorial board was set up. Every individual on the board went through rigorous rounds of assessment to prove their worth. After which they invested a large part of their time researching and compiling the most relevant data for our readers.

The editorial board has been involved in producing this book since its inception. They have spent rigorous hours researching and exploring the diverse topics which have resulted in the successful publishing of this book. They have passed on their knowledge of decades through this book. To expedite this challenging task, the publisher supported the team at every step. A small team of assistant editors was also appointed to further simplify the editing procedure and attain best results for the readers.

Apart from the editorial board, the designing team has also invested a significant amount of their time in understanding the subject and creating the most relevant covers. They scrutinized every image to scout for the most suitable representation of the subject and create an appropriate cover for the book.

The publishing team has been an ardent support to the editorial, designing and production team. Their endless efforts to recruit the best for this project, has resulted in the accomplishment of this book. They are a veteran in the field of academics and their pool of knowledge is as vast as their experience in printing. Their expertise and guidance has proved useful at every step. Their uncompromising quality standards have made this book an exceptional effort. Their encouragement from time to time has been an inspiration for everyone.

The publisher and the editorial board hope that this book will prove to be a valuable piece of knowledge for researchers, students, practitioners and scholars across the globe.

List of Contributors

Fung Jacky Wen and Po Sheun Chung
Department of Electronic Engineering, City University of Hong Kong, China

Atsushi Motogaito and Kazumasa Hiramatsu
Graduate School of Engineering, Mie University the Center of Ultimate Technology on nano-Electronics, Mie University, Japan

Gao Kan, Chen Haifeng, Liu Bing-Fang and Xu Qun-Yuan
Beijing Institute for Neuroscience, Beijing Center of Neural Regeneration and Repairing, Key Laboratory for Neurodegenerative Diseases of the Ministry Education Capital Medical University Beijing, China

Despina Fragouli, Roberto Cingolani and Athanassia Athanassiou
Center for Biomolecular Nanotechnologies, Istituto Italiano di Tecnologia, Italy

Christine C. Dupont-Gillain, Cristèle J. Nonckreman, Yasmine Adriaensen and Paul G. Rouxhet
Institute of Condensed Matter and Nanosciences - Bio & Soft Matter, Université Catholique de Louvain, Belgium

Guotao Duan and Weiping Cai
Key Laboratory of Materials Physics, Institute of Solid State Physics, Chinese Academy of Sciences, Anhui, China

Shuyan Gao
College of Chemistry and Environmental Science, Henan Normal University, Xinxiang, China

Changjian Lin
State Key Laboratory for Physical Chemistry of Solid Surfaces and College of Chemistry and Chemical Engineering, Xiamen University, China

Naoto Koshizaki
Nanosystem Research Institute (NRI) National Institute of Advanced Industrial Science and Technology (AIST), Japan

Atsushi Sekiguchi
Litho Tech Japan Corporation, Japan

Yue Li
Key Laboratory of Materials Physics, Institute of Solid State Physics, Chinese Academy of Sciences, Anhui, China
Nanosystem Research Institute (NRI) National Institute of Advanced Industrial Science and Technology (AIST), Japan

Masaki Yanagisawa
Sumitomo Electric Industries, Ltd. Japan

Keita Sakai
Canon Inc. Japan

SunHyung Lee and Katsuya Teshima
Faculty of Engineering, Shinshu University, Japan

Takahiro Ishizaki
National Institute of Advanced Industrial Science and Technology, Japan

Nagahiro Saito and Osamu Takai
EcoTopia Science Institute, Nagoya University, Japan

Zhong Chen
School of Materials Science and Engineering, Nanyang Technological University, Singapore

Yuekun Lai
State Key Laboratory for Physical Chemistry of Solid Surfaces and College of Chemistry and Chemical Engineering, Xiamen University, China
School of Materials Science and Engineering, Nanyang Technological University, Singapore

Index

B

Beam Splitter, 2

Bessel Function, 2-3

Binary Diffractive Lens, 20-30, 49

Binary Structure, 1, 22-25

Bioengineering, 65

C

Circular Dammann Grating, 1-3, 6, 8, 10, 12, 14, 16, 18-19

Circular Spot Rotation, 2-3, 18-19

Colloidal Lithography, 65, 80-82, 103-104, 106, 108, 110, 112, 114, 116-120, 122-126, 128, 130-136, 138, 140-142, 144-146, 231, 234

Colloidal Monolayer, 82, 103-107, 112-119, 122-137, 140-145

Colloidal Particles, 65-67, 70-75, 78-81

Convex Lens, 26-27, 30, 49

Crystal Damage, 150, 154

D

Dammann Grating, 1-3, 6, 8, 10, 12, 14, 16, 18-19

De-protection Reaction, 94-95, 98-99

Deposition Assisted Colloidal, 82, 103-106, 108, 110, 112, 114, 116, 118, 120, 122, 124, 126, 128, 130, 132, 134, 136, 138, 140, 142, 144-146

Diffraction, 1-3, 8-11, 18-23, 26-30, 48-53, 58-64, 106-107, 120-121, 149-151, 154, 156, 161-162, 220

Diffraction Angles, 1, 154, 156

Diffraction Grating, 10-11, 20-21, 154

Diffractive Lens, 20-30, 49

Dopaminergic Neurons, 31, 38-43

Duty Cycle, 16-19

E

Electron Beam, 10-11, 14-15, 20-26, 28, 65, 86-87, 149, 161, 183, 210, 214, 235

Electron Beam Lithography, 14-15, 20, 22-24, 26, 28, 149, 161, 183, 210, 214, 235

Extreme Ultraviolet, 83, 163

F

Field Emitters, 82, 103-105, 136-137, 144-145

Fluid Degradation, 167, 169, 182

Focal Length, 2-3, 8-9, 18-30, 49

G

Glass Substrates, 66-67

Grating Period, 1, 150, 154, 156-159

Grid Patterns, 38-43

H

Hankel Transform, 2-3, 6-7

High-index Immersion, 163-165, 167, 180-182

Hippocampal Cells, 38-39, 48

I

Immunocytochemical Staining, 40-41, 44-47

Immunofluorescent Image, 42-47

Immunostaining, 42-43

Intrinsic Birefringence, 164-166

L

Laser Diodes, 149, 161-162

Laser Irradiation, 50-55, 58-59, 62-63, 130-131, 167, 169-170, 173

Light Sensitive Polymer, 12-13

Lithographic Techniques, 10-11, 82, 103, 183

Lithography, 3, 7, 9-11, 13-15, 17, 19-29, 31-49, 51-53, 55, 57-65, 67, 69, 71, 73, 75, 77-101, 103-147, 149-150, 160-165, 190-191, 196-197, 200, 202, 210-211, 213-214, 218, 220, 227, 231-235

Lithography Simulation, 83-84, 86-88, 90, 92, 94, 96, 98, 100

Luminous Intensity, 26-30, 49

M

Micro Contact Printing, 31-33

Micro-fabrication, 31-32, 34, 36, 38, 40, 42, 44, 46-47

Microfluidic Devices, 30, 49-51, 62-63, 82, 103, 144-145

Microfluidic Technique, 38-41, 44-47
Microfluidics, 30-31, 48-51, 62-63
Microstencil Mold, 35-36

N

Nano-biotechnology, 82, 103
Nanophotonics, 82, 103
Nanostructured Arrays, 82, 103-114, 116-118, 120, 122-126, 128, 130-134, 136-138, 140-146
Nanostructured Surfaces, 72-73, 76-79
Neural Cells, 31, 38-41, 46-48
Neuroscience, 31, 36-39, 48

O

Open-frame Exposure System, 84-85, 92-93
Optical Films, 21-28
Optical Systems, 2-3, 83
Optoelectronic Devices, 82, 103, 213
Organic Monolayers, 183-186
Oxygen Pressures, 106-109, 122-123, 132-135

P

Parameter Values, 96-99
Photochromic Polymers, 30, 49-52, 54, 56, 58, 60, 62-63
Physical Deposition, 82, 103-106, 108, 110, 112, 114, 116, 118, 120, 122, 124, 126, 128, 130-132, 134, 136, 138, 140, 142, 144-146
Plasma Etching, 65, 124-125
Poly Methyl Methacrylate, 14-15
Prolith Lithography Simulator, 86-87, 92-93

R

Refractive Index, 10-11, 14-15, 20-23, 26-27, 30, 49, 52-53, 58-63, 163, 165, 167, 174, 181-182

S

Soft Lithography, 31-34, 36, 38-42, 44, 46-48, 52-53, 82, 103, 150, 186, 211, 233
Space Communication, 2-3, 18-19
Spin Curve, 12-15
Spiropyran, 50-51, 62-63
Spot Rotation Method, 2-3, 6-7, 18-19
Subsequent Etching, 124-125, 156
Surface Science, 82, 103, 211
Synapse, 36-37, 44-47

V

Virtual Lithography Evaluation, 83-85

W

Wavelength, 1-3, 8-11, 20-25, 28-29, 58-63, 83-85, 90-91, 94-95, 104-105, 149-150, 154, 156-159, 163, 168, 183, 230
Wettability, 48-50, 52-53, 58, 62-63, 81, 130-132, 214-216, 219, 221, 224, 231, 233-235

www.ingramcontent.com/pod-product-compliance
Lightning Source LLC
LaVergne TN
LVHW080632210926
848485LV00003B/251